Introduction to
Random Processes
in Engineering

Introduction to Random Processes in Engineering

A. V. BALAKRISHNAN

WILEY-INTERSCIENCE

A JOHN WILEY & SONS, INC., PUBLICATION

For general information on our other products and services please contact our Customer Care
Department within the U.S. at 877-762-2974, outside the U.S. at 317-572-3993 or fax 317-572-4002.

Wiley also publishes its books in a variety of electronic formats. Some content that appears in print,
however, may not be available in electronic format.

Library of Congress Cataloging-in-Publication is available.

ISBN 0-471-74502-2

10 9 8 7 6 5 4 3 2 1

CONTENTS

PREFACE

This book is intended for a senior-undergraduate or first-year-graduate level course in Random Processes in Communication/Control Engineering. It develops the theoretical framework fundamental to the processing of random signals and data, centered on the concepts of Stationarity, Correlation, and Spectrum, including the Ergodic Principle and the Sampling Principle, and principles of digital computer Simulation of Random Processes.

A course in probability, standard in most engineering schools, is a primary prerequisite. Additionally, a course in Systems and Signals (which includes some Fourier analysis, Series, and Transforms) and some elementary Linear Algebra, in lieu of a formal course, would be helpful. Proficiency in background areas such as Integral Calculus and Complex Variables is desirable. For ease of recall, a whole chapter—Review—is devoted to gathering together almost all the formulas used in the sequel.

In contrast to the current vogue in engineering texts which provide a quick once-over vague smorgasboard of topics, this is more a "math" book in that it emphasizes precision and eschews inaccurate descriptive phrases and in fact uses the theorem/proof style wherever possible, drawing attention to mathematical fine points. In addition, there is a conscious logical sequencing of chapters, each building systematically on the previous chapters. Indeed, with the widespread use of software with canned programs, it is more important than ever that the student understand the theoretical principles on which (hopefully) the programs are based so as to discern better the many underlying implicit assumptions and thereby be able to judge better the results generated— beyond the "bells and whistles." This is particularly so in dealing with random data.

On the other hand, the mathematics is kept in balance. Thus no knowledge of measure theory, essential to any axiomatic treatment of probability, is assumed, so that even a formal definition of a random variable is not invoked.

But we do, for example, emphasize the notion of sample paths. Ambiguities are minimized without getting mired in mathematical technicalities. Convergence of random variables is handled gently. Only mean square convergence is used for the most part, but we do call attention to sample path convergence, even if no proofs are provided except for use of the Chebychev inequality occasionally.

As a departure from the textbooks of the 1960s and 1970s, which treated almost exclusively one-dimensional processes, all processes are allowed to be multidimensional (multivariate) because of the universal use of State-Space models. No detailed knowledge of State-Space theory is required, however. The emphasis is moreover on discrete-time processes and systems, and it is possible to omit entirely the sections on continuous-time systems if necessary.

A few words regarding the organization of the chapters are appropriate.

We begin in Chapter 1 where we make a valiant attempt at a precise definition of what a "random process" is within the self-imposed limitation on mathematical level. Once the definition is made, subsequent demonstrations of random processes are made to be consistent with this definition, unlike most texts at this level. We also introduce the notion of sample paths, which we illustrate with the help of "trivial" processes. We define mean and covariance functions, the bread and butter in what is to follow. Chapter 2 is devoted to (second-order) stationary processes: the notion of spectral density (matrix) and its relation to the stationary covariance function. After some preliminary motivating examples we prove Bochner's theorem, albeit not in its full generality but adequate for our purposes.

The covariance and spectral density concepts are immediately put to use in the next two chapters devoted to describing the output response of linear systems to random inputs. We begin with discrete-time processes and systems in Chapter 3, because of their importance in digital computer processing. Both Weighting Pattern and State-Space models are considered with emphasis on the latter, and steady-state properties of the output process are derived. Finite memory systems are discussed in some detail because of their importance as a possible means of approximation, particularly as it would appear that we are less likely to be limited by available memory size. As a natural by-product to the study of systems excited by white noise, the important notion of signal generation models is introduced, and the role of physical realizability and rational spectra is noted. Because of its use in time-series analysis, ARMA models are discussed briefly and the equivalence to State-Space models is explained.

Chapter 4 extends these considerations to continuous-time systems noting the escalation in mathematical complexity. Thus the notion of the integral of a random process has to be clarified first, and we limit our consideration to processes with continuous mean and covariance functions so that the Riemann integral suffices. As in the discrete-time case we study both State-Space models and Weighting Function models, and the asymptotic or steady-state response is of major interest. One feature of the presentation here is a precise definition of "white noise" in contrast to the vague "delta-correlation" notion. Response of systems excited by white noise leads naturally to Signal Generation models—both State-Space and Weighting Function. As in the discrete-time case, the role of physical realizability and rational spectral density is noted.

Crucial in processing physical data, where as a rule only one sample path, however long, is available, the statistical averages such as mean and covariance have now to be evaluated as time averages; the basic paradigm here is the Ergodic Principle and is treated in Chapter 5. The important notion of correlation time is explained and a fairly general theorem embodying the Ergodic Principle is proved, with the warning that ultimately its use is a matter of faith—or experience.

Chapter 6 is devoted to the Sampling Principle of Shannon with special reference to random processes—this is of course basic to any digital computer processing embedded in the ubiquitous A-D converter. Estimates for the sampling or aliasing error are derived. The Karhunen–Loeve expansion is also treated here as another technique for faithful conversion of continuous-time data into discrete-time data.

Nowadays any optimal system design (a Kalman filter is an example) in which random signals are involved has to be preceded by careful digital computer simulation of system performance. In particular, random signals with prescribed properties have to be generated. Chapter 7 provides an introduction to the principles and techniques of digital computer simulation of random signals, even including truth models which may be given in continuous time. Here all the material developed in the previous chapters comes into play. All three types of signal simulation models—the Rice model, the Kalman model, and the finite-memory weighting-function models—are discussed.

Recently, digital computer processing of two-dimensional images has gained considerably in scope. The necessary theory here is that of Random Fields— random processes where the parameter set is now a Euclidean space of two or more dimensions. This is treated in Chapter 8, where we see that the crucial ideas are ready extensions of the notions developed in the previous chapters for the one-parameter case. A typical example is the Structure Function. On the other hand, because there is no natural time arrow in space, there are no State-Space Signal Generation models, and simulation techniques differ accordingly.

The final chapter, Chapter 9, is devoted to the most important single application area: filtering signals in noise, whether in communications, radar, or digital control. To keep within the bounds of an introductory treatment, only the linear theory is covered—and even then choices had to be made on topics as well as scope. Thus we have de-emphasized continuous parameter processes because of the added mathematical complexity, but even more because any implementation using digital computer software will automatically be limited to discrete-time data. A canonical filtering model for Gaussian random variables leads to the Wiener–Hopf equation and to Conditional Expectation in a variational (minimum mean square error) setting. We use it to calculate the conditional Gaussian density, and we generalize to Martingales later to be able to handle the limiting case of infinite data samples. An application to a tracking model is used as the route to on-line or recursive estimation for time-varying models. Parameter estimation is viewed as the "maximum-ignorance" case in Bayesian models.

Notes and comments following each chapter are not intended to be exhaustive of topics or literature. They are kept short in the hope that the student will read them and be tempted to read some of the original papers or trace the historical development. One is struck by the proliferation of new books in the 1960–1970 period and the slack in the 1980–1990 period, except for second and third editions. There are sound reasons for this, but we felt it best to break up the references into two classes—what may be called "classics" and books of more recent vintage, 1990+.

Problems

Problems are included for every chapter except the Review chapter. Perhaps a departure from traditional engineering texts, only a few problems are of the "plug-in" type—plugging different numbers into the same formulas. Also, there are no engineering "application" problems, since they take long to describe and even then simplifications that are almost deceiving have to be made to render them solvable. The problems are not necessarily ordered in terms of difficulty. But they should be considered an integral part of the text and often provide a succinct introduction to new kindred areas.

Some Helpful Hints for Teaching

For a real beginning class such as at the senior level, perhaps all material on multivariate processes can be eschewed, and of course this would mean giving up State-Space models but including both continuous time and discrete-time models albeit one-dimensional. After all, this was the case in the 1960–1970s era. However, it may be necessary to explain at least the notation, since the bulk of the material is in the matrix language, so that the student may specialize to the 1×1 case.

Another extreme is to omit all sections on continuous-time processes—this would mean Chapter 4 primarily; certainly e^{At} can be avoided in this way, including the attendant mathematical complications in defining continuous-time random processes.

It is better not to begin with the Review, since this can be boring. It is best to dip into the material when needed in the later chapters, or to provide a brief recall session on the topic involved as a prelude to its use.

Acknowledgement

It is a pleasure to acknowledge the help of many colleagues in the preparation of this book, including, in particular, Professor Simon Haykin, McMasters University, Canada.

The work was supported in part under NASA Dryden Flight Research Center grant no. NCC 2-374.

A. V. Balakrishnan

Los Angeles, California

REVIEW

This is a review chapter which collects together much (but not all!) of the prerequisite material on Linear Algebra and Probability and related topics—the necessary background. For students already familiar with the subject matter it may help to clarify the notation used (such as the inner product). Of course the review is just that—spotty and terse—and cannot replace a full course on the subject. In particular, we do not list all definitions, emphasizing only concepts and results which are essential in the sequel. Relevant textbooks are listed in the References for fuller treatment. There is no stage at which the student can be certain of proficiency in all the prerequisites. The rule is to simply "plunge in" and seek help only when unable to keep afloat.

LINEAR ALGEBRA

Linear Algebra—especially the algebra of rectangular matrices—is an essential tool, especially as we deal with State Space models. This is a marked departure from the earlier treatments of random processes of the 1950s and 1960s (e.g., reference 6).

Rectangular Matrices

A rectangular matrix—we generally use capital latters to denote them—is a rectangular array of complex numbers—mostly real, but not always. Thus

$$A = \{a_{ij}\}, \qquad i = 1, \ldots, n, j = 1, \ldots, m,$$

1

is an $n \times m$ matrix. When $m = 1$ we refer to it as a "column" vector, or simply vector when no confusion is possible—and a "row" vector if $n = 1$. In this case we generally use lowercase letters: $a, b, \ldots$. We denote by $\mathbf{E}^n$ the space of $n \times 1$ vectors with the usual rules of addition and scalar multiplication. We assume familiarity with rudimentary notions such as "linear independence" and "bases". The rules of addition and scalar multiplication make the class of $n \times m$ matrices equivalent to $\mathbf{E}^{nm}$. We shall use the notation $A + B$ only when A and B are both $n \times m$. If A is an $n \times m$ matrix as in (1.1) we define the "adjoint".

$$A^* = \{\bar{a}_{ji}\}, \qquad j = 1, \ldots, m, i = 1, \ldots, n,$$

which is then an $m \times n$ matrix. It is customary in engineering texts to define a "transpose"

$$A^T = \{a_{ji}\}$$

so that the "adjoint" is then a "conjugate transpose": $\overline{A^T} = A^*$. We shall stay with the term "adjoint," even if the matrices will often be real-valued.

Dot product—Inner Product   "Dot product" of two $n \times 1$ vectors a and b is defined as:

$$a \cdot b = \sum_1^n a_k \bar{b}_k.$$

We prefer the term "inner product" which we define for any two $n \times m$ rectangular matrices A and B:

$$[A, B] = \sum_{i=1}^n \sum_{j=1}^m a_{ij} \bar{b}_{ij}. \tag{1.1}$$

Then

$$\overline{[A, B]} = [B, A]$$

$$[A, A] \geq 0.$$

The *norm* of a rectangular matrix A:

$$\|A\| = \sqrt{[A, A]}.$$

Thus defined,

   (i) $A = 0$ if and only if $\|A\| = 0$,
   (ii) $\|A + B\| \leq \|A\| + \|B\|$ (triangle inequality),
   (iii) Schwarz inequality

$$|[A, B]| \leq \|A\| \|B\|,$$

   (iv) $\|A\| = \|A^*\|$.

Product We can define the product of two rectangular matrixes $A: n \times m$ and $B: m \times p$ as

$$AB = \left\{ \sum_{j=1}^{m} a_{ij} b_{jk} \right\}, \qquad i = 1, \ldots, n, k = 1, \ldots, p, \qquad (1.2)$$

which is thus an $n \times p$ matrix. The multiplication notation will be used only when the matrices are "linked" in this fashion. If $n = m$, the matrix I

$$I = \{ \delta_j^i \}, \qquad i = 1, \ldots, n, j = 1, \ldots, n,$$

is called the identity matrix since

$$IA = A.$$

We shall use

$$I_{n \times n}$$

when the dimension is not obvious from the context; otherwise, simply I.

$$(AB)^* = B^* A^*$$

$$\|AB\| \leq \|A\| \|B\|.$$

If A is $n \times m$ then A^* is $m \times n$ so that

$$AA^* \quad \text{is} \quad n \times n.$$

Indeed, if A and B are $n \times m$, we can define the product

$$AB^*, \quad \text{which is } n \times n.$$

Square Matrices

If A is a square matrix $n = m$, we define the trace (shortened to "Tr"):

$$\text{Tr } A = \sum_{1}^{n} a_{ii} (= \text{sum of diagonal terms}).$$

Note that for $n \times m$ matrices A and B we have

$$\|A\|^2 = \text{Tr } AA^* = \text{Tr } A^*A$$

$$[A, B] = \text{Tr } AB^*$$

$$\text{Tr } AB^* = \text{Tr } B^*A$$

(dimensions not necessarily same on both sides!). The determinant of a square matrix is denoted

$$\det A.$$

For a complex number λ, $\det (\lambda I - A)$ (where I is the identity matrix) is a polynomial of degree n, with the coefficient of $\lambda^n = 1$.

Cayley–Hamilton Theorem

$$A^n + \sum_{k=0}^{n-1} a_k A^k = 0, \tag{1.3}$$

where

$$\det (\lambda I - A) = \lambda^n + \sum_{k=0}^{n-1} a_k \lambda^k.$$

Matrices as Linear Transformations

Let A be a real-valued $n \times m$ matrix. For each x in $\mathbf{E}^m$ we can define (by the rules of multiplication of rectangular matrices)

$$y = Ax$$

mapping $\mathbf{E}^m$ into $\mathbf{E}^n$. This is a linear mapping (transformation):

$$A(\alpha x + \beta z) = \alpha Ax + \beta Az, \qquad x, z \in \mathbf{E}^m, \alpha, \beta \text{ scalars}.$$

Then

$$[Aa, b] = \operatorname{Tr} Aab^* = \operatorname{Tr} b^*Aa = \operatorname{Tr} (A^*b)^*a$$
$$= \operatorname{Tr} a(A^*b)^* = [a, A^*b].$$

Define the "operator norm" of A by

$$\|A\|_0 = \max \|Ax\|, \qquad \|x\| \le 1.$$

Then

$$\|A\|_0 = \max \left(\frac{|[Ax, y]|}{\|x\| \|y\|} \right), \qquad x = \mathbf{E}^m, y \in \mathbf{E}^n$$

$$\|Ax\| \le \|A\|_0 \|x\|$$

$$\|AB\|_0 \le \|A\|_0 \|B\|_0$$

$$\|A + B\|_0 \le \|A\|_0 + \|B\|_0$$

$$\|A\|_0 = \|A^*\|_0, \qquad \|\alpha A\|_0 = |\alpha| \|A^*\|_0$$

$$\|A\|_0 \le \|A\|, \qquad \|A\| \le \sqrt{mn} \|A\|_0.$$

If $A = \{a_{ij}\}$, then

$$|a_{ij}| \leq \|A\|_0 \quad \text{for every } j$$

$$\|A\|_o \leq \sqrt{mn}\left(\max_{i,j} |a_{ij}|\right).$$

The rank of A is defined as the dimension of the "range" space:

$$[Ax, x \in \mathbf{E}^m] \quad \text{in } \mathbf{E}^n.$$

Square Matrices: Eigenvalues

A: $n \times n$ real: (real-valued) matrix maps $\mathbf{E}^n$ into $\mathbf{E}^n$.

The roots of the determinantal equation

$$0 = \det(\lambda I - A) = \lambda^n + \sum_0^{n-1} a_k \lambda^k$$

are called "eigenvalues"; there are thus n of them. Some of them may be complex even if A is real. Hence we have to allow complex-valued scalars. Given any eigenvalue λ_i, we can find a vector e_i (of norm equal to unity) with possibly complex entries such that

$$Ae_i = \lambda_i e_i, \quad \|e_i\| = 1.$$

Every such vector is called an "eigenvector." Eigenvectors corresponding to distinct eigenvalues are linearly independent. A matrix is said to be *simple* if the eigenvectors form a basis for $\mathbf{E}^n$—for instance, when all the eigenvalues are distinct.† Note that if $\{e_i\}$ is a coordinate basis in $\mathbf{E}^n$—the only nonzero component of e_i is the ith which is equal to 1—we have

$$A = \{[Ae_i, e_j]\}, \quad i, j = 1, \ldots, n$$

$$\text{Tr } A = \sum_1^n [Ae_i, e_i]. \tag{1.4}$$

We say that x is orthogonal to y if

$$[x, y] = 0.$$

An "orthonormal" basis is any set of n vectors $\{e_i\}$ in $\mathbf{E}^n$ such that they are

† On a digital computer all eigenvalues may be considered distinct because all numbers are only approximate!

mutually orthogonal and of unit norm, and (1.4) continues to hold for such a basis.

The "spectrum" of a matrix is the set of eigenvalues $\{\lambda_i\}$.

$$r = \max_i |\lambda_i| \tag{1.5}$$

is called the spectral radius. We note that

$$r \leq \|A\|_0.$$

Equality need not hold in general. The eigenvalues of A^* are the conjugates of the eigenvalues of A.

Self-Adjoint Matrices A matrix is "self-adjoint" if it is its own adjoint (and in that case it has to be square). For a self-adjoint matrix A we have

$$\overline{[Ax, x]} = [x, Ax] = [A^*x, x] = [Ax, x]$$

and hence

$$[Ax, x] \text{ is real for every } x \text{ in } \mathbf{E}^n.$$

We already know that the eigenvalues must be real. Eigenvectors corresponding to distinct eigenvalues are orthonormal. The eigenvectors corresponding to the same eigenvalues can be orthonormalized and thus the eigenvectors form an orthonormal basis for $\mathbf{E}^n$. Also,

$$\|A\|_0 = \text{spectral radius}.$$

Let $\{e_i\}$ denote the orthonormalized eigenvectors. Then A has the representation

$$A = \sum_1^n \lambda_i e_i e_i^* \tag{1.6}$$

$$[Ax, y] = \sum_1^n \lambda_i [e_i, x][y, e_i], \qquad x, y \in \mathbf{E}^n.$$

Also

$$\lambda \|x\|^2 \leq [Ax, x] = \mu \|x\|^2, \tag{1.7}$$

where

$$\lambda = \min_i \lambda_i$$

$$\mu = \max_i \mu_i.$$

Orthogonal Projections A self-adjoint matrix P is called an orthogonal projection (or more simply a projection) if

$$P^2 = P.$$

Note that the nonzero eigenvalues are all equal to unity and that

$$\text{rank } P = \text{Tr } P.$$

Covariance Matrices   A self-adjoint matrix A with the property that

$$[Ax, x] \geq 0, \qquad \text{for all } x \text{ in } E^n,$$

is said to be "nonnegative definite"—a self-adjoint, nonnegative definite matrix is called a "covariance" matrix. It follows from (1.6) that the eigenvalues $\{\lambda_i\}$ are nonnegative. Where possible we use the letters R or Λ to denote covariance matrices. For any covariance matrix Λ we can define a positive square root $\sqrt{\Lambda}$ by

$$\sqrt{\Lambda} x = \sum_1^n \sqrt{\lambda_i}\, e_i [e_i, x], \tag{1.8}$$

which is also a covariance matrix. Furthermore,

$$[\Lambda x, y] = [\sqrt{\Lambda} x, \sqrt{\Lambda} y]$$

and hence

$$|[\Lambda x, y]| \leq \|\sqrt{\Lambda} x\|\, \|\sqrt{\Lambda} y\|$$

and since

$$\|\sqrt{\Lambda} x\|^2 = [\Lambda x, x]$$

we have that

$$|[\Lambda x, y]| \leq \sqrt{[\Lambda x, x]}\, \sqrt{[\Lambda y, y]}. \tag{1.8a}$$

Given N^2 matrices $\{\Lambda_{ij}\}$, $i, j = 1, \ldots, N$, such that each has the same dimension $n \times n$. Suppose

$$\Lambda_{ij}^* = \Lambda_{ji}$$

and Λ_{ii} is a covariance matrix for each i. Then the "compound" matrix $(N_n \times N_n)$

$$\Lambda = \{\Lambda_{ij}\}$$

is also a covariance matrix.

If the covariance matrix Λ is nonsingular, then the inverse Λ^{-1} is also a covariance matrix.

Given any two $n \times n$ covariance matrices Λ_1 and Λ_2, we have

$$\Lambda_1 = \{\lambda_{ij}\} \quad \text{and} \quad \Lambda_2 = \{\mu_{ij}\},$$

and the matrix

$$\{\lambda_{ij}\mu_{ij}\}$$

is a covariance matrix. This is a consequence of (1.6). Let us append a brief proof. Let $\{e_k\}$, $\{f_k\}$ denote the orthonormalized eigenvectors of Λ_1 and Λ_2, respectively. Then

$$\Lambda_1 = \sum_1^n \lambda_k e_k e_k^*, \quad \Lambda_1 e_k = \lambda_k e_k, \|e_k\| = 1$$

$$\Lambda_2 = \sum_1^n \mu_k f_k f_k^*, \quad \Delta_2 f_k = \mu_k f_k, \|f_k\| = 1.$$

Then

$$\lambda_{ij} = \sum_1^n \lambda_k e_{k,i} e_{k,j}, \quad \text{where } e_k = \text{col}\{e_{ki}\}, \lambda_k \geq 0,$$

$$\mu_{ij} = \sum_1^n \mu_k f_{k,i} f_{k,j}, \quad \text{where } f_k = \text{col}\{f_{ki}\}, \mu_k \geq 0.$$

Let

$$R = \{\lambda_{ij}\mu_{ij}\}.$$

Then for $\alpha = \text{col}\{a_i\}$ we have

$$[R\alpha, \alpha] = \sum_1^n \sum_1^n a_i \lambda_{ij} \mu_{ij} \bar{a}_j = \sum_{m=1}^n \sum_{m=1}^n \lambda_k \mu_m \left(\sum_1^n \sum_1^n a_i e_{ki} e_{kj} f_{mi} f_{mj} \bar{a}_j \right)$$

$$= \sum_1^n \sum_1^n \lambda_k \mu_m \left| \sum_1^n a_i e_{ki} f_{mi} \right|^2 \geq 0.$$

Given any linearly independent vectors, x_i, $i = 1, \ldots, N$, $x_i \in \mathbf{E}^n$,

$$M = \{[x_i, x_j]\}$$

is an $N \times N$ nonsingular covariance matrix and (1.7) holds for any x in $\mathbf{E}^m$. M is called a "moment" matrix.

Given two $n \times n$ self-adjoint matrices A and B we say

$$A \geq B \tag{1.9}$$

if $(A - B)$ is a covariance matrix. This induces a "(partial) ordering" in the class of self-adjoint matrices: $A \geq A$, $A \geq B$, and $B \geq C \Rightarrow A \geq C$.

Sequences of Matrices

Let $\{A_k\}$ be a sequence of $n \times n$ matrices. We say A_k converges to A, that is

$$A_k \to A,$$

if every component $(a_{ij})_k$ of A_k converges to the component (a_{ij}) of A. In other words,

$$[A_k e_i, e_j] \to [A e_i, e_j],$$

where $\{e_i\}$ is the coordinate basis. We have

$$A_k \to A \quad \text{iff} \quad \|A_k - A\|_0 \to 0$$
$$\text{iff} \quad \|A_k - A\| \to 0.$$

Matrix Exponential We can thus define

$$e^A = \sum_{k=0}^{\infty} \frac{A^k}{k!}.$$

The series on the right converges since

$$\sum_{N}^{N+p} \frac{\|A^k\|}{k!} \leq \sum_{N}^{N+p} \frac{\|A\|^k}{k!} \to 0 \qquad \text{as } N \to \infty,$$

and in particular we have that

$$\|e^A\| \leq \sum_{0}^{\infty} \frac{\|A\|^k}{k!} = e^{\|A\|}.$$

Also for any real number t, $-\infty < t < \infty$, we have

$$\frac{d}{dt} e^{At} = A e^{At} = e^{At} A,$$

from which we have also

$$\int_0^t A e^{As}\, ds = e^{At} - I$$

or

$$\int_0^t e^{As}\, ds = A^{-1}[e^{At} - I]$$

if A is nonsingular. Note that

$$\det e^A = e^{\operatorname{Tr} A}.$$

In fact

$$e^A e_i = e^{\lambda_i} e_i,$$

where

$$A e_i = \lambda_i e_i.$$

PROBABILITY

Distributions; Characteristic Functions

Distributions/Densities   The student is expected to be familiar with the elementary notions such as distributions and density of 1×1 or real-valued random variables. We allow the density functions to contain delta-functions, and will not need to consider more general distributions. Here we list some common probability density functions, denoted generically by $p(\cdot)$.

1. A uniform random variable is one with uniform density:

$$p(x) = \frac{1}{b - a}, \qquad a \leq x \leq b$$

$$= 0, \qquad\qquad \text{otherwise.} \tag{2.1}$$

For example: random phase angle, distributed uniformly in $[0, 2\pi]$.

2. Gaussian density (m, σ):

$$p(x) = \frac{1}{\sqrt{2\pi}\,\sigma} \exp\left[-\frac{1}{2} \frac{(x - m)^2}{\sigma^2} \right] \tag{2.2}$$

mean m, variance σ^2. See Figure 1 for a plot.

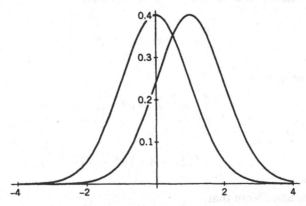

Figure 1. Example 2: Gaussian density. Mean 0, variance 1; mean 1, variance 1.

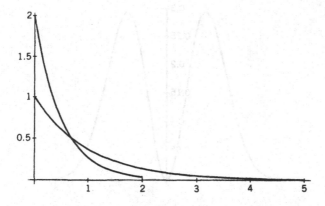

Figure 2. Example 3: Exponential density. $\lambda = 1$ and $\lambda = 2$.

3. Exponential:
$$p(x) = \lambda e^{-\lambda x}, \qquad x > 0, \lambda > 0$$
$$\qquad\quad = 0, \qquad\qquad x < 0. \tag{2.3}$$

Note the discontinuity at the origin. See Figure 2 for a plot.
  4. Rayleigh:
$$p(x) = \frac{x}{\sigma^2} \exp\left(-\frac{x^2}{2\sigma^2}\right), \qquad x > 0$$
$$\qquad\quad = 0, \qquad\qquad\qquad\quad x < 0. \tag{2.4}$$

See Figure 3 for a plot. Distribution function:
$$\Phi(x) = \left[1 - \exp\left(-\frac{x^2}{2\sigma^2}\right)\right], \qquad x \geq 0$$
$$\qquad\quad = 0, \qquad\qquad\qquad\qquad x < 0. \tag{2.5}$$

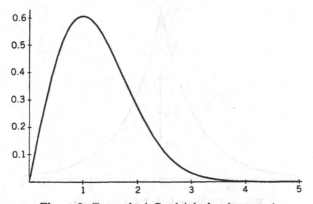

Figure 3. Example 4: Rayleigh density. $\sigma = 1$.

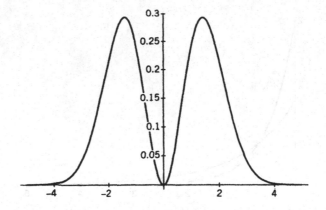

Figure 4. Example 5: Maxwell density. $\sigma = 1$.

5. Maxwell:

$$p(x) = \frac{1}{\sigma^3 \sqrt{2\pi}} x^2 \exp\left(-\frac{x^2}{2\sigma^2}\right).$$ (2.6)

See Figure 4 for a plot.

    6. Laplace:

$$p(x) = \tfrac{1}{2} e^{-|x|}.$$ (2.7)

See Figure 5 for a plot.

    7. Cauchy:

$$p(x) = \frac{1}{\pi} \frac{1}{1 + x^2}.$$ (2.8)

See Figure 6 for a plot.

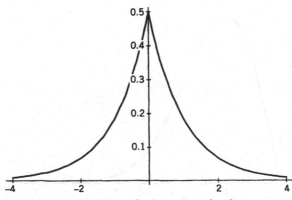

Figure 5. Example 6: Laplace density.

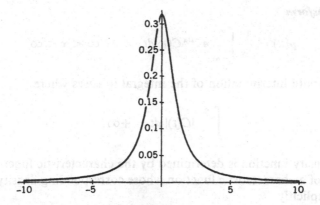

Figure 6. Example 7: Cauchy density.

Characteristic Functions

$$C(t) = \int_{-\infty}^{\infty} e^{itx} p(x)\, dx, \qquad -\infty < t < \infty. \tag{2.9}$$

Moment generation:

$$\left. \frac{d^n}{dt^n} C(t) \right|_{t=0} = (i)^n \int_{-\infty}^{\infty} x^n p(x)\, dx \tag{2.10}$$

if

$$\int_{-\infty}^{\infty} |x|^n p(x)\, dx < \infty.$$

Properties

$$C(0) = 1, \qquad |C(t)| \le 1$$

$$\overline{C(t)} = C(-t)$$

$$C(t) \text{ is continuous in } -\infty < t < \infty$$

$$C(\cdot) \text{ is ``positive definite''}$$

$$\sum_{1}^{N} \sum_{1}^{N} a_i C(t_i - t_j) \bar{a}_j \ge 0$$

for any N, any $\{t_i\}$, $\{a_i\}$.
The matrix

$$\{C(t_i - t_j)\}, \qquad i, j = 1, \dots, N,$$

is self-adjoint and nonnegative definite but not necessarily real-valued.

Inverse Transform

$$p(x) = \frac{1}{2\pi} \int_{-\infty}^{\infty} e^{-itx} C(t)\, dt, \qquad -\infty < x < \infty \qquad (2.11)$$

with appropriate interpretation of the integral in cases where

$$\int_{-\infty}^{\infty} |C(t)|\, dt = +\infty.$$

Thus the density function is determined by the characteristic function. Here is an example of a characteristic function whose corresponding density cannot be expressed explicitly:

$$C(t) = \exp(-|t|^{\alpha}), \qquad 0 < \alpha < 1/2.$$

Of course this is a pathological example in that no moments are finite! The characteristic function of all the seven densities as well as the moments should be calculated by the student as part of the review.

The characteristic function of the Laplace density is given by

$$\frac{1}{2} \int_{-\infty}^{\infty} e^{itx} e^{-|x|}\, dx = \frac{1}{2} \left[\frac{1}{1 + it} + \frac{1}{1 - it} \right] = \frac{1}{1 + t^2}.$$

Taking the inverse transform we obtain

$$\tfrac{1}{2} e^{-|x|} = \frac{1}{2\pi} \int_{-\infty}^{\infty} e^{-itx} \frac{1}{1 + t^2}\, dt,$$

and thus we see the characteristic function of the Cauchy density is given by

$$e^{-|t|}, \qquad -\infty < t < \infty.$$

Note that this function is not differentiable at the origin, and in fact we know that the Cauchy density does not have moments of any order—a pathology.

The characteristic function of the exponential density (as in reference 3),

$$C(t) = \frac{\lambda}{\lambda - it}. \qquad (2.12)$$

Here

$$\int_{-\infty}^{\infty} |C(t)|\, dt = \infty,$$

and the inverse Fourier transform has to be interpreted as a contour or Cauchy integral to yield the density function.

Notation: Expectation   If ζ denotes a random variable we use

$$E[\zeta] \qquad (E = \text{expectation})$$

to denote the first moment

$$\int_{-\infty}^{\infty} xp(x)\,dx.$$

Note that

$$E[f(\zeta)] = \int_{-\infty}^{\infty} f(x)p(x)\,dx.$$

In particular we can write

$$C(t) = E[\exp it\zeta].$$

For ζ Gaussian, mean m, and variance σ^2 we have

$$C(t) = E[\exp(it\zeta)] = e^{imt} \cdot e^{-\sigma^2 t^2/2}. \tag{2.13}$$

Vector/Multidimensional Random Variables

For our purpose we need to work with random variables which are $n \times 1$ vectors—referred to as "n-variate" (or "multivariate" generally) in the statistical literature and as "multidimensional" or "multiple" in engineering. We shall not need to distinguish or call attention to dimension: All variables will be generally taken to be $n \times 1$, where n is not necessarily equal to one—random variables with "range" in E^n, in other words.

The distribution function of the $n \times 1$ random variable ζ is given by

$$\Phi(x_1, \ldots, x_n) = \Pr\left[[\zeta, e_i] \le x_i, i = 1, \ldots, n\right],$$

where $\{e_i\}$ is the coordinate basis in E^n:

$$\int_{-\infty}^{x_1} \cdots \int_{-\infty}^{x_n} p(y_1, \ldots, y_n)\,dy_1 \cdots dy_n.$$

where $p(\cdot)$ is then the density. The simplest such density is the "product density"

$$p(y_1, \ldots, y_n) = p_1(y_1) \cdots p_n(y_n),$$

where $p_i(\cdot)$, $i = 1, \ldots, n$, are one-dimensional density functions.

For us the most important multidimensional density is the Gaussian

$$G(x) = \frac{1}{(\sqrt{2\pi})^n} \frac{1}{|\Lambda|^{1/2}} \exp\left\{-\tfrac{1}{2}[\Lambda^{-1}(x - m), x - m]\right\} \tag{2.14}$$

where m is the mean (an $n \times 1$ vector) and Λ, $n \times n$, is the covariance matrix

$$\Lambda = E[(\zeta - m)(\zeta - m)^*] \qquad (2.15)$$

and it is assumed that Λ is nonsingular.

The characteristic function of an n-dimensional density function $p(\cdot)$ is defined by

$$C(t_1, \ldots, t_n) = \int_{\mathbf{E}^n} e^{i[t, x]} p(x) |dx|, \qquad (2.16)$$

where

$p(x)$ is shorthand for $p(x_1, \ldots, x_n)$

$|dx|$ is shorthand for $dx_1 \cdots dx_n$

t denotes the column vector

$$\begin{vmatrix} t_1 \\ \vdots \\ t_n \end{vmatrix}, \quad t \in \mathbf{E}^n.$$

Again analogous to (2.10) we can use the characteristic function to generate the moments—all the "mixed" moments. For instance, using ζ_i to denote the components of ζ, we obtain

$$E[\zeta_i \zeta_j \zeta_k \zeta_l] = \frac{\partial^4}{\partial t_i \partial t_j \partial t_k \partial t_l} C(t_1, \ldots, t_n) \bigg|_{t=0}. \qquad (2.17)$$

The characteristic function of the n-dimensional Gaussian is given by

$$C(t_1, \ldots, t_n) = e^{i[t, m]} \exp -\tfrac{1}{2}[\Lambda t, t], \qquad (2.18)$$

which is thus well-defined even when Λ is singular. Analogous to (2.11), given the characteristic function, we can take the inverse Fourier transform to yield the density function

$$p(x_1, \ldots, x_n) = \frac{1}{(2\pi)^n} \int_{\mathbf{E}^n} e^{-i[t, x]} C(t) |dt| \qquad (2.19)$$

with the same proviso as in (2.11). Again, we may specify a random variable by specifying the characteristic function in place of the distribution (or density).

The characteristic function corresponding to the product density is the product of the characteristic functions

$$C(t_1, \ldots, t_n) = C_1(t_1) \cdots C_n(t_n).$$

Conversely, if

$$E[e^{i\sum_1^N [t_k, \zeta_k]}]$$

is the product

$$E[e^{i[t_1, \zeta_1]}] \cdots E[e^{i[t_N, \zeta_N]}],$$

then the variables $\{\zeta_i\}$ are independent.

For the Gaussian (2.14), with $m = 0$ and $n \geq 4$, we can use (2.17) to calculate the "four product"

$$
\begin{aligned}
E(x_1 x_2 x_3 x_4) &= \frac{\partial^4}{\partial t_1 \partial t_2 \partial t_3 \partial t_4} \exp \left. -\frac{1}{2} \sum_1^n \sum_1^n \lambda_{ij} t_i t_j \right|_{t=0} \\
&= \lambda_{12}\lambda_{34} + \lambda_{13}\lambda_{24} + \lambda_{14}\lambda_{23},
\end{aligned}
\tag{2.20}
$$

where

$$\Lambda = \{\lambda_{ij}\}.$$

This is best seen by expanding in a Taylor–MacLaurin series

$$\exp\left(-\tfrac{1}{2}[\Lambda t, t]\right) = 1 - \tfrac{1}{2}[\Lambda t, t] + \tfrac{1}{8}[\Lambda t, t]^2 + \cdots \tag{2.21}$$

where

$$[\Lambda t, t]^2 = \sum_1^n \sum_1^n \sum_1^n \sum_1^n \lambda_{ij}\lambda_{kl} t_i t_j t_k t_l,$$

and it allows us to calculate all moments in terms of $\{\lambda_{ij}\}$. It is immediate in particular that all "odd" moments (mixed moments of odd order) are zero.

Functions of Random Variables

Let the random variable η be defined as a function of the variable ζ:

$$\eta = F(\zeta), \qquad \zeta \in E^n, \eta \in E^m.$$

We often need to determine the distribution of η given the distribution of ζ. A general technique for this is to use the characteristic function

$$C_\eta(t) = E[e^{i[t, \eta]}] = E[e^{i[t, F(\zeta)]}], \qquad t \in E^m, \tag{2.22}$$

and hence the distribution of η is specified even if implicitly.

When the function $F(\cdot)$ is sufficiently smooth, we may calculate the density function without having to invert the Fourier transform. Thus for the case $n = m$, suppose that the Jacobian

$$J = \frac{\partial(\eta_1, \ldots, \eta_n)}{\partial(\zeta_1, \ldots, \zeta_n)} \tag{2.23}$$

of the transformation $F(\cdot)$ is nonsingular. Then the density function of η, $p_\eta(\cdot)$, may be calculated by

$$p_\eta(y) = p_\zeta(F^{-1}(y)) \frac{1}{|\det J|}, \tag{2.24}$$

where $p_\zeta(\cdot)$ is the density function of ζ. See p. 293 in reference 5 for more on this. As a simple example, consider the case $n = m = 1$,

$$\eta = A \sin \phi$$

where A is a constant and ϕ is uniform in $[0, 2\pi]$. Then

$$C_\eta(t) = E[e^{it \sin \phi}] = \frac{1}{2\pi} \int_0^{2\pi} e^{itA \sin \phi} \, d\phi = J_0(At), \tag{2.25}$$

where $J_0(\cdot)$ is the zeroth-order Bessel function. We can use either

$$p_\eta(y) = \frac{1}{2\pi} \int_{-\infty}^{\infty} e^{-ity} J_0(At) \, dt$$

or, directly,

$$C_\eta(t) = \frac{1}{2\pi} \left(\int_{-\pi/2}^{\pi/2} + \int_{\pi/2}^{3\pi/2} e^{iAt \sin \phi} \, d\phi \right)$$

$$= \frac{1}{2\pi} \cdot 2 \cdot \int_{-\pi/2}^{\pi/2} e^{iAt \sin \phi} \, d\phi$$

and make the change of variable $y = A \sin \phi$ in the integral

$$\frac{1}{\pi} \int_{-A}^{A} e^{iyt} \frac{1}{\sqrt{A^2 - y^2}} \, dy$$

and hence "read off" that

$$p_\eta(y) = \frac{1}{\sqrt{A^2 - y^2}}, \qquad -A < y < A$$

$$= 0, \qquad\qquad\qquad \text{otherwise.}$$

Note that this density is not continuous at $\pm A$!

We can extend this to the case where A is also random, independent of ϕ, and is Rayleigh with density given by (2.4). Thus

$$\eta = \begin{vmatrix} A \sin \phi \\ A \cos \phi \end{vmatrix} = F(\zeta)$$

$$J = \begin{vmatrix} A \cos \phi & \sin \phi \\ -A \sin \phi & \cos \phi \end{vmatrix}$$

and the determinant is equal to A. To find $F^{-1}(y)$, where

$$y = \begin{vmatrix} y_1 \\ y_2 \end{vmatrix},$$

we need to "solve"

$$y_1 = A \sin \phi, \quad y_2 = A \cos \phi.$$

The solution is

$$A = \sqrt{y_1^2 + y_2^2}, \quad \tan \phi = \frac{y_1}{y_2}.$$

Hence (2.24) yields

$$p_\eta(y) = \frac{1}{2\pi} \left(\frac{\sqrt{y_1^2 + y_2^2}}{\sigma^2} \right) \left[\exp \frac{-(y_1^2 + y_2^2)}{2\sigma^2} \right] \frac{1}{\sqrt{y_1^2 + y_2^2}}$$

$$= \frac{1}{2\pi\sigma^2} \exp \left[\frac{-(y_1^2 + y_2^2)}{2\sigma^2} \right]$$

$$= \left[\frac{1}{\sqrt{2\pi}\,\sigma} \exp \left(\frac{-y_1^2}{2\sigma^2} \right) \right] \left[\frac{1}{\sqrt{2\pi}\,\sigma} \exp \left(\frac{-y_2^2}{2\sigma^2} \right) \right]. \qquad (2.26)$$

Hence η here is zero-mean Gaussian with covariance matrix

$$\begin{vmatrix} \sigma^2 & 0 \\ 0 & \sigma^2 \end{vmatrix}.$$

In particular, $A \sin \phi$ and $A \cos \phi$ are independent $(0, \sigma)$ Gaussians. We can also use

$$C_\eta(t) = \int_0^\infty \frac{1}{2\pi} \int_0^{2\pi} e^{iAt_1 \sin \phi + iAt_2 \cos \phi} \, d\phi \, p(A) \, dA$$

with $p(\cdot)$, given by (2.4),

$$= \int_0^\infty J_0(A\sqrt{t_1^2 + t_2^2}) p(A) \, dA$$

(looking this up in integral tables [15, p. 717])

$$= \exp \frac{-\sigma^2}{2} (t_1^2 + t_2^2).$$

Conversely, suppose x and y are independent zero-mean Gaussians each with the same variance σ^2, and we define

$$r = \sqrt{x^2 + y^2}.$$

Then

$$E[e^{itr}] = \frac{1}{2\pi\sigma^2} \int_{-\infty}^{\infty} \int_{-\infty}^{\infty} e^{it\sqrt{x^2+y^2}} e^{(-(x^2+y^2))/2\sigma^2} \, dx \, dy,$$

and by changing to polar coordinates in the integral, we get

$$= \int_0^{\infty} e^{itr} \left(e^{-r^2/(2\sigma^2)} \frac{r}{\sigma^2} \right) dr$$

from which we can read off the density function (in parentheses).

Linear Transformation of a Gaussian Is Gaussian   Finally let ζ be $n \times 1$ Gaussian with mean m and covariance matrix Λ. Let

$$\eta = A\zeta,$$

where A is an $m \times n$ matrix. Then

$$\begin{aligned}
C_\eta(t) &= E[e^{i[A\zeta, t]}] \\
&= E[e^{i[\zeta, A^*t]}] \\
&= e^{i[m, A^*t]} \exp\left(-\tfrac{1}{2}[\Lambda A^*t, A^*t]\right) \\
&= e^{i[Am, t]} \exp\left(-\tfrac{1}{2}[A\Lambda A^*t, t]\right), \qquad t \in E^m,
\end{aligned}$$

from which we read off that η is Gaussian with mean Am and covariance $A\Lambda A^*$.

Uncorrelated Gaussians Are Independent   Let ζ, η be 1×1 "uncorrelated" Gaussians; that is,

$$E[\zeta\eta] - E[\zeta]E[\eta] = 0.$$

Then the covariance matrix can be written

$$\begin{vmatrix} \sigma_1^2 & 0 \\ 0 & \sigma_2^2 \end{vmatrix}.$$

But

$$C(t_1, t_2) = E[e^{i(t_1\zeta + t_2\eta)}] = e^{it_1 E[\zeta] + it_2 E[\eta]} \exp\left[-\tfrac{1}{2}(\sigma_1^2 t_1^2 + \sigma_2^2 t_2^2)\right]$$
$$= C_1(t_1)C_2(t_2)$$

proving the independence. The extension to the $n \times 1$ case is immediate.

Convergence of Random Variables

Convergence in the Mean   A sequence $\{\zeta_n\}$ of random variables is said to converge to a random variable ζ "in the mean of order p" if

$$E[\|\zeta_n - \zeta\|^p\} \to 0 \qquad \text{as } n \to \infty.$$

We are most interested in the case $p = 2$, and then refer to it as "convergence in the mean square" or "mean square convergence."

Convergence in Probability   We say that the sequence $\{\zeta_n\}$ converges "in probability" to the random variable ζ if, for any $\varepsilon > 0$,

$$\Pr\left[\|\zeta - \zeta_n\| > \varepsilon\right] \to 0$$

as $n \to \infty$. Convergence in the mean square implies convergence in probability by virtue of the Chebyshev inequality:

$$\Pr\left[\|\zeta_n - \zeta\| > \varepsilon\right] \le \frac{1}{\varepsilon^2} E[\|\zeta_n - \zeta\|^2].$$

Sample Path Convergence or Convergence with Probability One   For this notion it is necessary to introduce the sample space Ω of the sequence $\{\zeta_n\}$ of random variables. Thus for each ω in Ω we have a sequence $\{\zeta_n(\omega)\}$, a "sample path" (see also Chapter 1, p. 44). The sequence is said to converge with probability one ("almost surely") if the subset of Ω on which the sequence does *not* converge has probability zero. We refer to this also as "pathwise convergence."

Series   Let $\{\zeta_n\}$ be a sequence of random variables, mutually independent with zero mean, with convariance matrix Λ_n,

$$E[\|\zeta_n\|^2] = \sigma_n^2 = \text{Tr } \Lambda_n.$$

Then the infinite series

$$S = \sum_1^\infty \zeta_k$$

converges in the mean square sense if

$$\sum_1^\infty \sigma_n^2 < \infty.$$

We can prove this readily. For letting S_n denote the nth partial sum:

$$S_n = \sum_1^n \zeta_k.$$

We see that

$$E\|S_{n+N} - S_n\|^2 = \sum_{n+1}^{n+N} \sigma_k^2 \le \sum_{n+1}^\infty \sigma_k^2 \to 0 \qquad \text{as } n \to \infty.$$

If in addition, each ζ_k is Gaussian so is the sum S. Indeed, calculating the characteristic function, we have

$$C(t) = E[e^{i[t,\,S]}] = e^{-1/2\sum_1^\infty [\Lambda_k t,\, t]}$$
$$= e^{-1/2[\Lambda t,\, t]},$$

where

$$\Lambda = \sum_1^\infty \Lambda_k.$$

The series

$$\sum_1^\infty \Lambda_k$$

converges since

$$\text{Tr} \sum_1^N \Lambda_k = \sum_1^N \sigma_k^2$$

converges.

FOURIER SERIES

We collect together elementary facts which we shall need to use concerning Fourier series.

Let $f(\cdot)$ be a 1×1 real- or complex-valued function defined in $[-1/2, 1/2]$ such that

$$\int_{-1/2}^{1/2} |f(\lambda)|^2 \, d\lambda < \infty. \tag{3.1}$$

Then we have the Fourier series expansion

$$f(\lambda) = \sum_{-\infty}^\infty a_n e^{-2\pi i n\lambda}, \qquad -1/2 < \lambda < 1/2, \tag{3.2}$$

where the Fourier coefficients $\{a_n\}$ are given by

$$a_n = \int_{-1/2}^{1/2} f(\lambda)e^{2\pi in\lambda}\, d\lambda.\tag{3.3}$$

If $f(\cdot)$ is real-valued we have

$$a_{-n} = \bar{a}_n,$$

and if $f(\cdot)$ is real-valued and symmetric ($f(\lambda) = f(-\lambda)$) we obtain

$$a_{-n} = a_n$$

and the coefficients are real. Conversely, given $\{a_n\}$ such that

$$\sum_{-\infty}^{\infty} |a_n|^2 < \infty\tag{3.4}$$

we can construct

$$f(\lambda) = \sum_{-\infty}^{\infty} a_n e^{-2\pi in\lambda}, \qquad -1/2 < \lambda < 1/2\tag{3.5}$$

and we may extend the definition to all x, $-\infty < x < \infty$, by making it periodic with period 1. For our purposes, we usually have

$$\sum_{-\infty}^{\infty} |a_n| < \infty,\tag{3.6}$$

which of course implies (3.4). If (3.6) does not hold, the equality in (3.2), (3.5) will need to be interpreted appropriately. See reference 11 for the details.

Parseval's Formula   Let $f(\lambda)$ and $g(\lambda)$ be two functions, $-\infty < \lambda < \infty$, both periodic with period 1 and satisfying (3.1). Then if

$$f(\lambda) = \sum_{-\infty}^{\infty} a_n e^{-2\pi in\lambda}, \qquad a_n = \int_{-1/2}^{1/2} f(\lambda)e^{2\pi in\lambda}\, d\lambda,$$

$$g(\lambda) = \sum_{-\infty}^{\infty} b_n e^{-2\pi in\lambda}, \qquad b_n = \int_{-1/2}^{1/2} g(\lambda)e^{2\pi in\lambda}\, d\lambda$$

$$\int_{-1/2}^{1/2} f(\lambda)\overline{g(\lambda)}\, d\lambda = \sum_{-\infty}^{\infty} a_n \bar{b}_n.\tag{3.7}$$

In particular we have Parseval's theorem:

$$\int_{-1/2}^{1/2} |f(\lambda)|^2 \, d\lambda = \sum_{-\infty}^{\infty} |a_n|^2, \tag{3.8}$$

from which actually we can deduce (3.7).

Convolution Theorem With $f(\cdot)$ and $g(\cdot)$ as above, define

$$c_n = \sum_{-\infty}^{\infty} a_{n-m} b_m = \sum_{-\infty}^{\infty} b_{n-m} a_m. \tag{3.9}$$

The $\{c_n\}$ are the Fourier coefficients of the function

$$h(\lambda) = f(\lambda)g(\lambda), \quad -1/2 < \lambda < 1/2, \tag{3.10}$$

which can be deduced from (3.7). Also

$$d_n = a_n b_n \tag{3.11}$$

are the Fourier coefficients of the "convolution"

$$h(t) = \int_{-1/2}^{1/2} f(t-s)g(s) \, ds, \quad -1/2 < t < 1/2. \tag{3.12}$$

(Remember that $f(\cdot)$ and $g(\cdot)$ are periodic with period 1.)

Extension to the Multidimensional Case We can easily extend these to matrix-valued functions. Thus let $f(\lambda)$ and $g(\lambda)$ be $n \times m$ functions periodic in $-\infty < \lambda < \infty$ with period 1, and satisfy

$$\int_{-1/2}^{1/2} \|f(\lambda)\|^2 \, d\lambda < \infty, \quad \int_{-1/2}^{1/2} \|g(\lambda)\|^2 \, d\lambda < \infty.$$

The Fourier "coefficients" (3.3) are also $n \times m$ matrices. Parseval's theorem then becomes

$$\int_{-1/2}^{1/2} \|f(\lambda)\|^2 \, d\lambda = \sum_{-\infty}^{\infty} \|a_n\|^2 \tag{3.13}$$

$$\int_{-1/2}^{1/2} f(\lambda)g(\lambda)^* \, d\lambda = \sum_{-\infty}^{\infty} a_n b_n^*.$$

The convolution theorem reads

$$c_k = \sum_{-\infty}^{\infty} a_m b_{m-k}^* = \sum_{-\infty}^{\infty} a_{m+k} b_m^*, \tag{3.14}$$

where the c_k, which are $n \times n$ matrices, are the Fourier coefficients of the $n \times n$ function

$$h(\lambda) = f(\lambda)g(\lambda)^*, \qquad -1/2 < \lambda < 1/2. \tag{3.15}$$

Similarly,

$$d_k = a_k b_k^* \tag{3.16}$$

are the Fourier "coefficients" of

$$h(t) = \int_{-1/2}^{1/2} f(s)g(s-t)^* \, ds, \qquad -1/2 < \lambda < 1/2 \tag{3.17}$$

$$= \int_{-1/2}^{1/2} f(s+t)g(s)^* \, ds, \qquad -1/2 < \lambda < 1/2, \tag{3.18}$$

remembering that the functions have period 1.

FOURIER TRANSFORMS

Let $W(t)$, $-\infty < t < \infty$, be a real- or complex-valued function such that

$$\int_{-\infty}^{\infty} |W(t)|^2 \, dt < \infty. \tag{4.1}$$

Then we can define the Fourier transform (note the appearance of 2π and change of sign—this is more convenient for the needs of system theory)

$$\psi(f) = \int_{-\infty}^{\infty} e^{-2\pi i f t} W(t) \, dt, \qquad -\infty < f < \infty \tag{4.2}$$

(to be interpreted appropriately if

$$\int_{-\infty}^{\infty} |W(t)| \, dt = \infty).$$

Note that

$$\overline{\psi(f)} = \psi(-f)$$

if $W(\cdot)$ is real-valued, and

$$\psi(f) \text{ is real if } W(t) = W(-t).$$

Inverse Transform We can "invert" (4.2) to yield

$$W(t) = \int_{-\infty}^{\infty} e^{2\pi i\lambda t} \psi(\lambda)\, d\lambda, \qquad -\infty < t < \infty, \tag{4.3}$$

to be interpreted appropriately if

$$\int_{-\infty}^{\infty} |\psi(\lambda)|\, d\lambda = \infty.$$

Parseval's Theorem Parseval's theorem is defined as

$$\int_{-\infty}^{\infty} |\psi(\lambda)|^2\, d\lambda = \int_{-\infty}^{\infty} |W(t)|^2\, dt. \tag{4.4}$$

Convolution Theorem Also, if $H(\cdot)$ is another function also satisfying (4.1), and we define

$$\phi(f) = \int_{-\infty}^{\infty} e^{-2\pi i t f} H(t)\, dt, \qquad -\infty < f < \infty,$$

the Fourier transform of the convolution

$$h(t) = \int_{-\infty}^{\infty} F(t-s)H(s)\, ds = \int_{-\infty}^{\infty} F(s)H(t-s)\, ds, \qquad -\infty < t < \infty, \tag{4.5}$$

is the product of the Fourier transforms

$$\psi(\lambda)\phi(\lambda). \tag{4.6}$$

Also, Parseval's theorem generalizes to

$$\int_{-\infty}^{\infty} \psi(\lambda)\overline{\phi(\lambda)}\, d\lambda = \int_{-\infty}^{\infty} W(t)\overline{H(t)}\, dt. \tag{4.7}$$

Multidimensional Domains First we consider the case where the domain is multidimensional, for use in probability. Thus let $F(t)$, $t \in \mathbf{E}^n$, be a 1×1 real- or complex-valued function such that

$$\int_{\mathbf{E}^n} |F(t)|^2\, |dt| < \infty. \tag{4.8}$$

Then the Fourier transform is defined by

$$\hat{F}(\lambda) = \int_{\mathbf{E}^n} e^{i[t, \lambda]} F(t) \, |dt|, \qquad \lambda \in \mathbf{E}^n, \tag{4.9}$$

and the inverse Fourier transform is

$$F(t) = \frac{1}{(2\pi)^n} \int_{\mathbf{E}^n} e^{-i[t, \lambda]} \hat{F}(\lambda) \, |d\lambda|, \qquad t \in \mathbf{E}^n. \tag{4.10}$$

Parseval's theorem extends readily and so does the convolution theorem.

Multidimensional Functions Next we need to consider the case where the functions are $n \times m$ rectangular matrices but the domain is $\mathbf{E}^1$. The results here parallel those for Fourier series.

Thus let $W(t)$ and $H(t)$ denote $n \times m$ matrix functions, $-\infty < t < \infty$, such that

$$\int_{-\infty}^{\infty} \|W(t)\|^2 \, dt < \infty, \qquad \int_{-\infty}^{\infty} \|H(t)\|^2 \, dt < \infty, \tag{4.11}$$

or, equivalently, we may use operator norms:

$$\int_{-\infty}^{\infty} \|W(t)\|_0^2 \, dt < \infty.$$

We define the Fourier transform by

$$\psi(f) = \int_{-\infty}^{\infty} e^{-2\pi i f t} W(t) \, dt, \qquad -\infty < f < \infty,$$

$$\phi(f) = \int_{-\infty}^{\infty} e^{-2\pi i f t} H(t) \, dt, \qquad -\infty < f < \infty.$$

Note that

$$\psi(f)^* = \psi(-f)$$

if $W(\cdot)$ is self-adjoint. Then the inverse transform yields

$$W(t) = \int_{-\infty}^{\infty} e^{2\pi i f t} \psi(f) \, df, \qquad -\infty < t < \infty,$$

with the same provisos for interpretation if

$$\int_{-\infty}^{\infty} \|\psi(f)\| \, df = \infty.$$

The basic identity then is

$$\int_{-\infty}^{\infty} W(t)H(t)^* \, dt = \int_{-\infty}^{\infty} \psi(f)\phi(f)^* \, df. \tag{4.12}$$

From (4.2), setting $W(\cdot) = H(\cdot)$ we have Parseval's theorem:

$$\int_{-\infty}^{\infty} W(t)W(t)^* \, dt = \int_{-\infty}^{\infty} \psi(f)\psi(f)^* \, df. \tag{4.13}$$

The convolution theorem now takes the form

$$\int_{-\infty}^{\infty} W(t)H(t-s)^* \, dt = \int_{-\infty}^{\infty} e^{2\pi i f s}\psi(f)\phi(f)^* \, df, \qquad -\infty < s < \infty. \tag{4.14}$$

In particular,

$$\int_{-\infty}^{\infty} W(t)W(t-s)^* \, dt = \int_{-\infty}^{\infty} e^{2\pi i f s}\psi(f)\psi(f)^* \, df, \qquad -\infty < s < \infty \tag{4.15}$$

$$= \int_{-\infty}^{\infty} W(t+s)W(t)^* \, dt, \qquad -\infty < s < \infty. \tag{4.16}$$

SOME FREQUENTLY USED INEQUALITIES

Schwarz Inequality

$$\left| \sum_1^N a_i b_i \right| \le \sum_1^N |a_i b_i| \le \left(\sqrt{\sum_1^N |a_i|^2} \right)\left(\sqrt{\sum_1^N |b_i|^2} \right) \tag{5.1}$$

$$\int_{E^n} |f(t)g(t)| \, d|t| \le \left(\sqrt{\int_{E^n} |f(t)|^2 \, d|t|} \right)\left(\sqrt{\int_{E^n} |g(t)|^2 \, d|t|} \right). \tag{5.2}$$

Let $p(\cdot)$ denote a probability density. Then

$$\int_{E^n} |f(t)g(t)|p(t) \, d|t| \le \left(\sqrt{\int_{E^n} |f(t)|^2 p(t) \, d|t|} \right)\left(\sqrt{\int_{E^n} |g(t)|^2 p(t) \, d|t|} \right). \tag{5.3}$$

In particular, if ζ and η are $n \times m$ random variables, we obtain

$$|E[\zeta, \eta]| \le E|[\zeta, \eta]|$$

$$\le \sqrt{E[\|\zeta\|^2]} \sqrt{E[\|\eta\|^2]} \tag{5.4}$$

$$E[\|\zeta\|] \le \sqrt{E[\|\zeta\|^2]}. \tag{5.5}$$

Hölder Inequality Let

$$\| f \|^2 = \int_{\mathbf{E}^n} |f(t)|^2 \, d|t|$$

$$\| g \|^2 = \int_{\mathbf{E}^n} |g(t)|^2 \, d|t|.$$

Then

$$\| f + g \| \le \| f \| + \| g \|. \tag{5.6}$$

Let ζ and η be $n \times m$ random variables. Then

$$(E[\| \zeta + \eta \|^2])^{1/2} \le \sqrt{E[\| \zeta \|^2]} + \sqrt{E[\| \eta \|^2]}. \tag{5.7}$$

Chebyshev Inequality For any 1×1 random variable ζ, the probability is given by

$$\Pr\,[|\zeta| > \varepsilon] \le \frac{E[|\zeta|^2]}{\varepsilon^2}. \tag{5.8}$$

We can do better for a standard $(0, 1)$ Gaussian:

$$\Pr\,[|\zeta| > \varepsilon] \le \frac{2}{\varepsilon} \frac{\exp\,(-\varepsilon^2/2)}{\sqrt{2\pi}}. \tag{5.9}$$

In particular,

$$\Pr\,[|\zeta| > 3] < 3 \cdot 10^{-3},$$

which is a useful estimate in practice.

ALMOST PERIODIC SEQUENCES

A sequence $\{S_n\}$, $n \in \mathbf{I}$, defined by

$$S_n = \operatorname{Re} \sum_{k=1}^{M} A_k e^{2\pi i \lambda_k n}, \qquad -1/2 \le \lambda_k \le 1/2. \tag{6.1}$$

where the $\{A_k\}$ may be complex, is not necessarily periodic in n, for arbitrary $\{\lambda_k\}$. We call such a sequence "almost periodic." We define its "autocorrelation" function by

$$R(m) = \lim_{N \to \infty} \frac{1}{N} \sum_{n=1}^{N} S_n S_{n+m}, \qquad m \in \mathbf{I} \tag{6.2}$$

$$= \sum_{1}^{M} \frac{|A_k|^2}{2} \cos 2\pi \lambda_k m$$

as is readily verified. We see that the autocorrelation function is well defined even if M is infinite so long as

$$\sum_1^\infty |A_k|^2 < \infty.$$

We note that $R(m)$ can be represented as

$$R(m) = \int_{-1/2}^{1/2} e^{2\pi i\lambda m} p(\lambda) \, d\lambda, \qquad (6.3)$$

where

$$p(\lambda) = \frac{1}{2}\left(\sum_1^N \frac{|A_k|^2}{2}\left(\delta(\lambda - \lambda_k) + \delta(\lambda + \lambda_k)\right)\right), \qquad -1/2 \le \lambda_k \le 1/2, \quad (6.4)$$

where $\delta(\cdot)$ denotes the delta function.

GRADIENT

Let $f(\cdot)$ denote a function defined on R^n with range in R^m. The gradient of $f(\cdot)$ (we add the phrase "with respect to x" if unclear from context) is a generalization of the notion of derivative and is an $m \times n$ matrix function, denoted ∇f, and sometimes $\nabla_x f$, defined by

$$\nabla f(x)h = \frac{d}{d\lambda} f(x + \lambda h)\bigg|_{\lambda=0}, \qquad h \in R^m. \qquad (7.1)$$

As an illustration, consider the simplest such function, the "quadratic function":

$$f(x) = [Rx, x] - 2[z, x], \qquad x \in R^m, \qquad (7.2)$$

where R is a covariance matrix and z is a fixed element of R^m. To find the gradient we take the derivative

$$\frac{d}{d\lambda} f(x + \lambda h) = \frac{d}{d\lambda} \{[R(x + \lambda h), (x + \lambda h)] - 2[z, x + \lambda h]\}$$

$$= 2[R(x + \lambda h), h] - 2[z, h].$$

Hence

$$\nabla f(x)h = 2[Rx - z, h] = 2(Rx - z)^*h$$

or

$$\nabla f(x) = 2(Rx - z)^*.$$

being in this case a $1 \times m$ matrix.

The gradient is most often used in finding maxima or minima of numerical valued functions, since the gradient must vanish at a point of maximum or minimum. Thus in the case of the illustrative example above, the function $f(\cdot)$ has a maximum or minimum at

$$\nabla f(x) = 0$$

or at the point x satisfying

$$Rx = z.$$

PROBLEMS

Note: These problems are *not* necessarily exhaustive or representative!

1. Find the first and second moments of the seven density functions on pp. 10–13. Plot the Gaussian density with the same mean and variance on the same graph in each case.

2. Find all the moments of a Gaussian with zero mean and variance σ^2.

3. Show that the variance of the sum of independent variables is the sum of the variances.

4. Show that for a one-dimensional random variable ζ we obtain

$$E[|\zeta|]^2 \leq E[\zeta^2].$$

5. Given the second moment matrix

$$\begin{vmatrix} 1 & -1/2 & -1/2 \\ -1/2 & 1 & -1/2 \\ -1/2 & -1/2 & 1 \end{vmatrix}$$

write down the corresponding 3×1 Gaussian (zero-mean) characteristic function. What about the density function?

6. Show that the characteristic function of the uniform density

$$p(x) = 1, \qquad |x| \leq 1/2$$
$$= 0, \qquad |x| > 1/2$$

is given by

$$C(t) = \frac{\sin t/2}{t/2}.$$

Hence show that

$$\lim_{L \to \infty} \int_0^L \frac{\sin t}{t}\, dt = \frac{\pi}{2}.$$

Plot the left side as a function of L, $0 \le L < \infty$. Use Parseval's theorem to calculate:

$$\int_0^\infty \left(\frac{\sin t}{t}\right)^2 dt.$$

7. Find the density function corresponding to the characteristic function:

$$C(t) = \frac{1}{\lambda + it}, \qquad -\infty < t < \infty, \lambda > 0.$$

8. Let z denote a complex number. Calculate

$$\int_{-\infty}^\infty e^{zt} \frac{1}{\sqrt{2\pi}} e^{-t^2/2} dt.$$

9. Let ϕ denote an angle in radians uniformly distributed between 0 and 2π. Show that the variables

$$\cos \phi \quad \text{and} \quad \sin \phi$$

are uncorrelated. Are they independent?

10. Let ζ be a nonnegative random variable such that $\log \zeta$ is Gaussian with zero mean and unit variance. Calculate all the moments of ζ.

11. Let $\{\zeta_n\}$ be a sequence of real-valued random variables which converge in the mean square to the variable ζ. Show that

$$E[|\zeta_n - \zeta|] \to 0$$

and that

$$E[\zeta_n] \to E[\zeta]$$
$$E[\zeta_n^2] \to E[\zeta^2].$$

Show that if ζ_n is Gaussian for every n, then ζ is also Gaussian.
   Hint: For the last part we obtain

$$E[|e^{it\zeta_n} - e^{it\zeta}|^2] = E[|1 - e^{it(\zeta - \zeta_n)}|^2]$$

$$= 4E\left[\sin^2 \frac{t(\zeta - \zeta_n)}{2}\right] \le t^2 E[(\zeta - \zeta_n)^2]$$

$$\to 0 \qquad \text{as } n \to \infty.$$

12. Let A be Rayleigh with

$$E[A^2] = 2\sigma^2.$$

Calculate the density function of $\log A$ and show that it is not Gaussian.

13. Show that if λ_k, $k = 1, \ldots, N$, are real numbers, then

$$\sum_1^N \sum_1^N a_k e^{i(\lambda_k - \lambda_j)} \bar{a}_j = \left| \sum_1^N a_k e^{i\lambda_k} \right|^2 \geq 0$$

for any possible complex numbers a_k, $k = 1, \ldots, N$.

14. Let M and Q be $n \times n$ covariance matrices, possibly singular. Let I denote the $n \times n$ Identity matrix. Show that

$$(I + MQ)$$

is nonsingular.

15. Let C be a real-valued $m \times n$ matrix. Show that the $(m + n) \times (m + n)$ matrix.

$$\begin{vmatrix} O_{m \times m} & iC \\ -iC^* & O_{n \times n} \end{vmatrix}$$

is *not* a covariance matrix, unless C is zero.

16. Let

$$P(f) = \begin{vmatrix} 1 & e^{-2\pi i f \Delta} \\ e^{2\pi i f \Delta} & 1 \end{vmatrix}, \qquad -\infty < f < \infty, \ \Delta \text{ nonnegative.}$$

Find the eigenvalues and eigenvectors of $P(f)$ for each f. Show that

$$\sqrt{P(f)} = \frac{\sqrt{2} P(f)}{2}.$$

17. Show that

$$F_N(\lambda) = \frac{1}{\sqrt{1 + N}} \left(\sum_0^N e^{2\pi i k \lambda} \right), \qquad -1/2 \leq \lambda \leq 1/2$$

is such that

$$|F_N(\lambda)|^2 \to 0, \qquad \lambda \neq 0$$

$$|F_N(0)|^2 = (N + 1)$$

$$\int_{-1/2}^{1/2} |F_N(\lambda)|^2 \, d\lambda = 1$$

$$\lim_{N \to \infty} \int_{-1/2}^{1/2} |F_N(\lambda)|^2 h(\lambda) \, d\lambda = h(0)$$

for any continuous function $h(\cdot)$. Thus $\{|F_N(\lambda)|^2\}$ is a "δ-function sequence" over $-1/2 \le \lambda \le 1/2$. Calculate

$$\int_{-1/2}^{1/2} \int_{-1/2}^{1/2} \int_{-1/2}^{1/2} \int_{-1/2}^{1/2} F_N(\lambda_1 + \lambda_3)\overline{F_N(\lambda_2 + \lambda_4)}$$

$$\cdot F_N(\lambda_1 - \lambda_3)\overline{F_N(\lambda_2 - \lambda_4)}\, d\lambda_1\, d\lambda_2\, d\lambda_3\, d\lambda_4.$$

NOTES AND COMMENTS

This chapter should cover all the mathematics used without further explanation in the ensuing chapters, although surely here and there still lurks unnoticed material not included here (not unlike the "help" commands in computer software). The list of references is not intended to be exhaustive, but hopefully it is representative, here as elsewhere.

There seems to be no accepted universal notation for denoting inner product. Various books [1–4] use

$$\langle \ \rangle, \quad (\ ), \quad (\mid), \quad \text{etc.}$$

We like none of these for various reasons and have adopted [,]. Also we shall use $\mathbf{E}^n$ to denote n-dimensional vector space over complex scalars and $\mathbf{R}^n$ to denote the real subspace or the Eucliean space.

REFERENCES

Linear Algebra: Pure Mathematics Classic Treatise
1. P. H. Halmos. *Finite-Dimensional Vector Spaces*. Van Nostrand, 1958.

Recent Publications on Linear Algebra
2. P. Lowman and J. Stokes. *Introduction to Linear Algebra*. Harcourt Brace Jovanovich, 1991.
3. S. K. Berberian. *Linear Algebra*. Oxford Science Publications, 1992.
4. L. W. Johnson, R. D. Riess, and J. T. Arnold. *Introduction to Linear Algebra*. Addison-Wesley, 1993.

Probability: Stochastic Processes Pure Mathematics Classic Treatises
5. H. Cramer. *Mathematical Methods of Statistics*. Princeton University Press, 1946.
6. M. Parzen. *Stochastic Processes*. Holden-Day, 1962.

Recent Publications on Probability
7. D. Z. Peebles. *Probability, Random Variables and Random Signal Principles*. McGraw-Hill, 1991.
8. S. M. Ross. *Introduction to Probability and Statistics for Engineers and Scientists*. John Wiley and Sons, 1987.

9. R. W. Hamming. *The Art of Probability.* Addison-Wesley, 1991.

Fourier Series and Integrals: Mathematical Classic Treatises

10. E. C. Titchmarsh. *Introduction to the Theory of Fourier Integrals.* Clarendon Press, 1948.

11. R. V. Churchill. *Fourier Series and Boundary Value Problems.* McGraw-Hill, 1963.

Recent Publications on Systems and Signals

12. R. E. Ziemer, W. H. Tranter, and D. R. Fanin. *Signals and Systems: Continuous and Discrete.* Macmillan, 1993.

13. R. N. Bracewell. *The Fourier Transform and Its Applications.* McGraw-Hill, 1986.

14. N. Levan. *Signals and Systems.* Optimization Software Publications, 1992.

A Comprehensive Classic Treatise on Integral Tables

15. I. S. Gradshkeyn and I. M. Ryzhik. *Table of Integrals, Series and Products.* Academic Press, 1965.

RANDOM PROCESSES: BASIC CONCEPTS, PROPERTIES

1.1 RANDOM PROCESSES: BASIC DEFINITIONS, CONCEPTS

We begin with a formal definition:

Definition 1.1: A *random process* is an indexed family of random variables.

The catch in this definition is that we need to define first what a random variable is. What the student is familiar with is what we may call the "vague" definition (even if it is also the usual "dictionary" definition): A random variable is a variable whose values (real numbers, for our purposes) are taken randomly—in accord with a distribution law. This definition will have to suffice for us here since the precise definition will require a level of mathematics (measure theory) not assumed.† Thus, specifying a random variable is synonymous with specifying the distribution. Hence for us specifying a random process means specifying the joint distribution of variables corresponding to any finite subset of the index set T. As we shall see below in the examples, such a specification need not be explicit. Although by definition the index set can be quite arbitrary, the term "random process" is usually employed only when T is not finite, and in particular when $T = \mathbf{R}^1$, the set of all real numbers with the usual arithmetic operations, denoting a "time" variable. When $T = \mathbf{I}$, the set of all integers positive and negative, we use the term "time series" or simply "random sequence." We may regard the index set as a "parameter" set, and hence the often-used names "discrete-parameter process" for random sequences and

---

† It may be comforting to the student to know that most of the original workers in the area (e.g., references 2 and 3) were innocent of it as well and yet made great advances in the engineering applications.

"continuous-parameter process" when T is a "continuum" namely $\mathbf{R}^1$ (or a sub-interval thereof).

The term "process" is intended to suggest "evolution." However, T can well indicate a spatial variable where there is no arrow for "future." The case $T = \mathbf{R}^n$, $n > 1$, is distinguished by the name "random field"; and $n = 3$ finds application, for instance, in modeling fluid turbulence [9] and in geophysical data generally.

We shall be concerned only with $T = \mathbf{R}^1$ and $T = \mathbf{I}$, except in Chapter 8. Following custom for functional notation, our notation for a continuous-parameter process will be

$$x(t), \qquad -\infty < t < \infty$$

or an interval specified otherwise, and the sequence notation will be

$$x_n, \qquad n \in \mathbf{I}$$

or a subset specified otherwise, for random sequences.

Let us examine some illustrative examples (not chosen for practical significance.

1.1.1 Examples: Continuous-Parameter Processes

Example 1.1.1.

$$x(t) = A \sin (wt + \phi) \qquad (1.1.1)$$

where the amplitude A and (angular) frequency w are fixed, and ϕ is a random variable uniformly distributed in $(0, 2\pi)$. Thus we have a sine wave with random phase. Consistent with our definition of a random process, we can specify finite-dimensional joint distributions. It is convenient in this example to work instead with characteristic functions. Thus the characteristic function of $x(t_k)$, $k = 1, \ldots, N$, is

$$C_N(\lambda_1, \ldots, \lambda_N) = E\left(\exp \left[i \sum_1^N \lambda_k x(t_k) \right] \right) = E(\exp [iR \sin (\phi + \theta)]) \quad (1.1.2)$$

where

$$R = A \sqrt{\left(\sum_1^N \lambda_k \cos wt_k \right)^2 + \left(\sum_1^N \lambda_k \sin wt_k \right)^2}$$

$$= A \sqrt{\sum_1^N \sum_1^N \lambda_k \lambda_j \cos w(t_k - t_j)} \qquad (1.1.3)$$

$$\theta = \tan^{-1} \left(\frac{\sum_1^N \lambda_k \sin wt_k}{\sum_1^N \lambda_k \cos wt_k} \right).$$

Hence

$$C(\lambda_1, \ldots, \lambda_N) = \frac{1}{2\pi} \int_0^{2\pi} \exp\left[iR \sin(\phi + \theta)\right] d\phi$$

$$= \frac{1}{2\pi} \int_0^{2\pi} \exp\left(iR \sin \phi\right) d\phi$$

$$= J_0(R), \tag{1.1.4}$$

and J_0 is the Bessel function of order zero.

Example 1.1.2. We may generalize Example 1.1.1 to randomize the amplitude as well, still keeping the frequency fixed. Our particular choice of the distribution of A is the Rayleigh (cf. pages 10–15); the density function is given by

$$p(x) = \frac{x}{\sigma^2} \exp\left(\frac{-x^2}{2\sigma^2}\right), \qquad x \geq 0$$

$$= 0, \qquad x < 0.$$

We also assume that A is independent of ϕ. Thus

$$x(t) = A \sin(wt + \phi),$$

which is convenient to rewrite as

$$= x \cos wt + y \sin wt,$$

where

$$x = A \sin \phi, \qquad y = A \cos \phi.$$

As we have seen on pages 17–20, these are zero-mean independent Gaussians with the same variance σ^2. It is easy to calculate the characteristic functions:

$$C_N(\lambda_1, \ldots, \lambda_N) = E\left\{\exp\left[i \sum_1^N \lambda_k x(t_k)\right]\right\}$$

$$= E\left\{\exp\left[i\left(x \sum_1^N \lambda_k \cos wt_k\right)\right]\right\} E\left[\exp i\left(y \sum_1^N \lambda_k \sin wt_k\right)\right]$$

$$= \exp\left[\frac{-\sigma^2}{2} \sum_1^N \sum_1^N \lambda_j \lambda_k \cos w(t_k - t_j)\right], \tag{1.1.5}$$

showing in particular that the joint density is Gaussian. We call a process Gaussian if all joint distributions are Gaussian. Thus this is a Gaussian process. On the other hand, the process in Example 1.1.1 is *not* Gaussian.

Example 1.1.3. We may generalize this process to

$$x(t) = \sum_1^N A_k \sin(w_k t + \phi_k), \tag{1.1.6}$$

where the w_k are fixed frequencies but the A_k and ϕ_k are mutually independent random variables and the density function of A_k is given by

$$p(x) = \frac{x}{\sigma_k^2} \exp\left(\frac{-x^2}{2\sigma_k^2}\right), \qquad x > 0$$

$$= 0, \qquad\qquad\qquad x < 0$$

and each ϕ_k is uniform in $(0, 2\pi)$. We can rewrite (1.1.6) as

$$x(t) = \sum_1^N (x_k \cos w_k t + y_k \sin w_k t), \qquad x_k = A_k \sin \phi_k, y_k = A_k \cos \phi_k,$$

where the terms are independent and each is Gaussian. Hence $x(\cdot)$ is a Gaussian process. Note that if the $\{w_k\}$ are all rational numbers, then (1.1.6) will still be periodic in t, however large the period.

Example 1.1.4. As a final generalization of these examples, we may define

$$x(t) = F(t, \omega), \tag{1.1.7}$$

where ω is a multivariate random variable. In Example 1.1.3, for instance,

$$\omega = \{A_k, \phi_k\}, \qquad k = 1, \ldots, N.$$

Note that the process is specified as soon as the (finite-dimensional) random variable ω is. We shall call such processes "trivial" processes (as opposed to "nontrivial" where ω has to be infinite-dimensional, as in modeling "naturally occurring" processes).

Example 1.1.5. Consider the process $y(t)$, $t \geq 0$, defined by the differential equation

$$\frac{dy}{dt} + \mu y(t) = x(t), \qquad \mu > 0, \tag{1.1.8}$$

where $x(\cdot)$ is the process defined by (1.1.1). Recall that (1.1.8) has the solution

$$y(t) = e^{-\mu t} y(0) + \int_0^t e^{-\mu(t-s)} x(s)\, ds.$$

In our case we have

$$y(t) = e^{-\mu t}y(0) + \int_0^t e^{-\mu(t-s)}A \sin(ws + \phi)\, ds. \qquad (1.1.9)$$

The first term is a function of t and $y(0)$, and the second term a function of t and ϕ, and hence we have the form

$$y(t) = F(t, y(0), \phi), \qquad t \geq 0.$$

For each choice of $y(0)$ and ϕ we obtain a "trajectory" in t, $t \geq 0$. If $y(0)$ is taken to be a random variable, then the joint distribution $y(0)$ and ϕ will need to be specified. In most models they are taken to be independent, but this is not necessary to define the process. Note that the process $y(\cdot)$ is a trivial process, with

$$\omega = \{y(0), \phi\}.$$

If $x(\cdot)$ is defined as in Example 1.1.2 or 1.1.3 and we take $y(0)$ to be Gaussian and independent of the process $x(\cdot)$—that is, $y(0)$ is independent of $x(t_1), \ldots, x(t_N)$ for arbitrary choices $\{t_i\}$ and N (and in this case it is enough if $y(0)$ is independent of $\{A, \phi\}$ or $\{A_k, \phi_k\}$ respectively)—we see that the process $y(\cdot)$ is actually Gaussian. We note that we can carry out the integration in (1.1.9) for these examples. Thus from Example 1.1.3 we have

$$y(t) = e^{-\mu t}y(0) + \sum_1^N (x_k f_k(t) + y_k g_k(t)), \qquad (1.1.10)$$

where

$$x_k = A_k \cos \phi_k, \qquad y_k = A_k \sin \phi_k$$

$$f_k(t) = \frac{\mu \sin w_k t - w_k \cos w_k t + w_k e^{-\mu t}}{\mu^2 + w_k^2}$$

$$g_k(t) = \frac{\mu \cos w_k t + w_k \sin w_k t - \mu e^{-\mu t}}{\mu^2 + w_k^2}.$$

Example 1.1.6. A nontrivial process. Defining a nontrivial process takes more doing. We shall concoct a Gaussian process since Gaussian distributions are relatively easy to specify. Once again, we specify characteristic functions:

$$C_N(\lambda_1, \ldots, \lambda_N) = E\left\{\exp\left[i\sum_1^N \lambda_k x(t_k)\right]\right\} = \exp\left(-\frac{1}{2}\sum_1^N \sum_1^N \lambda_j \lambda_k \gamma_{jk}\right),$$

where

$$\gamma_{jk} = \exp(-\mu|t_j - t_k|), \qquad \mu > 0. \qquad (1.1.11)$$

This is the characteristic function of a zero-mean Gaussian if we can show that the $N \times N$ matrix

$$M = \{\gamma_{jk}\}_{jk=1,\ldots,N}$$

is nonnegative definite. But this is immediate since the function

$$\exp(-\mu|t|), \qquad -\infty < t < \infty$$

is a characteristic function, as we have seen in the Review Chapter. Note that the characteristic function of any subset of variables $m < N$ is

$$C_m(\lambda_1, \ldots, \lambda_m) = E\left\{\exp\left[i\sum_1^m \lambda_k x(t_k)\right]\right\} = C_N(\lambda_1, \ldots, \lambda_m, 0, \ldots, 0).$$

In other words, the distribution of $x(t_1), \ldots, x(t_m)$ can be obtained as the marginal distribution from the distribution of the variables $x(t_1), \ldots, x(t_N)$, $m < N$. We call this the "consistency" property of the class of distributions being specified. We note that we can generalize definition (1.1.11) to

$$\gamma_{ik} = C(t_j - t_k),$$

where $C(\cdot)$ is a characteristic function (of a single variable) or a positive multiple thereof.

It is of interest that as $\mu \to \infty$ in (1.1.11) we obtain in the limit a process $x(t)$, $-\infty < t < \infty$, in which for any $\{t_i\}$, $i, \ldots, N$, $x(t_i)$, $i, \ldots, N$, are independent zero-mean Gaussians, the joint density being defined as the product of the individual densities. More on this process later, in Chapter 4.

The student will note that what we have defined in this example is a consistent family of distributions. That we can then actually define a (nonfinite) family of random variables with these distributions requires formal proof and was in fact given by A. N. Kolmogorov in a now classic, justly celebrated work in 1930 [1]. Unfortunately the proof is beyond our scope.

1.1.2 Examples: Discrete-Parameter Processes

Next let us consider illustrative examples of random sequences. We shall naturally take advantage of the continuous-time examples.

Example 1.1.7. Our first example is a canonical way of generating sequences. Periodic sampling: Let $x(t)$, $-\infty < t < \infty$, be a random process. Define

$$x_n = x(n\Delta), \tag{1.1.12}$$

where $\Delta > 0$ is the sampling period. Then x_n defines a discrete-parameter process or random sequence whose properties are readily derived from those of the parent process. Applying this to Example 1.1.2, we have

$$x_n = A \sin(w\Delta n + \phi), \qquad n \in I, \qquad (1.1.13)$$

yielding a discrete-parameter process in which all joint distributions are Gaussian, and thus we have a discrete-parameter Gausssian process. Note that we can write

$$\sin(w\Delta n + \phi) = \sin(2\pi\lambda n + \phi),$$

where

$$\frac{w\Delta}{2\pi} = m + \lambda, \qquad -1/2 \le \lambda \le 1/2; m \text{ integer}.$$

Similarly, periodic sampling of the Gaussian process of Example 1.1.3 yields a discrete-parameter Gaussian process:

$$x_n = \sum_1^N A_k \sin(2\pi\lambda_k n + \phi_k), \qquad -1/2 \le \lambda_k \le 1/2. \qquad (1.1.14)$$

Example 1.1.8. Trivial processes. Define

$$x_n = F(n, \omega),$$

where ω denotes a finite-dimensional random variable. This is a trivial process analogous to Example 1.1.4.

Example 1.1.9. Analogous to the differential equation in Example 1.1.5, we may consider a "difference equation" to define a process $\{y_n\}$ by

$$y_n = \rho y_{n-1} + x_{n-1}, \qquad 0 < \rho, \qquad (1.1.15)$$

where

$$x_n = A \sin(2\pi\lambda_0 n + \phi). \qquad (1.1.16)$$

Taking the starting index as zero, we have for $n \ge 0$:

$$y_n = \rho^n y_0 + \sum_0^{n-1} \rho^{n-1-j} x_j = \rho^n y_0 + \sum_0^{n-1} \rho^{n-1-j} A \sin(2\pi\lambda_0 j + \phi). \qquad (1.1.17)$$

We see that this is again a trivial process:

$$y_n = F(n, \omega),$$

where

$$\omega = \{y_0, A, \phi\}.$$

Again if y_0 is a random variable, we have to specify the joint distribution of y_0 and ϕ (and A, if A is a random variable). Taking

$$x_n = \sum_1^N A_k \sin (2\pi n\lambda_k + \phi_k) \tag{1.1.18}$$

with A_k and ϕ_k as in Example 1.1.3, we see that if in addition we take y_0 to be Gaussian independent of $\{A_k, \phi_k\}$, we have a Gaussian process. For $\{x_n\}$ given by (1.1.18) we can get a "closed form" for y_n analogous to (1.1.10).

Example 1.1.10 (IID Sequence; Bernoulli Sequence; White-Noise Sequence). Let

$$x_n, \quad n = 0, 1, \ldots$$

be a sequence of independent random variables each with the same (identical) distribution. This is referred to as an IID sequence for short. When $x_n = 0$ or 1 with probability 1/2 each, we call it a Bernoulli sequence. When the common distribution is Gaussian, we call it a white-noise sequence.

Even though it may look more plausible to the reader, Example 1.1.10 involves a nonfinite number of random variables and what we have defined is actually only finite-dimensional joint distributions. We have therefore to invoke the Kolmogorov theorem [1] referred to earlier for a formal proof. This formality is not necessary if we are only dealing with a finite number of variables at a given time, which is usually the case, the prominent exception being when we consider convergence with "probability one." These ideas will become clearer in context, later. In Example 1.1.8, the problem is shifted to the original continuous-parameter process.

And now we go on to consider what is a crucial notion in the theory.

1.1.3 Sample Paths: Continuous-Parameter Processes

The notion of sample paths generalizes the sample space notion for (finite-dimensional) random variables. Let us begin with continuous-parameter processes. Consider first Example 1.1.1. Note that we have actually a function of two variables: t and ϕ. For each t we have a random variable. On the other hand we can fix ϕ and let t "run" to yield a waveform or a function of time; we call it a "sample path."† Each sample path is a "realization" of the process. See Figure 1.1. See Figure 1.2 for sample paths of Example 1.1.2, and see Figure 1.3 for Example 1.1.3.

The "outcome" of any repeatable "experiment" (standard phraseology in

---

† The name "sample path" derives from physics where the most celebrated example of randomness is the motion of particles suspended in a light fluid first observed by the botanist John Brown in 1829. Each particle executes its own sample path jostled around by others. See reference 9 for more on this and Brownian motion.

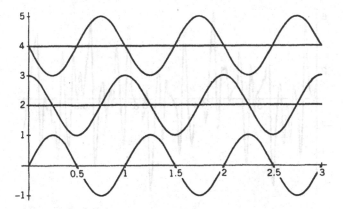

Figure 1.1. Sample paths: Example 1.1.1. $x(t) = \sin(2\pi t + \phi)$.

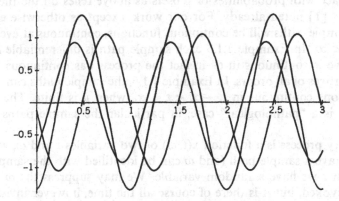

Figure 1.2. Sample paths: Example 1.1.2. $x(t) = A \sin(2\pi t + \phi)$.

texts) as it unfolds in time is a sample path. The ensemble of all such paths is the "sample space" of the process. Generalizing the sine wave example to any trivial process, for each t we have a random variable; and for each ω we get a sample path as a function of time. Let Ω denote the sample space. Let B denote the subset of sample paths for which

$$a_i < x(t_i) < b_i, \qquad i = 1, \ldots, N. \tag{1.1.19}$$

Then the proportion of these paths to the total—the probability of the "event" B—is the probability that (1.1.19) holds for the random variables $x(t_i)$, $i = 1, \ldots, N$.

This idea extends of course to nontrivial processes as well except that ω is now no longer finite-dimensional. What we are really given is a consistent family of finite-dimensional distributions. That we can still construct sample paths and

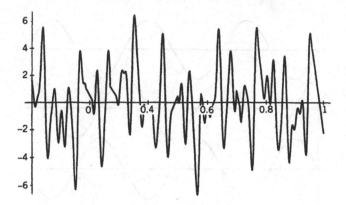

Figure 1.3. Sample paths: Example 1.1.3 with $N = 8$; $w_7 = 303.82$, $w_6 = 198.46$, $w_5 = 136.11$, $w_4 = 96.01$, $w_3 = 74.05$, $w_2 = 41.04$, $w_1 = 15.385$, $w_0 = 0$.

a sample space with probabilities of subsets as above relies on the theorem of Kolmogorov [1] noted already. For our work, except as otherwise explicitly noted, all sample paths will be continuous functions, continuous at every point t, $-\infty < t < \infty$. In Example 1.1.5 each sample path is differentiable in t and the derivative is continuous in t—in fact the process has continuous "sample path" derivatives of all orders. In Example 1.1.6, the sample paths can be taken to be functions continuous in t, $-\infty < t < \infty$, when μ is finite. The limiting case $\mu = \infty$ is a "pathological" case; in particular the sample paths are not continuous.

Thus every process is a function $x(t, \omega)$ of two variables t and ω, where for each ω we have a sample path (and ω can be identified with the sample path) and for each t we have a random variable. We may suppress the ω variable when not invoked, but it is there of course all the time, however invisible!

1.1.4 Sample Paths: Discrete-Parameter Processes

The definition of sample paths is the same as for continuous-parameter processes, except that they are sequences and notions of continuity or differentiability are irrelevant. Thus every process can be expressed as a function of two variables and written

$$x_n(\omega),$$

where for each ω we get a sample path sequence and for each n we get a random variable. Where not specifically necessary we shall ignore the sample path variable ω. See Figure 1.4 for sample paths of the process (1.1.13), and see Figure 1.5 for discretized version of (1.1.6) with $\Delta = 1/50$.

Finally let us emphasize that the examples in this section, however illustrative, do not even come close to exhausting the many ways in which random processes arise or can be specified.

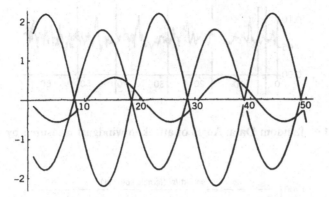

Figure 1.4. Sample paths of process (1.1.13). $A \sin (2\pi n\lambda + \phi)$, $2\pi\lambda = 0.30771$. Values interpolated.

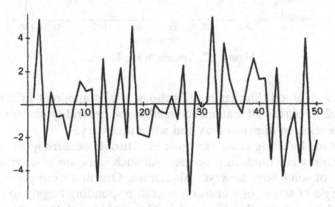

Figure 1.5. Discretized version of (1.1.6) with $\Delta = 1/50$. Values interpolated by linear segments.

1.1.5 Vector-Valued Processes

So far we have considered only processes in which for each value of t in T, $x(t)$ is a one-dimensional random variable. In many applications, and nowadays with digital computer processing, it is necessary to allow $x(t)$ to be multidimensional. More specifically we shall consider the case where for each t, $x(t)$ is an $n \times 1$ rectangular matrix ("column vector") with the associated matrix algebra. This is no more than efficient "bookkeeping" at one level, but the advantages are far more than merely notational. But there is little new conceptually in terms of the basic concepts; only the algebra. Thus the characteristic function of N variables $x(t_i)$, $i = 1, \ldots, N$, is given by

$$C(\lambda_1, \ldots, \lambda_N) = E\left(\exp \left\{ \sum_1^N i[\lambda_k, x(t_k)] \right\} \right),$$

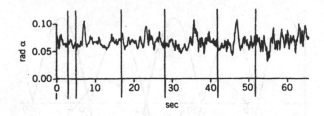

Figure 1.6. Random Data: Angle of attack in windgust measured by vane.

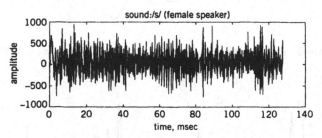

Figure 1.7. Speech Wave Form.

where each λ_k is $n \times 1$. Similar modifications will be apparent in use. In what follows we shall phrase all statements valid in the generality of the vector case and call attention to dimension as and when necessary.

We conclude by citing some examples of naturally occurring (as opposed to computer generated) random processes. All such data are of course screened by a sensor of some sort, however ubiquitous. Our first example is telemetry data: the angle of attack of a cruising aircraft responding largely to wind gust, as measured by a vane. See Figure 1.6. The analogue data is sampled at 50 samples/second, but the rate is so high that it is virtually "continuous-time."

The second example (Figure 1.7) is a speech wave form sampled at the rate of 4.82×10^4 samples/second. The third example, given in Figure 1.8, shows laser beam scintillation (log amplitude) due to wind turbulence sampled at 2.2×10^4 samples/second. The final example (Figure 1.9) is a time series: annual rainfall in inches in the Los Angeles basin from 1877 to 1994. Note the difference in time scale from the previous examples. We may model each as a sample path of a random (discrete-time) process.

PROBLEMS

1.1.1. Calculate the joint characteristic function of $x(t_1), \ldots, x(t_N)$ in Example 1.1.3.

1.1.2. Let

$$x(t) = \sin \zeta t, \qquad -\infty < t < \infty,$$

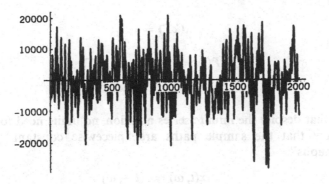

Figure 1.8. Laser beam (log amplitude) scintillation due to wind turbulence.

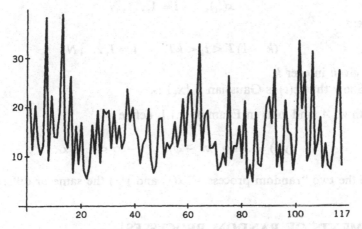

Figure 1.9. Annual rainfall in inches in the Los Angeles Basin: 1877–1994.

where

ζ is a Gaussian variable with mean m and variance σ^2.

Sketch some sample paths of the process. Sketch some paths (describe the function) which are not sample paths. Describe some sample paths which have small probability of occurring.

1.1.3. Reversing Example 1.1.7, we consider defining (interpolating!) a continuous-parameter process from a random sequence. Thus let $\{x_k\}$ denote a 1×1 random sequence and define

$$x(t) = \sum_{-\infty}^{\infty} x_k P(t - kT), \qquad t \neq kT$$

and

$$x(kT) = x_{k-1},$$

where

$$P(t) = 1, \qquad 0 < t < T,$$

$$= 0, \qquad \text{otherwise.}$$

Note that despite the infinite series notation, no limits need to be taken!
Show that the sample paths are "piecewise constant" and "left continuous":

$$x(t, \omega) = x(t-, \omega).$$

Find the distribution (or characteristic function) of the variables

$$x(t_i), \qquad i = 1, \dots, N$$

where

$$(k-1)T < t_i < kT, \qquad i = 1, \dots, N$$

for given integer k.
Show that $x(t)$ is Gaussian if $\{x_k\}$ is.

1.1.4. With w, A, and ϕ as in Example 1.1.1, define

$$y(t) = A \sin(wt - \phi), \qquad -\infty < t < \infty.$$

Are the two "random processes" $x(\cdot)$ and $y(\cdot)$ the same or different?

1.2 MOMENTS OF RANDOM PROCESSES

As with random variables, most of the properties of random processes are determined by the moments. We are primarily interested in the first and second moments of processes.

Let $x(t)$, $t \in T$ be an $n \times 1$ random process. We assume that

$$E[\|x(t)\|^2] < \infty \qquad \text{for each } t \text{ in } T. \qquad (1.2.1)$$

We define

$$m(t) = E(x(t)). \qquad (1.2.2)$$

This is the "mean"—an $n \times 1$ function of t, defined on T. The 2nd-moment matrix

$$R(t_1, t_2) = E((x(t_1) - m(t_1))((x(t_2) - m(t_2))^*) \qquad (1.2.3)$$

defines an $n \times n$ matrix function of two variables t_1 and t_2 ($R(\cdot, \cdot)$ is defined

on $T \times T$ and is called the "covariance function" of the process.) We note that
$$R(t_1, t_2) = E(x(t_1)x(t_2)^*) - m(t_1)m(t_2)^*$$
$$R(t_1, t_2)^* = R(t_2, t_1), \qquad (1.2.4)$$
and in particular
$$R(t, t)$$
is nonnegative definite for each t.

Let us specialize the dimension to be 1×1. Then we note that for any choice of $t_i \in T$, $i = 1, \ldots, N$ and complex constants a_i, $i = 1, \ldots, N$ we have
$$\sum_1^N \sum_1^N a_i R(t_i, t_j) \bar{a}_j \geq 0. \qquad (1.2.5)$$

This is immediate from
$$0 \leq E\left[\left(\left|\sum_1^N a_i(x(t_i) - m(t_i))\right|\right)^2\right]$$
$$= E\left[\left(\sum_1^N a_i(x(t_i) - m(t_i))\right)\overline{\left(\sum_1^N a_j(x(t_j) - m(t_j))\right)}\right]$$
$$= E\left(\sum \sum a_i(x(t_i) - m(t_i))(x(t_j) - m(t_j))\bar{a}_j\right)$$
$$= \sum \sum a_i R(t_i, t_j)\bar{a}_j.$$

In particular of course
$$R(t, t) \geq 0. \qquad (1.2.6)$$

Note also that the function of two variables $R(t_1, t_2)$ is "positive definite" if (1.2.4) and (1.2.5) hold; or equivalently the $N \times N$ matrix
$$\Lambda = \{R(t_i, t_j)\}$$
is nonnegative definite. Thus the covariance function is positive definite.

Let us generalize this result to the $n \times 1$ vector case. Here we have that for any complex constants a_i, vectors $x_i \in E^n$, and any choice t_i, $i = 1, \ldots, N$,
$$\left(\sum_1^N a_i x_i^*(x(t_i) - m(t_i))\right)\left(\sum_1^N a_i x_i^*(x(t_i) - m(t_i))\right)^*$$
is nonnegative and hence so is the average
$$E\left[\left(\sum_1^N a_i x_i^*(x(t_i) - m(t_i))\right)\left(\sum_1^N a_i x_i^*(x(t_i) - m(t_i))\right)^*\right]$$
$$= \sum_1^N \sum_1^N a_i [R(t_i, t_j)x_i, x_j] \bar{a}_j \geq 0,$$

or the $N \times N$ matrix

$$\Lambda = \{[R(t_i, t_j)x_i, x_j]\}$$

is nonnegative definite. We say that an $n \times n$ matrix function of two variables t_1 and t_2 is positive definite if

$$R(t_i, t_j)^* = R(t_j, t_i)$$

and for any $x_i \in E^n$, the $N \times N$ matrix

$$\Lambda = [R(t_i, t_j)x_i, x_j] \tag{1.2.6}$$

is nonnegative definite. Hence the covariance *function* is "positive definite"—note the emphasis on "function"; for each t_i, and t_j, $R(t_i, t_j)$ is *not* necessarily positive definite! In other words, (1.2.4) and (1.2.5) hold in the vector case as well, if interpreted properly.

Let $R(t_1, t_2)$, $t_1 \in T$, $t_2 \in T$, be a (real-valued)† positive definite function. Then for any t_i, $i = 1, \ldots, N$, the matrix

$$\Lambda = \{R(t_i, t_j)\}, \qquad i, j = 1, \ldots, N$$

is a symmetric nonnegative definite matrix. Hence

$$C(t) = \exp\left(-\tfrac{1}{2}[\Lambda t, t]\right), \qquad t \in R^N$$

is the characteristic function of an N-variable Gaussian with zero mean and covariance matrix Λ. More importantly, these finite-dimensional Gaussian distributions are consistent, as the definition of the characteristic functions shows. Hence by the Kolmogorov theorem [1] we can define a zero-mean Gaussian process $x(t)$, $t \in T$, with this function as the covariance function. We can define any function $m(\cdot)$ as the mean by defining

$$y(t) = m(t) + x(t).$$

For this reason we may call any positive definite function a "covariance" function.

The importance of the mean and covariance functions is thus that given these two functions we can find a Gaussian process with these mean and covariance functions. Since most of our work will involve the first- and second-order moments, the resulting theory is often labeled "second-order theory."

---

† All the random variables we consider will be real-valued (all components will be real in the vector case). Hence unless otherwise stated, all covariance functions will be real-valued (all components will be real-valued in the matrix case).

1.2.1 Continuous-Parameter Processes: Continuity in the Mean Square

When the index set is a continuum—and to be specific let us consider the case where T is an interval: $a \leq t \leq b$—we shall naturally be concerned about the smoothness—differentiability, continuity—of the mean and covariance functions and what this implies in turn for the process.

We say that a process $x(t)$, $a \leq t \leq b$†, is continuous in the mean of order p at the point t if

$$E[\| x((t + \Delta) - x(t)\|^p] \to 0 \qquad \text{as } |\Delta| \to 0.$$

And, more simply, it is continuous in the mean of order p if this holds for every t in $a \leq t \leq b$. We are mainly concerned with $p = 1, 2$. For $p = 2$, we also use the term "continuity in the mean square." By Schwarz inequality

$$E[\| x(t + \Delta) - x(t)\|]^2 \leq E[\| x(t + \Delta) - x(t)\|^2]$$

we see that continuity in the mean of order two implies continuity in the mean of order one. Hence let us look at the former. We have:

Theorem 1.2.1. A process $x(t)$, $a \leq t \leq b$, is continuous in the mean of order two if and only if

(i)
$$m(t) = E(x(t))$$
is continuous in $a \leq t \leq b$.

(ii) The (matrix) covariance function $R(t_1, t_2)$ is continuous in the rectangle $a \leq t_1, t_2 \leq b$.

Proof. Let

$$x(t) = \begin{vmatrix} x_1(t) \\ \vdots \\ x_n(t) \end{vmatrix}.$$

Suppose $x(t)$ is continuous in the mean square. Then so is $x_i(t)$ for each i, $i = 1, \ldots, n$. Moreover,

$$E[x_i(t)^2]$$

is continuous in t for each i. This follows from the Hölder inequality (see (0.5.7))

$$E[x_i(t)^2]^{1/2} \leq E[|x_i(t) - x_i(s)|^2]^{1/2} + E[x_i(s)^2]^{1/2}$$

$$E[x_i(s)^2]^{1/2} \leq E[|x_i(t) - x_i(s)|^2]^{1/2} + E[x_i(t)^2]^{1/2}$$

† If $a = -\infty$ and b is finite, the interval is defined as $-\infty < t \leq b$; if a is finite but $b = \infty$, the interval is defined as $a \leq t < \infty$.

and hence

$$|E[x_i(t)^2]^{1/2} - E[x_i(s)^2]^{1/2}| \leq E[x_i(t) - x_i(s)^2]^{1/2}.$$

Thus

$$E[x_i(t)^2]^{1/2}$$

and hence

$$(E[x_i(t)^2]^{1/2})^2 = E[x_i(t)^2]$$

is continuous in t.

Since

$$E[\| x(t_1) - x(t_2)\|^2] = \| m(t_1) - m(t_2)\|^2 + E[\| \tilde{x}(t_1) - \tilde{x}(t_2)\|^2], \quad (1.2.7)$$

where

$$\tilde{x}(t) = x(t) - m(t),$$

and each term on the right side of (1.2.7) is nonnegative, it follows that $m(t)$ must be continuous and that $\tilde{x}(t)$ must be continuous in the mean of order two. To prove the continuity of $R(t_1, t_2)$, we note that we can write

$$R(t_1 + \Delta_1, t_2 + \Delta_2) - R(t_1, t_2)$$
$$= \{r_{ij}(t_1 + \Delta_1, t_2 + \Delta_2) - r_{ij}(t_1 + \Delta_1, t_2) + r_{ij}(t_1 + \Delta_1, t_2) - r_{ij}(t_1, t_2)\},$$

where

$$R(t, s) = \{r_{ij}(t, s)\}, \qquad i, j = 1, \ldots, n.$$

But (and this is the key calculation), taking each pair of terms in curly brackets in turn, we obtain

$$|r_{ij}(t_1 + \Delta_1, t_2) - r_{ij}(t_1, t_2)| = |E[(\tilde{x}_i(t_1 + \Delta_1) - \tilde{x}_i(t_1))x_j(t_2)]|,$$

which by the Schwarz inequality (5.4 on page 28) becomes

$$\leq E[\tilde{x}_j(t_2)^2]^{1/2} E[|\tilde{x}_i(t_1 + \Delta_1) - \tilde{x}_i(t_1)|^2]^{1/2} \to 0, \qquad \text{as } |\Delta_1| \to 0$$

and

$$|r_{ij}(t_1 + \Delta_1, t_2 + \Delta_2) - r_{ij}(t_1 + \Delta_1, t_2)| = |E[\tilde{x}_i(t_1 + \Delta_1)(\tilde{x}_j(t_2 + \Delta_2) - \tilde{x}_j(t_2))]|$$
$$\leq E[x_i(t_1 + \Delta_1)^2]^{1/2} E[|x_j(t_2 + \Delta_2) - x_j(t_2)|^2]^{1/2}$$

and because of the continuity of the function

$$E[x_j(t)^2]^{1/2}$$

the first factor is bounded as $\Delta_1 \to 0$ and the second factor goes to zero. Hence it follows that $R(t_1, t_2)$ is continuous.

Conversely, suppose that conditions (i) and (ii) are satisfied. Then $r_{ij}(t_1, t_2)$, $1 \leq i, j \leq n$, is continuous in t_1, t_2. Now

$$E[(\tilde{x}_i(t + \Delta) - \tilde{x}_i(t))^2]$$

$$= r_{ii}(t + \Delta, t + \Delta) - r_{ii}(t + \Delta, t) + r_{ii}(t, t) - r_{ii}(t, t + \Delta)$$

$$\leq |r_{ij}(t + \Delta, t + \Delta) - r_{ii}(t + \Delta, t)| + |r_{ii}(t, t) - r_{ii}(t, t + \Delta)|.$$

The second term clearly goes to zero as $|\Delta| \to 0$, from the continuity of $r_{ii}(t_1, t_2)$. The first term becomes

$$\leq |r_{ii}(t + \Delta, t + \Delta) - r_{ii}(t, t)| + |r_{ii}(t, t) - r_{ii}(t + \Delta, t)|$$

and then by the continuity condition on $r_{ii}(t_1, t_2)$, becomes

$$\to 0 \qquad \text{as } |\Delta| \to 0.$$

From

$$E[(x_i(t + \Delta) - x_i(t))^2] = (m_i(t + \Delta) - m_i(t))^2 + E[(\tilde{x}_i(t + \Delta) - \tilde{x}_i(t))^2]$$

the mean square continuity follows since $m(\cdot)$ is continuous. This concludes the proof.

Remark. It follows from the arguments in the proof that $R(t_1, t_2)$ is continuous in t_1, t_2, as soon as $r_{ii}(t_1, t_2)$ is continuous in t_1, t_2, for each i, $i = 1, \ldots, n$.

We note that $R(t_1, t_2)$ in Example 1.2.3 below is *not* continuous—indeed $R(t, t)$ is *not* continuous; neither are the sample paths, as we have seen. For most of our work, however, the covariance functions will be continuous, and the processes will be mean square continuous.

1.2.2 Examples

Let us calculate the covariance functions of the processes in our examples of Section 1.1.

Example 1.1.1.

$$R(t_1, t_2) = \frac{A^2}{2} \cos w(t_1 - t_2), \qquad \text{zero mean.}$$

Example 1.1.2.

$$x(t) = A \sin (wt + \phi) = x \cos wt + y \sin wt$$

$$E[x(t)] = E[x] \cos wt + E[y] \sin wt = 0$$

$$R(t_1, t_2) = E[(x \cos wt_1 + y \sin wt_1)(x \cos wt_2 + y \sin wt_2)]$$

$$= \sigma^2[\cos wt_1 \cos wt_2 + \sin wt_1 \sin wt_2]$$

$$= \sigma^2 \cos w(t_1 - t_2). \tag{1.2.8}$$

In particular we note that the function

$$r(t_1, t_2) = \cos w(t_1 - t_2)$$

is positive definite.

Example 1.1.3.

$$E[x(t)] = 0$$

$$R(t_1, t_2) = \sum_1^n \sigma_k^2 \cos w_k(t_1 + t_2). \tag{1.2.9}$$

Example 1.2.1. Let $T = (-\infty, \infty)$. Then we note that

$$m(t) = 0$$

$$R(t_i, t_j) = 0, \qquad i \neq j$$

$$= 1 \text{ (or identity matrix in vector case)}, \qquad t_i = t_j$$

defines a Gaussian random process in which $x(t_1)$ and $x(t_2)$ are independent for $t_1 \neq t_2$ and the covariance of $x(t)$ is the identity. If $T = I$, the same definition yields a white-noise sequence.

Example 1.2.2. Compare Problem 1.1.2.

$$\zeta \text{ Gaussian mean } m, \text{ variance } \sigma^2$$

$$x(t) = \sin \zeta t$$

$$E[x(t)] = E[\sin \zeta t] = \text{Im } E[e^{i\zeta t}].$$

But

$$E[e^{i\zeta t}] = e^{imt} \cdot e^{-\sigma^2 t^2/2}$$

(characteristic function of Gaussian density). Hence

$$E[x(t)] = (\sin mt)e^{-\sigma^2 t^2/2}.$$

Next using the identity

$$2 \sin A \sin B = \cos(A - B) - \cos(A + B)$$

we have

$$
\begin{aligned}
E[(x(t_1)x(t_2)] &= E[\tfrac{1}{2}(\cos \zeta(t_1 - t_2) - \cos \zeta(t_1 + t_2))] \\
&= \tfrac{1}{2}\,\mathrm{Re}\,[E[e^{i\zeta(t_1 - t_2)} - e^{i\zeta(t_1 + t_2)}]] \\
&= \tfrac{1}{2}[e^{-\sigma^2(t_1 - t_2)^2/2}\cos m(t_1 - t_2) - e^{-\sigma^2(t_1 + t_2)^2/2}\cos m(t_1 + t_2)].
\end{aligned}
$$

Hence

$$
\begin{aligned}
R(t_1, t_2) &= E[x(t_1)x(t_2)] - E[x(t_1)]E[x(t_2)] \\
&= \tfrac{1}{2}[(e^{-\sigma^2(t_1 - t_2)^2/2} - e^{-\sigma^2(t_1^2 + t_2^2)/2})\cos m(t_1 - t_2) \\
&\quad - (e^{-\sigma^2(t_1 + t_2)^2/2} - e^{-\sigma^2(t_1^2 + t_2^2)/2}\cos m(t_1 + t_2)]. \qquad (1.2.10)
\end{aligned}
$$

Example 1.2.3. Compare Problem 1.1.3.

$$E[x(t)] = \sum_{-\infty}^{\infty} E[x_k]P(t - kT).$$

Hence

$$
\begin{aligned}
E[x(t) &= E[x_k], \qquad (k - 1)T < t < kT \\
&= E[x_{k-1}], \qquad t = kT \\
&= 0, \qquad \text{since } E[x_k] \equiv 0 \text{ for every } k.
\end{aligned}
$$

$$R(t_1, t_2) = \sum_{-\infty}^{\infty}\sum_{-\infty}^{\infty} r_{jk}P_k(t_1 - kT)P_k(t_2 - jT),$$

$$= r_{m,n},$$

for t_1 and t_2 such that $mT < t_1 < (m + 1)R$ and $nT < t_2 < (n + 1)T$. If $t_1 = mT$ and $nT < t_2 < (n + 1)T$, we have

$$R(t_1, t_2) = r_{m-1, n}.$$

If $t_1, = mT$ and $t_2 = nT$, we have

$$R(t_1, t_2) = r_{m-1, n-1}.$$

Note that $R(t, t)$ is not continuous in t.

1.2.3 Equivalent Gaussian Processes

Finally a cautionary note concerning Gaussian processes with the same mean and covariance function as a given process: By way of illustration let us consider the (non-Gaussian) process defined in Example 1.1.1. The (second-order) equivalent Gaussian process would have variance

$$\frac{A^2}{2} \cos w(t_1 - t_2)$$

and hence sample paths of the form

$$r \cos (wt + \phi), \qquad 0 \le r < \infty,$$

where r is Rayleigh with density given by (2.4 on page 11) where

$$\sigma^2 = \frac{A^2}{2}.$$

The crucial point here is that the amplitudes of the sine waves are no longer constant as in the original process! Thus the sample paths of the equivalent Gaussian need not necessarily be the same as that of the original process.

PROBLEMS

1.2.1. Show that the sum as well as the product of two positive definite functions is a positive definite function.

  Hint: For sum, use definition. For product, use the result on covariance matrices in the Review Chapter. Or construct mutually independent Gaussian processes, each with one of the covariances, and consider the product and sum of the process.

1.2.2. If $R(t_1, t_2)$, $-\infty < t_1, t_2 < \infty$, is a covariance function, show that so is

$$aR(at_1, at_2)$$

for any real number $a > 0$.

1.2.3. Let T be the positive half-line: $0 \le t < \infty$. Show that

$$R(t_1, t_2) = \min (t_1, t_2), \qquad 0 \le t_1, t_2 < \infty$$

is a covariance function.

  Hint: Use induction. For $t_1 < t_2$, the matrix

$$\begin{vmatrix} t_1 & t_1 \\ t_1 & t_2 \end{vmatrix}$$

is clearly a covariance matrix. Assume that the result is true for t_i,

$i = 1, \ldots, N$. Let us prove it for t_i, $i = 1, \ldots, N, N + 1$. Arrange the $\{t_i\}$ in ascending order:

$$t_{i+1} \geq t_i,$$

then

$$\min t_i = t_1$$

and hence

$$\sum_1^{N+1} \sum_1^{N+1} a_i \min (t_i, t_j) \bar{a}_j = \sum_2^{N+1} \sum_2^{N+1} a_i \min (t_i, t_j) \bar{a}_j$$
$$+ |a_1|^2 t_1 + a_1(\bar{a}_2 + \cdots + \bar{a}_{N+1}) t_1$$
$$+ \bar{a}_1(a_2 + \cdots + a_{N+1}) t_1.$$

Now for $i, j \geq 2$

$$\min (t_i, t_j) - t_1 = \min (t_i - t_1, t_j - t_1)$$

and

$$\sum_2^{N+1} \sum_2^{N+1} a_i \min (t_i, t_j) \bar{a}_j - \sum_2^{N+1} \sum_2^{N+1} a_i t_1 \bar{a}_j$$
$$= \sum_2^{N+1} \sum_2^{N+1} a_i \min (t_i - t_1, t_j - t_1) \bar{a}_j,$$

which by the induction hypothesis is nonnegative. Hence

$$\sum_1^{N+1} \sum_1^{N+1} a_i \min (t_i, t_j) \bar{a}_j \geq t_1 \sum_1^{N+1} \sum_1^{N+1} a_j \bar{a}_j \geq 0.$$

1.2.4. Let $\{x_k\}$ be an $n \times 1$ white-noise process such that

$$E[x_k] = m$$

covariance matrix $E[(x_k - m)(x_k - m)^*] = \Lambda.$

Let

$$y_n = \frac{1}{n} \sum_1^n x_k, \qquad n \geq 1.$$

Find the mean and covariance function of the process $\{y_n\}$, $n \geq 1$.
    Answer:

$$\text{cov } R(m, n) = \min \left(\frac{1}{m}, \frac{1}{n} \right) \Lambda.$$

1.2.5. Let $W(t)$, $0 \leq t < \infty$ be a zero-mean Gaussian process with the covariance function in Problem 1.2.3. Show that for $s < t$,

$$E[(W(t) - W(s))^2] = t - s.$$

The process is called a "Wiener process." For $t_1 < t_2 < t_3$ show that $(W(t_2) - W(t_1))$ is independent of $(W(t_3) - W(t_2))$.

Hint: It is enough to show that the correlation is zero.

Let

$$Z(t) = W(e^{\mu t}) \exp\left(\frac{-\mu t}{2}\right), \qquad \mu > 0.$$

Calculate the mean and covariance functions of the process $Z(\cdot)$.

1.2.6. Let $W(t)$, $t \geq 0$, denote the Wiener process as defined in Problem 1.2.5. Let

$$x(t) = \sin(2\pi f_c t + W(t)), \qquad t \geq 0,$$

where f_c is a fixed frequency. Calculate the mean and covariance functions of the process $x(\cdot)$. Show that

$$\lim_{t \to \infty} E[x(t)] = 0.$$

Let $R(t_1, t_2)$ denote the covariance function. Show that

$$\lim_{L \to \infty} R(t + L, L) = \tfrac{1}{2}(\cos 2\pi f_c t)e^{-|t|/2}.$$

1.2.7. If the matrix covariance function $R(s, t)$, $-\infty < s, t < \infty$, is such that

$$\text{Tr } R(t, t) = 0, \qquad -\infty < t < \infty,$$

what conclusion can you reach about the process $x(\cdot)$ with this covariance function?

1.2.8. Calculate the mean and covariance functions of the process defined by (1.1.10).

1.2.9. Carry out the summation in (1.1.17) and express y_n in "closed form." Calculate the mean and covariance function of the process $\{y_n\}$, $n \geq 0$, for $\{x_n\}$ given by (1.1.18).

1.2.10. Let $\{x_k\}$, $k \in I$, denote a one-dimensional Gaussian process. Calculate the mean and covariance of the process

$$z_k = x_k x_{k+p},$$

where p is a fixed integer.

Hint: Use the formula for expectation of four products of Gaussians [cf. Eq. (2.20) on page 17].

Note that the process $\{z_k\}$ is *not* Gaussian.

1.2.11. Calculate the mean and covariance function of the process defined by (1.1.7). Find

$$\lim_{n \to \infty} E[y_n].$$

1.2.12. A process $y(t)$, $-\infty < t < \infty$, is called "lognormal" if it has the form

$$y(t) = e^{x(t)}$$

where $x(t)$, $-\infty < t < \infty$, is a Gaussian process. Calculate the mean and covariance function on the process $y(\cdot)$ in terms of those of the process $x(\cdot)$.

Hint: Use

$$E[e^\zeta] = e^m \cdot e^{\sigma^2/2}$$

for ζ Gaussian with mean m and variance σ^2.

1.2.13. Show that

$$e^{R(t_1, t_2)}, \qquad t_1, t_2 \in T$$

is a covariance function if $R(t_1, t_2)$ is.

Hint: Use Problem 1.2.12.

1.2.14. Show that the sample paths $x(t)$, $-\infty < t < \infty$, defined in (1.1.6) are actually periodic if

$$w_k = 2\pi f_k,$$

where the $\{f_k\}$ are rational, and determine the period. Similarly (1.1.14) is periodic if the $\{\lambda_i\}$ are rational. Any such sequence generated on a computer will be periodic, even if the period is large compared to the length of the sequence generated.

NOTES AND COMMENTS

The basic idea that a signal ("wave form") is "random" as opposed to "deterministic"—there would be no need to transmit it if it were deterministic in the sense that it would then be "known" to the receiver—is fundamental to communication system design and is due to C. E. Shannon, the originator of "Information Theory." His landmark paper of 1948 [3] has a legitimate claim as marking the beginning of modern communications. Prior to that, all signals were treated as "periodic" in the engineering literature. The extension to nonperiodic signals—"almost periodic"—was made by Wiener in the 1930's under the title of "abstract harmonic analysis" [6], but this mathematics was too complicated. For instance, if one defines an "almost periodic" signal as one which has finite power (finite "root mean square" value), it does not follow that the sum of two such signals has the same property!

It was natural that Shannon start with the sinusoid with random phase to illustrate random signals. S. O. Rice in his now historic article [2] models signals as sums of sinusoids with random amplitude and phase still describable in terms of "mean square" time averages. The relation to *stationary* random processes became apparent in the late 1940's largely through the work of Wiener and Khinchin (see reference 4). In other words, the language of stationary or "metrically transitive ergodic" processes (see reference 4) was much easier operationally than the "abstract harmonic analysis" of Wiener, even though the results were the same.

REFERENCES

Classic Papers
1. A. N. Kolmogorov. *Foundations of Theory of Probability*. Chelsea Publishing Co., 1950.
2. S. O. Rice. "Mathematical Analysis of Random Noise," *Bell System Technical Journal*, Vols. 23 and 24. Reprinted in: N. Wax, *Noise and Stochastic Processes*. Dover Publications, 1954.
3. C. E. Shannon. "A Mathematical Theory of Communication," *Bell Technical Journal*, Vol. 27 (1948). See also reprinted version in: *The Mathematical Theory of Communication*. C. E. Shannon and W. Weaver. University of Illinois Press, 1949.

Mathematical Treatises
4. J. L. Doob. *Stochastic Processes*. John Wiley and Sons, 1953.
5. P. Billingsley. *Probability and Measure*. John Wiley and Sons, 1979.
6. N. Wiener. *The Fourier Integral and Certain of Its Applications*. Dover Publications, 1933.

Engineering Texts
7. A. Papoulis. *Probability, Random Variables and Stochastic Processes*. McGraw-Hill, 1984.
8. R. E. Mortensen. *Random Signals and Systems*. John Wiley and Sons, 1986.

Application of Random Fields
9. A. S. Monin and A. M. Yaglom. *Statistical Fluid Mechanics*. M.I.T. Press, 1971.

2

STATIONARY RANDOM PROCESSES: COVARIANCE AND SPECTRUM

Perhaps the most important single concept in the theory of random processes is that of "stationarity." This concept applies only to the case $T = \mathbf{R}^1$ or $\mathbf{I}$, and this will be assumed whenever we speak of a stationary random process or simply stationary process.

Definition 2.1. A random process $(T = \mathbf{R}^1)$ is said to be "strictly stationary" if the joint distribution of

$$x(t_1), \dots, x(t_N)$$

is the same as that of

$$x(t_1 + t), \dots, x(t_N + t)$$

for every t, $-\infty < t < \infty$, and for any choice of t_i, $i = 1, \dots, N$.

We say for short that the finite-dimensional distributions "are invariant with respect to translation in time."

The definition for random sequences $(T = \mathbf{I})$ is similar: The joint distribution of

$$x_{n_1}, \dots, x_{n_N}$$

is the same as that of

$$x_{n_1 + m}, \dots, x_{n_N + m}$$

for all integers m (positive or negative) and all $\{n_i\}$, $i = 1, \dots, N$.

In many physical problems the phenomena observed are invariant with respect to translation in time—the definition of stationarity captures this property for the random process models. We can state this also in terms of the

63

characteristic function for vector-valued processes

$$E\left[\exp\left(i\sum_1^N [\lambda_k, x(t_k)]\right)\right] = E\left[\exp\left(i\sum_1^N [\lambda_k, x(t + t_k)]\right)\right], \qquad \lambda \in \mathbf{R}^n$$

for every t and any $\{t_i\}$, $i = 1, \ldots, N$. We can use this to see that the process in Example 1.1.1 is strictly stationary.

Suppose a process is "strictly stationary." Then let us see what this implies for the moments. For the first moment we have

$$m(t) = E[x(t)] = E[x(0 + t)] = E[x(0)]$$

since the distribution of $x(0)$ is the same as that of $x(0 + t)$, and hence

$$m(t) = m(0) = \text{constant}.$$

As for the covariance, $R(t_1, t_2)$, we have

$$R(t_1, t_2) = E[x(t_1)x(t_2)^*] - m(t_1)m(t_2)^*$$
$$= E[x(t_1 + t)x(t_2 + t)^*] - m(0)m(0)^*,$$

and taking

$$t = -t_2$$

we have that

$$R(t_1, t_2) = R(t_1 - t_2, 0),$$

or $R(t_1, t_2)$ depends only on the difference $(t_1 - t_2)$. In other words, there is a function of one variable $R(t)$, $-\infty < t < \infty$, such that

$$R(t_1, t_2) = R(t_1 - t_2). \tag{2.1}$$

Note that the function $R(\cdot)$ must have the property

$$R(t)^* = R(-t) \tag{2.2}$$

and (cf. 1.2.5a)) for x_i in E^n, $i = 1, \ldots, N$,

$$\Lambda = \{[R(t_i - t_j)x_i, x_j]\} \geq 0. \tag{2.3}$$

$R(\cdot)$ is thus a "positive definite function"—an $n \times n$ matrix function, in general. In addition, we shall always assume that it is continuous in $-\infty < t < \infty$. We call such a function also a "stationary covariance function," and sometimes we may be sloppy and omit the qualification "stationary" when it is clear from the context.

Suppose now we have a process for which

$$E[x(t)] = \text{constant} \tag{2.4}$$

$$R(t_1, t_2) = R(t_1 - t_2). \tag{2.5}$$

We call such a process "second-order stationary" (or "wide-sense stationary"). We note that if the process is Gaussian and that (2.4) and (2.5) hold, then the process is strictly stationary. Also of course given a process which is second-order stationary we can create a strictly stationary Gaussian process with the same mean and covariance function. Since we shall be concerned mostly with Gaussian processes, we shall omit the qualifications "strictly," "second-order," etc., and just use "stationary."

2.1 CONSTRUCTING COVARIANCE FUNCTIONS: CONTINUOUS-PARAMETER PROCESSES

2.1.1 Covariance Functions: Examples

Note that in the one-dimensional case every characteristic function, being positive definite as we have seen in the Review Chapter, is a covariance function even if not real-valued. Of course every characteristic function is a continuous function and we shall only need to deal with continuous stationary covariance functions.

Example 1.1.1 yields a stationary process which is *not* Gaussian. Examples 1.1.2 and 1.1.3 are stationary Gaussian processes with stationary covariance function

$$E[x(s)x(s + t)] = R(t) = \sum_{1}^{N} \sigma_k^2 \cos w_k t, \qquad -\infty < t < \infty. \tag{2.6}$$

See Figure 2.1, where $\sigma_k^2 = 1$ and w_k are as in Figure 1.3.

Example 1.1.6 also is a stationary Gaussian process, since the covariance function is

$$R(t) = \exp(-\mu|t|), \qquad \mu > 0. \tag{2.7}$$

See Figure 2.2. Note that the covariance functions (2.6) and (2.7) are continuous. However, there is one difference: For (2.7) we have

$$\int_{-\infty}^{\infty} |R(t)|\, dt < \infty,$$

but for (2.6) we obtain

$$\int_{-\infty}^{\infty} |R(t)|\, dt = \infty.$$

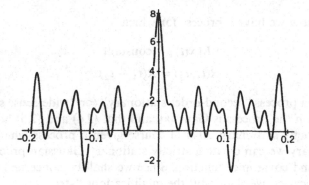

Figure 2.1. Correlation function of Example 1.1.3 with w_k as in Figure 1.3 and $\sigma_k^2 = 1$.

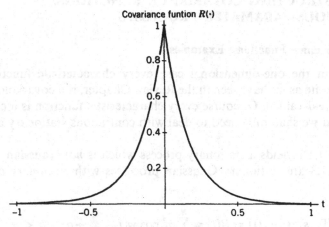

Figure 2.2. $R(t) = \exp(-\mu|t|)$, $\mu = 2\pi$.

Let us now indicate a general technique for constructing covariance functions.

Theorem 2.1. Let $P(f)$, $-\infty < f < \infty$, be a (one-dimensional) nonnegative function such that

$$\int_{-\infty}^{\infty} P(f)\, df < \infty. \tag{2.8}$$

Then

$$R(t) = \int_{-\infty}^{\infty} e^{2\pi i f t} P(f)\, df, \qquad -\infty < t < \infty$$

defines a covariance function, which is continuous at each t.

Proof. We can readily verify the requisite properties

$$\overline{R(t)} = \int_{-\infty}^{\infty} e^{-2\pi i f t} P(f)\, df = R(-t)$$

and

$$\sum_{1}^{N}\sum_{1}^{N} a_j R(t_j - t_k)\bar{a}_k = \int_{-\infty}^{\infty} \left(\sum_{1}^{N}\sum_{1}^{N} a_j e^{2\pi i (t_j - t_k) f} \bar{a}_k \right) P(f)\, df$$

$$= \int_{-\infty}^{\infty} \left| \sum_{1}^{N} a_k e^{2\pi i f t_k} \right|^2 P(f)\, df \geq 0.$$

Remark. We note that $R(t)$ is real-valued if we make $P(\cdot)$ a symmetric function of f:

$$P(f) = P(-f).$$

In this case we can write

$$R(t) = 2 \int_{0}^{\infty} \cos 2\pi f t P(f)\, df.$$

Also because of condition (2.8), $R(t) \to 0$ as $|t| \to \infty$.

Recall also that $P(\cdot)$ can be obtained by taking the inverse Fourier transform

$$P(f) = \int_{-\infty}^{\infty} e^{-2\pi i f t} R(t)\, dt,$$

with appropriate interpretation if

$$\int_{-\infty}^{\infty} |R(t)|\, dt = +\infty.$$

Every characteristic function is a covariance function, with $P(\cdot)$ as the probability density function. In particular we may generalize Theorem 2.1 to allow $P(\cdot)$ to be delta functions. Thus

$$P(f) = \left(\frac{\delta(f - f_c) + \delta(f + f_c)}{2} \right)$$

will yield

$$R(t) = \left(\frac{\exp 2\pi i f_c t + \exp -2\pi i f_c t}{2} \right)\sigma^2 = \sigma^2 \cos 2\pi f_c t$$

covering Example 1.1.2, and

$$P(f) = \sum_{1}^{N} \sigma_k^2 \left(\frac{\delta(f - f_k) + \delta(f + f_k)}{2} \right) \qquad (2.9)$$

takes care of Example 1.1.3. Of course (2.8) is no longer satisfied and we note that $R(t)$ does *not* converge to zero as $|t| \to \infty$.

We call $P(\cdot)$ the Power Spectral Density or simply Spectral Density because

$$R(0) = \int_{-\infty}^{\infty} P(f)\, df$$

yields the total power over all frequencies, $-\infty < t < \infty$, and $P(f)$ represents the power "per unit frequency." The integral

$$F(\lambda) = \int_{-\infty}^{\lambda} P(f)\, df, \qquad -\infty < \lambda < \infty$$

is known as the spectral "distribution," and the analogy to probability "distribution" is apparent.

Let us now go on to the matrix case.

Theorem 2.2. Let $P(f)$, $-\infty < f < \infty$, be an $n \times n$ matrix function which is self-adjoint and nonnegative definite:

$$P(f)^* = P(f)$$

$$P(f) \geq 0$$

and further

$$\int_{-\infty}^{\infty} \operatorname{Tr} P(f)\, df < \infty. \qquad (2.10)$$

Then

$$R(t) = \int_{-\infty}^{\infty} e^{2\pi i f t} P(f)\, df, \qquad -\infty < t < \infty \qquad (2.11)$$

defines an $n \times n$ (matrix) continuous covariance function. It is real-valued if

$$\overline{P(f)} = P(-f).$$

Proof. Again we verify

$$R(t)^* = \int_{-\infty}^{\infty} e^{-2\pi i f t} P(f)^* \, df = \int_{-\infty}^{\infty} e^{-2\pi i f t} P(f)\, df = R(-t),$$

and, to verify (2.3),

$$\sum_1^N \sum_1^N a_j [R(t_j, t_k)x_j, x_k] \bar{a}_k = \int_{-\infty}^{\infty} \left(\sum_1^N \sum_1^N a_j e^{2\pi i f t_j} [P(f)x_j, x_k] e^{-2\pi i f t_k} \right) df$$

$$\int_{-\infty}^{\infty} \left[p(f) \left(\sum_1^N a_j e^{2\pi i f t_j} x_j \right), \left(\sum_1^N a_k e^{2\pi i f t_k} x_k \right) \right] df$$

is nonnegative, since the integrand is for every f. The "technical"† condition (2.10) shows that $R(t)$ is defined for every t and is continuous. Note that

$$R(0) = \int_{-\infty}^{\infty} P(f)\, df > 0.$$

Also,

$$\overline{R(t)} = \int_{-\infty}^{\infty} e^{-2\pi i f t} \overline{P(f)}\, df = \int_{-\infty}^{\infty} e^{2\pi i f t} \overline{P(-f)}\, df$$

$$= R(t) \qquad \text{if } P(-f) = \overline{P(f)}.$$

Example 2.1. Let

$$P(f) = \frac{1}{k^2 + 4\pi^2 f^2}.$$

See Figure 2.3. Power spectral density is usually plotted in

$$\text{db} = (10 \log_{10} P(f)).$$

See Figure 2.4. The "3-db point" is the frequency at which the power is down 3 db from the maximum. For this example it is given by

$$k^2 + 4\pi^2 f^2 = 2k^2 \quad \text{or} \quad f = \frac{k}{2\pi}.$$

To calculate $R(t)$, let

$$\psi(f) = \frac{1}{k + 2\pi i f}, \qquad k > 0.$$

Then

$$P(f) = \psi(f)\overline{\psi(f)}.$$

† A "technical" condition is one which holds in the applications but must be specified for the mathematics.

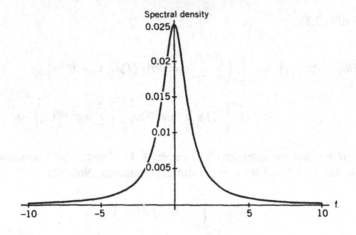

Figure 2.3. $P(f) = \dfrac{1}{k^2 + 4\pi^2 f^2}$, $k = 2\pi$.

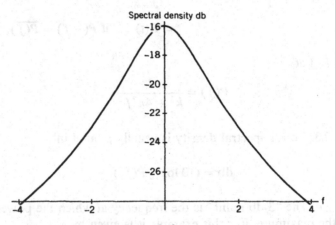

Figure 2.4. $-10 \log (k^2 + 4\pi^2 f^2)$, $k = 2\pi$, 3-db point at $f = 1$.

Now

$$\psi(f) = \int_0^\infty e^{-2\pi i f t} e^{-kt}\, dt$$

$$\overline{\psi(f)} = \int_0^\infty e^{2\pi i f t} e^{-kt}\, dt = \int_{-\infty}^0 e^{-2\pi i f t} e^{-k|t|}\, dt,$$

and hence we can calculate that

$$R(t) = \int_{-\infty}^\infty e^{2\pi i f t} |\psi(f)|^2\, df = \frac{e^{-k|t|}}{2k} \tag{2.12}$$

by the convolution theorem [cf. Eq. (4.16) on page 28]

$$\int_{-\infty}^{\infty} e^{2\pi i f t}\psi(f)\overline{\psi(f)}\,df = \int_0^{\infty} e^{-k(t+s)}e^{-ks}\,ds = \frac{e^{-kt}}{2k}, \qquad t\ge 0$$

$$= \int_{|t|}^{\infty} e^{-k(s-|t|)}e^{-ks}\,ds, \qquad t\le 0$$

$$= \int_0^{\infty} e^{-ks}e^{-k(s+|t|)}\,ds = \frac{e^{-k|t|}}{2k}.$$

This generalizes to the matrix case.

Let

$$P(f) = \psi(f)\psi(f)^*,$$

where

$$\psi(f) = (2\pi i f I - A)^{-1} = \int_0^{\infty} e^{-2\pi i f t}e^{At}\,dt, \qquad (2.13)$$

where A is a real-valued $(n\times n)$ stable matrix (real part of eigenvalues strictly less than zero). Here the stability of the matrix A assures condition (2.10). Note also that in this example $P(f)$ is nonsingular for every f. By the rules for products of Fourier transforms [cf. Eq.(4.16) on page 28] we have that

$$R(t) = \int_0^{\infty} e^{A(t+s)}e^{A^*s}\,ds = e^{At}\int_0^{\infty} e^{As}e^{A^*s}\,ds = e^{At}R(0), \qquad t\ge 0, \quad (2.14)$$

where

$$R(0) = \int_0^{\infty} e^{As}e^{A^*s}\,ds$$

$$R(-|t|) = \int_{|t|}^{\infty} e^{A(s-|t|)}e^{A^*s}\,ds$$

$$= \int_0^{\infty} e^{As}e^{A^*(|t|+s)}\,ds$$

$$= R(|t|)^* = R(0)e^{A^*|t|}. \qquad (2.15)$$

Example 2.2. Band-limited white noise. Let

$$P(f) = D \qquad -W < f < W, W > 0$$

$$= 0 \qquad |f| > W,$$

where D is $n\times n$, self-adjoint nonnegative definite. Then

$$R(t) = \int_{-\infty}^{\infty} e^{2\pi f t}P(f)\,df = D\int_{-W}^{W} e^{-2\pi i f t}\,df$$

$$= (2DW)\frac{\sin 2\pi Wt}{2\pi Wt} = R(0)\frac{\sin 2\pi Wt}{2\pi Wt} \qquad (2.16)$$

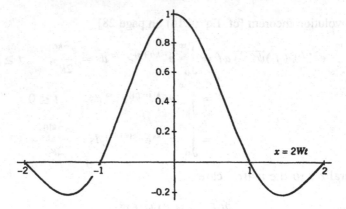

Figure 2.5. Curve for $R(t) = \dfrac{\sin(2\pi Wt)}{2\pi Wt}$ (2.16), $|t| \leq 2\left(\dfrac{1}{2W}\right)$.

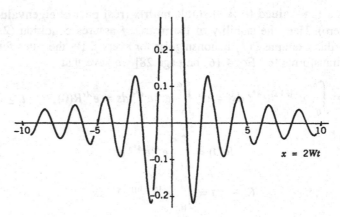

Figure 2.6. Curve for $R(t) = \dfrac{\sin(2\pi Wt)}{2\pi Wt}$ (2.16), $|t| \leq 10\left(\dfrac{1}{2W}\right)$.

since

$$2DW = R(0).$$

For $R(t)$ to be real, it is necessary that D be real. A zero-mean Gaussian process with this spectral density is called "band-limited white noise." A plot of (2.16) is given in Figures 2.5 and 2.6 for two different ranges of t.

Example 2.3. Let

$$P(f) = \left(\frac{\sin 2\pi f T}{2\pi f T}\right)^2, \qquad -\infty < f < \infty.$$

From Example 2.2 or otherwise,

$$\frac{\sin 2\pi f T}{2\pi f T} = \frac{1}{2T} \int_{-T}^{T} e^{-2\pi i f t} dt = \int_{-\infty}^{\infty} e^{-2\pi i f t} h(t) dt,$$

where

$$h(t) = \frac{1}{2T}, \qquad -T < t < T.$$

Hence, because $P(f)$ is the square of the absolute value of the Fourier transform of $h(\cdot)$, we may invoke the convolution theorem [cf. Eq. (4.5) on page 26] to obtain $R(\cdot)$ as the convolution

$$R(t) = \int_{-\infty}^{\infty} h(t-s)h(s)\, ds, \qquad -\infty < t < \infty$$

$$= \frac{1}{2T} \int_{-T}^{T} h(t-s)\, ds$$

$$= \frac{1}{2T} \int_{t-T}^{t+T} h(\sigma)\, d\sigma$$

$$= \frac{1}{2T} \left(1 - \frac{|t|}{2T} \right), \qquad 0 \le |t| \le 2T$$

$$= 0, \qquad |t| \ge 2T.$$

Example 2.4. Von Karman turbulence spectrum:

Here is a "physically" derived power spectral density (Von Karman's modification of the Kolmogorov formula for the spectral density of vertical wind velocity in turbulence), providing a standard example of a spectral density which is *not* a rational functional of the frequency

$$P(f) = \frac{2k\left(1 + \left(\frac{8}{3}\right)k^2(1.339)^2\omega^2\right)}{(1 + k^2(1.339)^2\omega^2)^{11/6}}, \qquad \omega = 2\pi f, \ -\infty < f < \infty,$$

where k is a scale parameter. Let

$$W(t) = \frac{1}{k}\left(\frac{-0.157}{k} t^{5/6} + 0.452 t^{-1/6} \right) \exp\left(\frac{-t}{1.339k} \right), \qquad 0 < t < \infty.$$

Then it is known that

$$P(f) = |\psi(f)|^2,$$

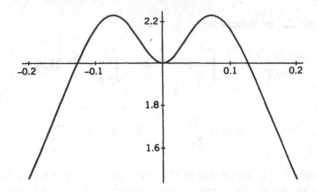

Figure 2.7. Von Karman spectral density in decibels versus frequency in Hertz; $k = 1$.

where

$$\psi(f) = \int_0^\infty e^{-2\pi i f t} W(t)\, dt.$$

However, we cannot express the covariance function in closed form. See Figure 2.7 for a plot of the spectral density in decibels ($10 \log_{10} P(f)$) for $k = 1$.

Here is another example of a nonrational spectral density:

$$P(f) = \frac{1}{(k^2 + 4\pi^2 f^2)^{v+(1/2)}}, \qquad v + 1/2 > 0.$$

The corresponding covariance function is available in closed form

$$R(t) = \frac{1}{\sqrt{\pi}} \left(\frac{t}{2k}\right)^v \frac{1}{\Gamma(v + 1/2)} K_v(kt), \qquad kt > 0,$$

where $K_v(\cdot)$ is the modified Hankel function (see reference 15 on page 35, for more on this function). For $v = 1/2$ we have

$$K_v(kt) = \sqrt{\pi/2kt}\, e^{-kt}, \qquad kt > 0.$$

2.2 CONSTRUCTING COVARIANCE FUNCTIONS: DISCRETE-PARAMETER PROCESSES

As in the continuous-parameter case, for strictly stationary random sequences we have

$$x_n, \qquad n \in I$$

$$E[x_n] = \text{constant}.$$

The covariance function is given by

$$R(m, n) = E[x(m)x(n)^*] - E[x_0]E[x_0]^*$$
$$= R(m - n, 0) = R(m - n),$$

where the (stationary) covariance sequence $\{R(n)\}$ satisfies

$$R(n)^* = R(-n) \tag{2.17}$$

$$\sum_1^N \sum_1^N a_i R(n_i - n_j)\bar{a}_j \geq 0. \tag{2.18}$$

The analog of Theorem 2.1 is now the following:

Theorem 2.3. Let $p(\lambda)$, $-1/2 < \lambda < 1/2$, be a nonnegative function such that

$$\int_{-1/2}^{1/2} p(\lambda)\, d\lambda < \infty.$$

Then

$$R(n) = \int_{-1/2}^{1/2} e^{2\pi i n \lambda} p(\lambda)\, d\lambda, \qquad n \in I$$

defines a covariance function. $R(\cdot)$ is real-valued if

$$p(\lambda) = p(-\lambda),$$

and in that case we can express $R(n)$ as the cosine transform

$$R(n) = 2 \int_0^{1/2} p(\lambda) \cos 2\pi n\lambda\, d\lambda.$$

Proof. The proof is similar to that of Theorem 2.1.

As in the continuous-parameter case, we may allow $p(\cdot)$ to contain δ-functions.

Finally Theorem 2.2 now becomes:

Theorem 2.4. Let $P(\lambda)$, $-1/2 < \lambda < 1/2$, be an $n \times n$ matrix function such that

$$P(\lambda)^* = P(\lambda)$$
$$P(\lambda) \geq 0$$

and

$$\int_{-1/2}^{1/2} \operatorname{Tr} P(\lambda)\, d\lambda < \infty.$$

Then

$$R(k) = \int_{-1/2}^{1/2} e^{2\pi ik\lambda} P(\lambda)\, d\lambda, \qquad k \in \mathbf{I} \tag{2.19}$$

defines an $n \times n$ matrix covariance function. It is real-valued if $\overline{P(\lambda)} = P(-\lambda)$.

Proof. The proof is again similar to that of Theorem 2.2.

Example 2.5. Discrete-time analog of Example 2.1. Let

$$\psi(\lambda) = (1 - \rho e^{-2\pi i\lambda})^{-1}, \qquad |\lambda| \le 1/2, 0 < \rho < 1$$

and let

$$
\begin{aligned}
p(\lambda) &= |\psi(\lambda)|^2 \\
&= \left(\sum_{0}^{\infty} \rho^k e^{-2\pi ik\lambda} \right) \left(\sum_{0}^{\infty} \rho^k e^{2\pi ik\lambda} \right)
\end{aligned}
$$

since

$$\sum_{0}^{\infty} \rho^k e^{-2\pi ik\lambda} = \frac{1}{1 - \rho e^{-2\pi i\lambda}}.$$

Hence by the Fourier convolution theorem [cf. Eq. (3.9) on page 24] we have

$$
\begin{aligned}
\int_{-1/2}^{1/2} e^{2\pi in\lambda} p(\lambda)\, d\lambda &= \sum_{0}^{\infty} \rho^{n+m} \rho^m, \qquad n > 0 \\
&= \sum_{0}^{\infty} \rho^m \rho^{m+|n|}, \qquad n < 0
\end{aligned}
$$

or

$$
\begin{aligned}
R(n) &= \frac{\rho^{|n|}}{1 - \rho^2} \\
&= R(0)\rho^{|n|}.
\end{aligned}
$$

Note that

$$p(\lambda) = \frac{1}{1 + \rho^2 - 2\rho \cos 2\pi\lambda} \tag{2.20}$$

and

$$\sum_{-\infty}^{\infty} \rho^{|n|} e^{-2\pi in\lambda} = \frac{1 - \rho^2}{1 + \rho^2 - 2\rho \cos 2\pi\lambda}.$$

See Figure 2.8 for a plot of $- 10 \log_{10} p(\lambda)(1 - \rho^2)$ for $\rho = 0.9$.

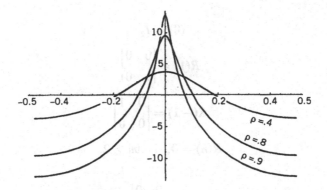

Figure 2.8. Curves for $10 \log_{10}\left(\dfrac{1 - \rho^2}{1 + \rho^2 - 2\rho \cos 2\pi\lambda}\right)$. Power spectral density in decibels.

We can readily generalize to the matrix case. Thus

$$P(\lambda) = \psi(\lambda)\psi(\lambda)^*, \tag{2.21}$$

where

$$\psi(\lambda) = (e^{2\pi i\lambda}I - A)^{-1},$$

where all the eigenvalues of A are strictly inside the unit circle in the complex plane (absolute values strictly less than one). Then

$$R(n) = A^n \sum_0^\infty A^k A^{*k} = A^n R(0), \tag{2.22}$$

where

$$R(0) = \sum_0^\infty A^k A^{*k}$$

$$R(-|n|) = R(0)(A^*)^{|n|}. \tag{2.23}$$

Example 2.6. Consider the 2×2 spectral density matrix

$$P(\lambda) = \begin{vmatrix} 1 & e^{2\pi i\lambda} \\ e^{-2\pi i\lambda} & 1 \end{vmatrix}, \qquad -1/2 < \lambda < 1/2.$$

We can verify that $P(\lambda)$ is self-adjoint and nonnegative definite for each λ. The corresponding covariance functions obtained from

$$R(n) = \int_{-1/2}^{1/2} e^{2\pi i\lambda n} p(\lambda) \, d\lambda$$

yields

$$R(0) = I_{2 \times 2}$$

$$R(1) = \begin{vmatrix} 0 & 0 \\ 1 & 0 \end{vmatrix}$$

$$R(-1) = \begin{vmatrix} 0 & 1 \\ 0 & 0 \end{vmatrix}$$

$$R(n) = 0, \qquad |n| \geq 2.$$

Define

$$A = \begin{vmatrix} 0 & 0 \\ 1 & 0 \end{vmatrix}.$$

Then

$$A^n = 0 \qquad \text{for } n \geq 2$$

(A is a "nilpotent" matrix). Thus $R(n)$ satisfies (2.22). In particular,

$$\psi(\lambda) = (e^{2\pi i \lambda} I - A)^{-1} = \begin{vmatrix} e^{-2\pi i \lambda} & 0 \\ e^{-4\pi i \lambda} & e^{-2\pi i \lambda} \end{vmatrix}$$

$$\psi(\lambda)^* = \begin{vmatrix} e^{2\pi i \lambda} & e^{4\pi i \lambda} \\ 0 & e^{2\pi i \lambda} \end{vmatrix}$$

and we can readily calculate that

$$P(\lambda) = \psi(\lambda)\psi(\lambda)^*.$$

2.3 CHARACTERIZATION OF COVARIANCE FUNCTIONS: SPECTRAL REPRESENTATION THEOREM

It turns out that the technique of constructing covariances we have described is in fact the only way to generate them. The general result is due to Bochner (see reference 1). We shall prove first a more restricted version of this result, which is adequate for our purposes. We begin with the continuous-parameter case.

2.3.1 Continuous-Parameter Processes

Theorem 2.5. Let $R(t)$, $-\infty < t < \infty$, be a one-dimensional continuous covariance function such that

$$\int_{-\infty}^{\infty} |R(t)| \, dt < \infty. \tag{2.24}$$

Then we can find a nonnegative continuous function $P(f)$, $-\infty < f < \infty$, such that

$$R(t) = \int_{-\infty}^{\infty} e^{2\pi i f t} P(f) \, df, \qquad -\infty < t < \infty, \qquad (2.25)$$

where in fact

$$P(f) = \int_{-\infty}^{\infty} e^{-2\pi i f t} R(t) \, dt. \qquad (2.26)$$

If $R(t)$ is real-valued, then $P(f) = P(-f)$.

Proof. By virtue of (2.24) we can define the Fourier transform as

$$\int_{-\infty}^{\infty} e^{-2\pi i f t} R(t) \, dt. \qquad (2.27)$$

It is enough to show that (2.27) is nonnegative for every f. We do this in a slightly "roundabout" way. Let

$$P_T(f) = \frac{1}{T} \int_0^T \int_0^T e^{2\pi i f t} R(t - s) e^{-2\pi i f s} \, ds \, dt.$$

Since $R(\cdot)$ is continuous, the integral can be approximated by finite sums of the form

$$\frac{1}{T} \sum_0^N \sum_0^N e^{2\pi i f t_k} R(t_k - t_j) e^{-2\pi i f t_j} \Delta t_k \Delta t_j,$$

but each of these sums is nonnegative. Hence

$$P_T(f) \geq 0.$$

Next we can transform the double integral into a single integral by the change of variable:

$$\tau = t - s$$
$$s = s.$$

Thus

$$P_T(f) = \int_{-T}^{T} R(\tau) \left(1 - \frac{|\tau|}{T} \right) e^{2\pi i f \tau} \, d\tau = \int_{-\infty}^{\infty} R_T(\tau) e^{2\pi i f \tau} \, d\tau, \qquad (2.28)$$

where

$$R_T(\tau) = R(\tau) \left(1 - \frac{|\tau|}{T} \right) \qquad -T < \tau < T$$
$$= 0 \qquad\qquad\qquad \tau > |T|. \qquad (2.29)$$

This function converges, as $T \to \infty$, to

$$R(\tau)$$

for each τ, $-\infty < \tau < \infty$. Also,

$$|R_T(\tau)| \leq |R(\tau)|, \qquad -\infty < \tau < \infty,$$

and we are given that

$$\int_{-\infty}^{\infty} |R(\tau)| \, d\tau < \infty.$$

Hence† as $T \to \infty$

$$P_T(f) \text{ converges to } \int_{-\infty}^{\infty} R(\tau) e^{2\pi i f \tau} \, d\tau$$

and for each f, the limit is nonnegative. Hence

$$P(f) = \int_{-\infty}^{\infty} R(\tau) e^{2\pi i f \tau} \, d\tau \geq 0,$$

where $P(f)$ is actually continuous in f because of (2.24). Now, taking inverse Fourier transforms, in (2.28), we have

$$\int_{-\infty}^{\infty} P_T(f) \, df = R(0) < \infty$$

and hence‡

$$\int_{-\infty}^{\infty} P(f) \, df \leq \lim \int_{-\infty}^{\infty} P_T(f) \, df = R(0) < \infty.$$

Thus we can take inverse Fourier transform to yield

$$R(t) = \int_{-\infty}^{\infty} e^{2\pi i f t} P(f) \, df$$

for every t, $-\infty < t < \infty$, as required.

Remark. Since $R(\cdot)$ is real and $R(t) = R(-t)$, it follows from (2.26) that $P(f) = P(-f)$.

---

† The mathematical technicality here is that we may invoke the Lebesgue bounded convergence theorem. Compare reference 2, p. 91.
‡ The technicality here is to use Fatou's lemma. Compare reference 2, p. 89.

There are cases where the covariance function does not satisfy (2.24)—see Example 1.1.3. To cover these cases it is necessary to allow $P(\cdot)$ in (2.25) to contain "δ-functions," as we have already seen [cf. (2.9)]. We shall not dwell on further generalization since for most of our work (2.24) will be satisfied, and the extension to include δ-functions will take care of *all* cases of interest to us.

The requirement that the covariance be continuous is essential for the theorem to hold. Here is a simple counterexample which is of some independent interest as well. Let

$$R(t) = 0, \qquad t \neq 0$$
$$= 1, \qquad t = 0.$$

This is a covariance function, but it cannot be represented as in (2.26)! For if so, we would have

$$R(0) = 1 = \int_{-\infty}^{\infty} P(f)\, df$$

while since (2.24) is satisfied,

$$P(f) = \int_{-\infty}^{\infty} e^{-2\pi i f t} R(t)\, dt = 0, \qquad -\infty < f < \infty,$$

which is a contradiction.

The function $P(f)$, $-\infty < f < \infty$, in (2.25) is called the spectral density of the process; often only the spectral density can be specified and the covariance function implicitly then as the Fourier transform. Also techniques are available for directly estimating the spectral density from a sample path. (See reference 1.)

Let us next state the matrix version of the theorem.

Theorem 2.6. Let $R(t)$, $-\infty < t < \infty$, be an $n \times n$ matrix covariance function

$$R(t) = \{r_{ij}(t)\}, \qquad i, j = 1, \ldots, n$$

where $r_{ij}(t)$ is continuous in t and

$$\int_{-\infty}^{\infty} |r_{ij}(t)|\, dt < \infty \qquad \text{for each } i, j.$$

Then there is a (spectral density) matrix function $P(f)$, continuous in f, $-\infty < f < \infty$, which is self-adjoint and nonnegative definite such that

$$R(t) = \int_{-\infty}^{\infty} e^{2\pi i f t} P(f)\, df$$

for every t, $-\infty < t < \infty$. If $R(t)$ is real-valued for every t, then

$$\overline{P(f)} = P(-f).$$

Proof. The proof is a straightforward extension of the one-dimensional case, considering $P_T(f)$ defined by (2.28) component by component.

It must be noted that $P(f) \neq P(-f)$ in general! On the other hand, if $R(t)$ is real,

$$\overline{P(f)} = P(-f),$$

as follows from

$$R(t) = \int_{-\infty}^{\infty} e^{-2\pi i f t} \overline{P(f)}\, df = \int_{-\infty}^{\infty} e^{2\pi i f t} P(-f)\, df.$$

Remark. The condition

$$\int_{-\infty}^{\infty} |r_{ij}(t)|\, dt < \infty \qquad \text{for every } i, j,$$

is equivalent to the condition that

$$\int_{-\infty}^{\infty} |[R(t)x, y]|\, dt < \infty$$

for every x, y in E^n and is also equivalent to the condition that

$$\int_{-\infty}^{\infty} \|R(t)\|_0\, dt < \infty.$$

The latter follows from the matrix norm inequalities on pages 4–9.

Example 2.7. Show that $R_T(\cdot)$ defined in (2.29) is a covariance function and that

$$P_T(f) = \int_{-\infty}^{\infty} P(f - \sigma) \frac{\sin^2 \pi \sigma T}{\pi^2 \sigma^2 T}\, d\sigma. \qquad (2.30)$$

Solution: Note first that from (2.28) $P_T(\cdot)$ is the Fourier transform of $R_T(\cdot)$.

The function

$$\left(1 - \frac{|\tau|}{T}\right), \qquad -T < \tau < T$$

$$0, \qquad\qquad \tau > T$$

has the Fourier transform (defined by Eq. (4.2) on page 25)

$$\frac{\sin^2 \pi f T}{\pi^2 f^2 T}, \qquad -\infty < f < \infty$$

and hence is a covariance function; and (2.30) follows by the Fourier convolution theorem [cf. Eq. (4.5) on page 26] applied to $R_T(\cdot)$. Also, since

$$\int_{-\infty}^{\infty} \frac{\sin^2 \pi f T}{\pi^2 f^2 T} \, df = 1$$

and as $T \to \infty$,

$$\frac{\sin^2 \pi f T}{\pi^2 f^2 T} = T\left(\frac{\sin \pi f T}{\pi f T}\right)^2 \to \infty \qquad \text{for } f = 0$$

$$\to 0 \qquad \text{for } f \neq 0$$

we see that

$$\frac{\sin^2 \pi f T}{\pi^2 f^2 T}$$

is a "δ-function sequence," and of course the integral in (2.30) converges to $P(f)$ as $T \to \infty$. A plot of the function

$$\frac{\sin^2 \pi f T}{\pi^2 f^2 T}$$

for various values of T is given in Figure 2.9.

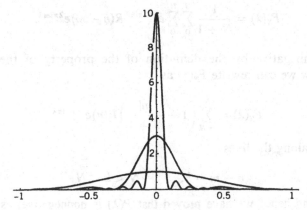

Figure 2.9. Curves for $P(f) = \dfrac{\sin^2 \pi f T}{\pi^2 f^2 T}$, $T = 1, 3, 10$.

2.3.2 Discrete-Parameter Processes

Let us now specialize to discrete-parameter processes and prove the analog of Theorem 2.5. Continuity is no longer relevant, but we shall retain the analog of (2.2).

Theorem 2.7. Let $R(n)$, $n \in I$ be a covariance function such that

$$\sum_{-\infty}^{\infty} |R(n)| < \infty. \tag{2.31}$$

Then

$$R(n) = \int_{-1/2}^{1/2} e^{2\pi i n \lambda} P(\lambda) \, d\lambda, \tag{2.32}$$

where

$$P(\lambda) \geq 0$$

and is continuous in $[-1/2, 1/2]$. In fact, $P(\lambda)$ is defined by the Fourier series

$$P(\lambda) = \sum_{-\infty}^{\infty} R(n) e^{-2\pi i n \lambda}. \tag{2.33}$$

Proof. Because of (2.31) we can define the Fourier series with $R(n)$ as coefficients to be

$$P(\lambda) = \sum_{-\infty}^{\infty} R(n) e^{2\pi i n \lambda}, \qquad -1/2 < \lambda < 1/2.$$

This function is continuous in $[-1/2, 1/2]$ and hence we have (2.32). However, we need to prove $P(\lambda)$ is nonnegative. We have to be somewhat roundabout for this purpose and let

$$P_N(\lambda) = \frac{1}{N+1} \sum_{0}^{N} \sum_{0}^{N} e^{-2\pi i n \lambda} R(n-m) e^{2\pi i m \lambda},$$

which is nonnegative by the definition of the property of the covariance function. Now we can rewrite $P_N(\cdot)$ as

$$P_N(\lambda) = \sum_{-N}^{N} \left(1 - \frac{|n|}{N+1}\right) R(n) e^{-2\pi i n \lambda} \tag{2.34}$$

by summing along the lines

$$n - m = k, \qquad -N < k < N.$$

for every λ and hence we have proved that $P(\lambda)$ is nonnegative, as required. If $R(\cdot)$ is real-valued, we have

But from (2.34) we have that

$$P_N(\lambda) \to P(\lambda)$$

$$P(-\lambda) = \sum_{-\infty}^{\infty} R(n)e^{-2\pi in\lambda} = \sum_{-\infty}^{\infty} R(-n)e^{2\pi in\lambda} = \sum_{-\infty}^{\infty} R(n)e^{2\pi in\lambda} = P(\lambda)$$

As in the continuous-parameter case, condition (2.31) is too restrictive. For instance, the covariance function in (1.1.14) does *not* satisfy it. To take care of this we allow δ-functions in the spectral density. Thus

$$P(\lambda) = \sum_{1}^{N} \sigma_k^2 \left(\frac{\delta(\lambda - \lambda_k) + \delta(\lambda + \lambda_k)}{2} \right)$$

corresponding to (1.1.14).

Let us next state the matrix version, analogous to Theorem 2.6.

Theorem 2.8. Let $R(k)$, $k \in I$, be an $n \times n$ matrix covariance function such that

$$\sum_{-\infty}^{\infty} |r_{ij}(k)| < \infty \qquad \text{for every } i, j$$

where

$$\{r_{ij}(k)\} = R(k).$$

Then there is a spectral density function, an $n \times n$ matrix function $P(\lambda)$, continuous in $-1/2 \leq \lambda \leq 1/2$ which is self-adjoint and nonnegative definite such that

$$R(k) = \int_{-1/2}^{1/2} e^{2\pi ik\lambda} P(\lambda) \, d\lambda, \qquad k \in I$$

and

$$P(\lambda) = \sum_{-\infty}^{\infty} R(k)e^{-2\pi ik\lambda}, \qquad -1/2 \leq \lambda \leq 1/2.$$

If $R(\cdot)$ is real-valued, then

$$\overline{P(\lambda)} = P(-\lambda).$$

Proof. The proof is again a straightforward extension of that of Theorem 2.7 and is omitted. We note that the condition for each i, j, namely

$$\sum_{-\infty}^{\infty} |r_{ij}(k)| < \infty,$$

is implied by

$$\sum_{-\infty}^{\infty} \|R(k)\|_0 < \infty.$$

As in the continuous-parameter case we see that the function

$$R(n) = \left(1 - \frac{|n|}{N + 1}\right), \qquad |n| \le N$$

$$= 0, \qquad\qquad\qquad |n| > N$$

defines a covariance function with spectral density

$$\sum_{-N}^{N} \left(1 - \frac{|n|}{N + 1}\right) e^{-2\pi i n\lambda} = \frac{1}{N + 1} \sum_{0}^{N} \sum_{0}^{N} e^{2\pi i \lambda(m - n)}$$

$$= \frac{1}{N + 1} \left|\sum_{0}^{N} e^{2\pi i n\lambda}\right|^2$$

$$= \frac{1}{N + 1} \left|\frac{1 - e^{2\pi i\lambda(N - 1)}}{1 - e^{2\pi i\lambda}}\right|^2$$

$$= \frac{1}{N + 1} \left(\frac{1 - \cos 2\pi\lambda(N + 1)}{1 - \cos 2\pi\lambda}\right)$$

$$= \frac{1}{N + 1} \frac{\sin^2 \pi\lambda(N + 1)}{\sin^2 \pi\lambda}.$$

Hence by the convolution theorem, the spectral density of

$$R_N(n) = R(n)\left(1 - \frac{|n|}{N + 1}\right), \qquad |n| \le N$$

$$= 0, \qquad\qquad\qquad |n| > N$$

is given by

$$P_N(\lambda) = \int_{-1/2}^{1/2} P(\lambda - \sigma)\left(\frac{\sin^2 \pi\sigma(N + 1)}{(N + 1)\sin^2 \pi\sigma}\right) d\sigma.$$

Note that

$$\int_{-1/2}^{1/2} \frac{\sin^2 \pi\sigma(N + 1)}{(N + 1)\sin^2 \pi\sigma} d\sigma = 1 \qquad (= R(0)!)$$

and as $N \to \infty$,

$$\frac{\sin^2 \pi\sigma(N+1)}{(N+1)\sin^2 \pi\sigma} \to \infty \qquad \text{for } \sigma = 0$$

$$\to 0 \qquad \text{for } \sigma \neq 0$$

or

$$\frac{\sin^2 \pi\sigma(N+1)}{(N+1)\sin^2 \pi\sigma}$$

is a δ-function sequence in N. (Compare the "Fejer kernel" in the theory of Fourier series—see reference 11 of the Review Chapter.)

Example 2.8. Discrete-parameter processes often arise by periodic "sampling" of continuous-parameter processes. Thus let

$$x_n = x(n\Delta), \tag{2.35}$$

where Δ is the sampling period, and $x(t)$, $-\infty < t < \infty$, is a (second-order) stationary stochastic process with mean zero and stationary covariance function $R(\cdot)$ and spectral density $P(\cdot)$. The covariance function of the process $\{x_n\}$ is

$$E[x(n\Delta)x(m\Delta)^*] = R(\overline{n-m}\Delta)$$

and is thus clearly (second-order) stationary also. Our main problem is then to determine the spectral density in terms of spectral density $P(\cdot)$ of the continuous-time process.

It is convenient to set

$$\Delta = \frac{1}{2W},$$

where $2W$ is the sampling "rate." We have

$$R(k\Delta) = \int_{-\infty}^{\infty} e^{2\pi i f k\Delta} P(f)\, df = \int_{-\infty}^{\infty} e^{\pi i f k/W} P(f)\, df. \tag{2.36}$$

The function

$$e^{\pi i f k/W}, \qquad -\infty < f < \infty$$

is periodic in f with period $2W$. Hence we may break up the range of integration of (2.36) into multiples of $2W$:

$$\int_{-\infty}^{\infty} e^{\pi i f k/W} P(f)\, df = \sum_{n=-\infty}^{\infty} \int_{(2n-1)W}^{(2n+1)W} e^{\pi i f k/W} P(f)\, df.$$

But each term

$$\int_{(2n-1)W}^{(2n+1)W} e^{\pi i f k/W} P(f)\, df$$

(exploiting the periodicity of $e^{\pi i f k/W}$)

$$= \int_{-W}^{W} e^{\pi i f k/W} P(f + 2nW)\, df.$$

Hence

$$R(k\Delta) = \sum_{n=-\infty}^{\infty} \int_{-W}^{W} e^{\pi i f k/W} P(f + 2nW)\, df. \tag{2.37}$$

Because $P(f)$ is nonnegative definite for each f and

$$\int_{-\infty}^{\infty} \text{Tr}\, P(f)\, df = \text{Tr}\, R(0) < \infty,$$

it is legitimate to change the order of summation and integration in (2.37) and hence

$$R(k\Delta) = \int_{-W}^{W} e^{\pi i f k/W} \sum_{n=-\infty}^{\infty} P(f + 2nW)\, df. \tag{2.38}$$

Finally, making a change of variable

$$\frac{f}{2W} = \lambda.$$

we have

$$R(k\Delta) = \int_{-1/2}^{1/2} e^{2\pi i n \lambda k} \left(\sum_{-\infty}^{\infty} P(2\lambda W + 2nW) \right)(2W)\, d\lambda.$$

Hence the spectral density of the process $\{x_n\}$ is given by

$$(2W) \sum_{-\infty}^{\infty} P(2\lambda W + 2nW), \qquad -1/2 \le \lambda \le 1/2, \tag{2.39}$$

even if the infinite series can rarely be expressed in closed form.
   For example, for

$$P(f) = \frac{1}{k^2 + 4\pi^2 f^2}$$

taking

$$2W = \frac{2\alpha k}{\pi}$$

we have

$$\sum_{-\infty}^{\infty} P(2\lambda W + 2nW) = \left(\frac{1}{k^2}\right) \sum_{-\infty}^{\infty} \frac{1}{1 + 16\alpha^2(n + \lambda)^2}.$$

For α sufficiently large, the term corresponding to $n = 0$ suffices. The case $\alpha = 1$ corresponds to taking W to be the "3 db-point" of $P(\cdot)$, with the frequency W defined by

$$P(W) = \frac{1}{2} P(0).$$

Finally, let us note that given any stationary discrete-time process $\{x_n\}$, $n \in \mathbf{I}$, with spectral density $p(\lambda)$, $-1/2 < \lambda < 1/2$, we can express it as

$$x_n = x(n),$$

where the continuous-parameter process $x(t)$, $-\infty < t < \infty$, has the spectral density

$$P(f) = p(f), \qquad -1/2 < f < 1/2,$$
$$= 0, \qquad |f| > 1/2.$$

This follows from (2.39) by taking

$$2W = 1$$

therein.

2.4 ELEMENTARY PROPERTIES OF STATIONARY COVARIANCE FUNCTIONS: CONTINUOUS-TIME MODELS

We list now some elementary yet oft-used properties of stationary covariance functions. We shall consider only the continuous-time case, but the results which do not involve differentiability will be valid for the discrete-time models as well.

Let $R(t)$, $-\infty < t < \infty$, denote an $n \times n$ stationary covariance function and let $P(\cdot)$ be the corresponding spectral density function. Then

1.

$$|\text{Tr } R(t)| \leq \text{Tr } R(0)$$
$$\|R(t)\|_0 \leq \|R(0)\|_0. \tag{2.40}$$

Proof. For each x, y in E_n, specializing (2.3), we have that the 2×2 matrix

$$\begin{vmatrix} [R(0)x, x] & [R(t)x, x] \\ [R(t)y, x] & [R(0)y, y] \end{vmatrix}$$

is nonnegative definite. Hence

$$|[R(t)x, y]| \leq \sqrt{[R(0)x, x]} \sqrt{[R(0)y, y]}.$$

In particular,

$$|[R(t)x, x]| \leq [R(0)x, x].$$

Let $\{e_i\}$ denote the coordinate basis vectors in E^n. Then

$$|[R(t)e_i, e_i]| \leq [R(0)e_i, e_i].$$

Hence

$$|Tr\ R(t)| = \left| \sum_1^n [R(t)e_i, e_i] \right| \leq \sum_1^n |[R(t)e_i, e_i]|$$

$$\leq \sum_1^n [R(0)e_i, e_i] = Tr\ R(0).$$

Next

$$\frac{|[R(t)x, y]|}{\|x\|\,\|y\|} \leq \frac{\sqrt{[R(0)x, x]}}{\|x\|} \frac{\sqrt{[R(0)y, y]}}{\|y\|} \leq \|R(0)\|_0,$$

and hence [cf. pages 4–9] it follows that

$$\|R(t)\|_0 \leq \|R(0)\|_0.$$

2. If $R(t)$ is once differentiable in a neighborhood of $t = 0$, then

$$Tr\ R'(0) = 0.$$

Proof. Since

$$[R(t)x, x] \leq [R(0)x, x],$$

zero will be a maximum point; and if the function is differentiable, the derivative must vanish there. Hence

$$\frac{d}{dt}[R(t)x, x] = 0, \qquad \text{at } t = 0.$$

Hence

$$\frac{d}{dt}\left(\sum_1^n [R(t)e_i, e_i] \right) = 0, \qquad \text{at } t = 0$$

or

$$Tr\ R'(0) = 0.$$

3. If $R(t)$ is twice differentiable at $t = 0$, then

$$R''(0) < 0$$

(strictly less than zero unless $R(t)$ is identically zero). Moreover,

$$-R''(0) = 4\pi^2 \int_{-\infty}^{\infty} f^2 P(f) \, df. \qquad (2.41)$$

Proof.

$$R''(0) = \lim_{t \to 0} \frac{R(2t) - 2R(0) + R(-2t)}{(2t)^2}.$$

But

$$R(2t) - 2R(0) + R(-2t) = \int_{-\infty}^{\infty} (e^{4\pi i f t} - 2 + e^{-4\pi i f t}) P(f) \, df$$

$$= \int_{-\infty}^{\infty} (e^{2\pi i f t} - e^{-2\pi i f t})^2 P(f) \, df$$

$$= (-1) \int_{-\infty}^{\infty} 4 \sin^2 2\pi f t P(f) \, df.$$

Hence

$$-R''(0) = \lim_{t \to 0} \int_{-\infty}^{\infty} \left(\frac{\sin 2\pi f t}{t} \right)^2 P(f) \, df.$$

But

$$\lim_{t \to 0} \left(\frac{\sin 2\pi f t}{t} \right)^2 = 4\pi^2 f^2.$$

This is enough to show that

$$\int_{-\infty}^{\infty} f^2 P(f) \, df < \infty$$

and hence

$$-R''(0) = \int_{-\infty}^{\infty} \lim_{t \to 0} \left(\frac{\sin 2\pi f t}{t} \right)^2 P(f) \, df = 4\pi^2 \int_{-\infty}^{\infty} f^2 P(f) \, df.$$

In particular, we see that

$$R''(0) \neq 0,$$

unless $R(t) = 0$ for all t.

4. Suppose

$$\int_{-\infty}^{\infty} |f|^n \operatorname{Tr} P(f) \, df < \infty.$$

Then

$$R(t) \text{ is } n\text{-times differentiable}$$

and

$$\frac{d^n}{dt^n} R(t) = i^n \int_{-\infty}^{\infty} (2\pi f)^n e^{2\pi i f t} P(f) \, df. \tag{2.42}$$

Proof. With $P(f)$ being self-adjoint and nonnegative definite, we obtain

$$|[P(f)x, y]| \leq \sqrt{[P(f)x, x]} \sqrt{[P(f)y, y]}$$
$$\leq (\operatorname{Tr} P(f)) \|x\|^2 \|y\|^2.$$

Hence

$$\int_{-\infty}^{\infty} |f|^n |[P(f)x, y]| \, df < \infty.$$

Thus in

$$[R(t)x, y] = \int_{-\infty}^{\infty} e^{2\pi i f t} [P(f)x, y] \, df$$

we may differentiate inside the integral sign n-times yielding (2.42).

In particular we see that the covariance function $R(\cdot)$ is twice differentiable at $t = 0$ if and only if

$$\int_{-\infty}^{\infty} f^2 \operatorname{Tr} P(f) \, df < \infty.$$

We shall mostly be concerned with spectral density functions which are rational† and hence we see that

$$\int_{-\infty}^{\infty} |f|^n P(f) \, df = \infty$$

for all $n >$ some n_0.

In the discrete-parameter case the differentiability question does not of course arise. We do have that

$$\int_{-1/2}^{1/2} |\lambda|^{2n} P(\lambda) \, d\lambda < \infty$$

for all n, although this is not of particular significance. What is of some significance is whether

$$\int_{-1/2}^{1/2} |\log \operatorname{Tr} P(\lambda)| \, d\lambda < \infty.$$

---

† "Rational" = ratio of polynomials in f. Hence n_0 = degree of denominator polynomial of highest degree.

In fact (see reference 1), in the 1×1 case if

$$\int_{-1/2}^{1/2} |\log P(\lambda)| \, d\lambda < \infty$$

we can write

$$P(\lambda) = |\psi(\lambda)|^2,$$

where

$$\psi(\lambda) = \sum_{0}^{\infty} w_k e^{2\pi i k \lambda}.$$

See Chapter 3 for what this representation implies.

5. Let

$$P(f) = \{p_{ij}(f)\}_{i,j=1,\ldots,n}, \qquad -\infty < f < \infty$$

be a covariance matrix for each f. Then

$$\int_{-\infty}^{\infty} P(f) \, df < \infty$$

if and only if

$$\int_{-\infty}^{\infty} \text{Tr} \, P(f) \, df < \infty.$$

Proof. The proof follows from

$$\left(\int_{-\infty}^{\infty} |p_{ij}(f)| \, df \right)^2 \leq \left(\int_{-\infty}^{\infty} p_{ii}(f) \, df \right) \left(\int_{-\infty}^{\infty} p_{jj}(f) \, df \right).$$

6. If the covariance function is real-valued, for each real-valued x in E^n we obtain

$$[P(f)x, x] = [P(-f)x, x], \qquad -\infty < f < \infty$$

and

$$\text{Tr} \, P(f) = \text{Tr} \, P(-f).$$

Proof.

$$[P(f)x, x] \geq 0$$

and

$$[P(f)x, x] = \int_{-\infty}^{\infty} e^{-2\pi i f t}[R(t)x, x] \, dt.$$

Hence the integral is real-valued and hence equal to the conjugate

$$\int_{-\infty}^{\infty} e^{2\pi i f t}\overline{[R(t)x, x]}\, dt.$$

But $R(t)$ is real-valued (all entries in the matrix are real) and because x is also real-valued we obtain

$$\overline{[R(t)x, x]} = [R(t)x, x].$$

Hence

$$[P(f)x, x] = \int_{-\infty}^{\infty} e^{2\pi i f t}[R(t)x, x]\, dt = [P(-f)x, x].$$

In particular if $\{e_i\}$ denote the coordinate basis vectors in $\mathbf{E}^n$, then

$$[P(f)e_i, e_i] = [P(-f)e_i, e_i]$$

and hence

$$\operatorname{Tr} P(f) = \operatorname{Tr} P(-f).$$

In particular, all the diagonal entries in $P(f)$ are nonnegative. These results also follow from

$$\overline{P(f)} = P(-f),$$

which holds as we have seen for real-valued covariance functions.

7. Let $R_1(t)$ and $R_2(t)$ be two stationary covariances $-\infty < t < \infty$. With the product of two covariance functions being a covariance function, we know that

$$R_3(t) = R_1(t)R_2(t)$$

defines a stationary covariance. The corresponding spectral density is the convolution of the spectral densities by the convolution theorem [cf. Eq. (4.6) on page 26]. Thus

$$P_3(f) = \int_{-\infty}^{\infty} e^{-2\pi i f t}R_3(t) = \int_{-\infty}^{\infty} P_1(f - \sigma)P_2(\sigma)\, d\sigma$$

$$= \int_{-\infty}^{\infty} P_2(f - \sigma)P_1(\sigma)\, d\sigma, \qquad -\infty < t < \infty,$$

where $P_3(\cdot)$, $P_2(\cdot)$ and $P_1(\cdot)$ are the spectral densities corresponding to $R_3(\cdot)$, $R_2(\cdot)$, and $R_1(\cdot)$ respectively. In the discrete-time analog,

$$R_3(k) = R_1(k)R_2(k)$$

is a stationary covariance if $R_1(\cdot)$ and $R_2(\cdot)$ are. The corresponding spectral density is given by

$$p_3(\lambda) = \int_{-1/2}^{1/2} p_1(\lambda - \sigma)p_2(\sigma)\,d\sigma = \int_{-1/2}^{1/2} p_2(\lambda - \sigma)p_1(\sigma)\,d\sigma,$$
$$-1/2 < \lambda < 1/2,$$

where, in the integrals, $p_1(\lambda)$ and $p_2(\lambda)$ are defined for *all* λ by making them periodic with period one.

PROBLEMS

2.1. If $R(t)$, $-\infty < t < \infty$, is a covariance function, so is

$$R(t) \cos 2\pi f_c t$$

for any frequency f_c. In fact the corresponding spectral density is given by

$$\frac{P(f - f_c) + P(f + f_c)}{2},$$

where $P(\cdot)$ is the spectral density corresponding to $R(\cdot)$. Generalize to

$$R(t) \cos^n (2\pi f_c t)$$

for any positive integer n.
  State and prove the discrete-parameter analog.

2.2. Let $x(t)$, $-\infty < t < \infty$, be a stationary Gaussian process with zero mean and covariance function $R(\cdot)$ and spectral density $P(\cdot)$. Let $y(t)$, $-\infty < t < \infty$, be a similar process with the same mean and covariance, independent of the process $x(\cdot)$. Calculate the covariance and spectral density of the process

$$z(t) = x(t) \cos 2\pi f_c t + y(t) \sin 2\pi f_c t.$$

Sketch the spectral density function.

2.3. Use Problem 2.2 to find the spectral density corresponding to

(a) $\qquad (\exp(-\mu|t|)) \cos 2\pi f_c t, \qquad -\infty < t < \infty.$

(b) $\qquad \left(\dfrac{\sin 2\pi W t}{2\pi W t}\right) \cos 2\pi f_c t, \qquad -\infty < t < \infty, \ W < f_c.$

2.4. Let $x(t)$, $-\infty < t < \infty$, denote a stationary Gaussian process with mean zero and spectral density $P(\cdot)$ and stationary covariance function $R(\cdot)$. Show that the covariance function of the lognormal process

$$y(t) = e^{x(t)}$$

is given by

$$e^{R(0)}[e^{R(t)} - 1].$$

Specialize to the case

$$R(t) = e^{-k|t|}, \qquad k > 0$$

and determine also the spectral density.
　　Answer:

$$P(f) = e \sum_{1}^{\infty} \frac{2k}{(n - 1!)(4\pi^2 f^2 + k^2 n^2)}, \qquad -\infty < f < \infty.$$

2.5. Show that the function

$$F(t) = \exp(-t^{2n}), \qquad -\infty < t < \infty$$

(even though the appearance is similar to $\exp(-t^2)$) cannot be a covariance function for any n bigger than one.
　　Hint: Calculate $F''(0)$.
　　Let

$$F(t) = \cos 2\pi t - \cos 2\pi \sqrt{1 + t^2}, \qquad -\infty < t < \infty.$$

Show that even though

$$-F''(0) > 0,$$

this is not a covariance function. Is $-F''(t)$, $-\infty < t < \infty$, a covariance function?

2.6. Show that the spectral density function of white-noise sequence is given by

$$P(\lambda) = \Lambda, \qquad -1/2 \leq \lambda \leq 1/2,$$

where

$$\Lambda \text{ is the white-noise covariance matrix.}$$

2.7. Define the sequence

$$R(n) = 1, \qquad n = 0, \pm 1$$
$$= 0, \qquad \text{otherwise.}$$

Is this a covariance function?

Define

$$R(0) = 6$$
$$R(\pm 1) = -4$$
$$R(\pm 2) = 1$$
$$R(n) = 0, \qquad |n| > 2.$$

Is this a covariance function?
Define

$$R(\pm n) = N - |n|, \qquad |n| \leq N$$
$$= 0, \qquad\qquad |n| > N.$$

Is this a covariance function?

2.8. Let A be an $n \times n$ matrix with all its eigenvalues strictly inside the unit circle. Define

$$R(m) = A^m D, \qquad m \geq 0$$
$$= DA^{*|m|}, \qquad m \leq 0$$

where D is $n \times n$. What conditions should D satisfy in order that $R(n)$ so defined yields a covariance function?

2.9. Let $\{\zeta_k\}$ be a 1×1 IID sequence with mean zero. Let $x(t)$, $-\infty < t < \infty$, be a zero-mean stationary Gaussian process. Define

$$x_n = x\left(\sum_1^n \zeta_k \right), \qquad n \geq 1.$$

Find the mean and covariance functions of the process $\{x_n\}$. This is a version of "random sampling" in contrast to the usual periodic sampling.
Hint:

$$E[x_n] = 0$$

$$E[x_m \; x_n] = E\left[R\left(\sum_{m+1}^n \zeta_k \right) \right], \qquad n > m,$$

where $R(\cdot)$ is the covariance function of the process $x(\cdot)$. The right side

$$= \int_{-\infty}^{\infty} E\left[e^{i\left(\sum_{m+1}^n \zeta_k\right) f} \right] P(f)\, df$$

$$R(m, n) = \int_{-\infty}^{\infty} C(f)^{n-m} P(f)\, df, \qquad n > m.$$

where $P(\cdot)$ is the spectral density corresponding to $R(\cdot)$ and $C(\cdot)$ is the characteristic function corresponding to the common distribution of $\{\zeta_k\}$.

2.10. Let $\{x_k\}$ denote a 1×1 stationary Gaussian process. Let $\{y_k\}$ denote the $n \times 1$ process

$$y_k = \begin{vmatrix} x_k \\ x_{k+1} \\ \vdots \\ x_{k+n-1} \end{vmatrix}.$$

Find the covariance function and spectral density of the process $\{y_k\}$.

2.11. Let $\{x_k\}$ denote an $n \times 1$ stationary Gaussian process. Let C be an $m \times n$ matrix, and let

$$z_k = Cx_k.$$

Find the covariance function and spectral density of the process $\{z_k\}$.

2.12. Let $\{x_k\}$ be an $n \times 1$ zero-mean stationary process with covariance function $R(k)$ such that $R(0)$ is singular. Express x_k as

$$x_k = L\zeta_k,$$

where $\{\zeta_k\}$ is an $m \times 1$ stationary process with

$$E[\zeta_k \quad \zeta_k^*] = I_{m \times m},$$

where m is the rank of $R(0)$; L is $n \times m$ and

$$L^*L = D > 0,$$

where D is $m \times m$ diagonal.

Hint: Use the eigenvectors of $R(0)$ corresponding to nonzero eigenvalues as a new basis.

2.13. Let $R_1(k)$ and $R_2(k)$, $k \in \mathbf{I}$, denote 1×1 stationary covariance functions. Show that

$$R_3(k) = R_1(k)R_2(k)$$

is a stationary covariance function and find the spectral density.

2.14. Let $\{x_k\}$ denote a zero-mean, 1×1, stationary Gaussian process. Define for fixed m the process

$$z_k = x_k x_{k+m}.$$

Show that the process $\{z_k\}$ is second-order stationary and calculate the mean and covariance functions as well as the spectral density. Specialize to the case where

$$E(x_k x_{k+p}) = \rho^{|p|}, \qquad 0 \le \rho < 1$$

and determine the spectral density explicitly.

2.15. Let $\{\Lambda_k\}$ be a sequence of covariance matrices (self-adjoint, nonnegative definite) of some dimension, say $n \times m$. Show that

$$\|\Lambda_k\| \to 0 \qquad \text{as } k \to \infty$$

as soon as

$$\mathrm{Tr}\,\Lambda_k \to 0, \qquad \text{as } k \to \infty,$$

and conversely.

Hint: Use

$$|[\Lambda_n e_i, e_j]| \le \sqrt{[\Lambda_n e_i, e_i]}\sqrt{[\Lambda_n e_j, e_j]}.$$

2.16. Let

$$P(f) = \frac{1}{(1 + 4\pi^2 f^2)^2}\begin{vmatrix} 4\pi^2 f^2 + 5 & -2 - 4\pi i f \\ -2 + 4\pi i f & 1 + 4\pi^2 f^2 \end{vmatrix}, \qquad -\infty < f < \infty.$$

Show that $\overline{P(f)} = P(-f)$. Is this a spectral density function? If so, how many times differentiable is the covarance function? Denoting the covariance function by $R(\cdot)$, calculate

$$\int_{-\infty}^{\infty} R(t)\, dt$$

and show that it is a covariance matrix. Repeat the problem with

$$P(f) = \frac{1}{(1 + 4\pi^2 f^2)^2}\begin{vmatrix} 1 & -2\pi i f \\ 2\pi i f & 4\pi^2 f^2 \end{vmatrix}.$$

Let

$$p(f) = P_{11}(f), \qquad -\infty < f < \infty$$

(the 11 component of $P(\cdot)$). Repeat both problems for $p(\cdot)$. Plot $p(\cdot)$ in decibels and compare with Figure 2.4. Mark the 3-db point.

Moral: Some terms in the covariance matrix function may be differentiable when others are not.

2.17. Let

$$A = \begin{vmatrix} 0 & 1 & 0 \\ 0 & 0 & 1 \\ -a_0 & -a_1 & -a_2 \end{vmatrix}.$$

Calculate

$$P(f) = (2\pi i f - A)^{-1} D (2\pi i f - A)^{*-1}, \qquad -\infty < f < \infty,$$

where

$$D = \begin{vmatrix} 0 & 0 & 0 \\ 0 & 0 & 0 \\ 0 & 0 & 1 \end{vmatrix}.$$

You may assume that all eigenvalues of A have strictly negative real parts. Show that

$$\int_{-\infty}^{\infty} f^{2k} \operatorname{Tr} P(f) df = \infty, \qquad k \geq 3.$$

Show that $\overline{P(f)} = P(-f)$, if A is real.

2.18. Show that

$$R(t) = e^{-bt}(\cos t + b \sin t), \qquad 0 \leq t$$
$$= R(-t), \qquad t \leq 0$$

is a stationary covariance function, for every $b > 0$, and determine the corresponding density. More generally, given $b > 0$, determine the range of values of ϕ, $|\phi| < \pi/2$ for which

$$R(t) = e^{-b|t|} \cos(|t| - \phi), \qquad -\infty < t < \infty$$

is a stationary covariance function, and the corresponding spectral density.

Hint: Calculate Fourier transform.
Answer:

$$|\tan \phi| \leq b.$$

Hint: Transforms of even functions being real-valued, we can readily verify that the Fourier transform of

$$e^{-b|t|} \cos(|t| - \phi)$$

is given by

$$\text{Re} \int_{-\infty}^{\infty} e^{(-2\pi i f t - b|t| + i|t| - i\phi)}\, dt$$

$$= \text{Re } e^{-i\phi}\left(\int_{0}^{\infty} e^{-2\pi i f t - bt + it}\, dt + \int_{0}^{\infty} e^{2\pi i f t - bt + it}\, dt\right)$$

$$= \text{Re } e^{-i\phi}\left(\frac{1}{b - i(1 + 2\pi f)} + \frac{1}{b - i(1 - 2\pi f)}\right)$$

$$= \text{Re } e^{-i\phi}\left(\frac{b + i(1 + 2\pi f)}{b^2 + (1 + 2\pi f)^2} + \frac{b + i(1 - 2\pi f)}{b^2 + (1 - 2\pi f)^2}\right)$$

$$= \text{Re } e^{-i\phi}\left(\frac{2b(b^2 + 1 + 4\pi^2 f^2) + i(2b^2 + 2 - 8\pi^2 f^2)}{(b^2 + (1 - 2\pi f)^2)(b^2 + (1 + 2\pi f)^2)}\right)$$

$$= \frac{(2b^2 + 2)(b\cos\phi + \sin\phi) + 8\pi^2 f^2(b\cos\phi - \sin\phi)}{(b^2 + (1 - 2\pi f)^2)(b^2 + (1 + 2\pi f)^2)}.$$

2.19. Examples of stationary covariance functions

(a) $R(t) = (1 + |t|)e^{-|t|}$

(b) $R(t) = e^{-|t|}\dfrac{\sin t}{t}$

(c) $R(t) = e^{-|t|/\sqrt{2}} \sin\left(\dfrac{\pi}{4} + \dfrac{|t|}{\sqrt{2}}\right).$

In each case calculate $R''(0)$. Verify that the corresponding spectral densities are

(a) $P(f) = \dfrac{4}{(1 + 4\pi^2 f^2)^2}$

(b) $P(f) = \text{Arctan}\left(\dfrac{2}{4\pi^2 f^2}\right)$

(c) $P(f) = \dfrac{2}{1 + (2\pi f)^4}.$

In each case calculate

$$\int_{-\infty}^{\infty} f^2 P(f)\, df.$$

Hint: $= \infty$ for case (b).

2.20. show that the process defined in Problem 1.1.2 is "asymptotically" second-order stationary; that is,

$$E[x(t)] \to 0 \qquad \text{as } t \to \infty$$

$$\lim_{L \to \infty} R(t_1 + L, t_2 + L) = R(t_1 - t_2).$$

Calculate the stationary covariance and the corresponding spectral density.

2.21. Let $\{x_n\}$ be a stationary stochastic process, $n \in I$ with covariance function

$$R(n) = \rho^{|n|}, \qquad 0 < \rho < 1.$$

Define $z_n = x_{pn}$, where p is a nonzero positive integer. Find the spectral density of the process $\{z_n\}$. What happens as $p \to \infty$?
    Answer:

$$\frac{(1 - \rho^{2p})}{1 + \rho^{2p} - 2\rho^p \cos 2\pi\lambda}.$$

2.22. Construct a Gaussian process with the same mean and covariance function as the process z_k defined in Problem 2.15 with $m = 0$ therein, and draw some sample paths of the Gaussian process. Indicate how they differ from the sample paths of the original process.

2.23. Let $\{N_k\}$ denote one-dimensional white noise with unit covariance. Calculate the mean and covariance function of the process

$$y_k = (N_k - 2N_{k-1} + N_{k-2})^2.$$

Determine also the spectral density.

2.24. Let

$$x(t) = \begin{vmatrix} x_1(t) \\ x_2(t) \end{vmatrix}, \qquad -\infty < t < \infty$$

be a 2×1 stationary Gaussian process with zero mean. Let

$$A(t) = \sqrt{x_1(t)^2 + x_2(t)^2}.$$

Show that $A(t)$, $-\infty < t < \infty$ is also a stationary process. Calculate the (first-order) density function of $A(t)$ given that

$$E[x(t)x(s)^*] = \begin{vmatrix} r_{11}(t-s) & r_{12}(t-s) \\ r_{12}(t-s) & r_{11}(t-s) \end{vmatrix}.$$

Calculate $E[A(t)^2 A(s)^2]$.

Answer: Density function

$$= \frac{r}{\sigma^2\sqrt{1-\rho^2}} I_0\left(\frac{r^2\rho}{2\sigma^2(1-\rho^2)}\right) \exp\left(\frac{-r^2}{2\sigma^2(1-\rho^2)}\right),$$

where

$$\sigma^2 = r_{11}(0), \; r_{12}(0) = \rho\sigma^2$$

$I_0(\cdot)$: Bessel function of order zero with imaginary argument

(reduces to Rayleigh when $\rho = 0$).

2.25. Let R be an $(n + p) \times (n + p)$ "covariance" matrix. Let it be partitioned as

$$R = \begin{vmatrix} R_{11} & R_{12} \\ R_{21} & R_{22} \end{vmatrix},$$

where

$$R_{11} \text{ is } n \times n$$
$$R_{22} \text{ is } p \times p$$
$$R_{12} \text{ is } n \times p$$
$$R_{21} \text{ is } R_{12}^*.$$

Show that

(a) $$\|R_{21}\|^2 = \|R_{12}\|^2 \leq (\text{Tr } R_{11})(\text{Tr } R_{22})$$

(b) $$\|R_{12}\|_0^2 = \|R_{21}\|_0^2 \leq \|R_{11}\|_0 \cdot \|R_{22}\|_0.$$

Hint: This is an extension of (2.40) and is an easy consequence of

$$|[R_{12}x, y]|^2 \leq [R_{11}x, x][R_{22}y, y], \qquad x \in E^p, y \in E^n,$$

itself deriving from the nonnegative definiteness property of R.

2.26. Show that if a stationary covariance $R(t)$ is twice continuously differentiable in $-\infty < t < \infty$, then

$$-R''(t), \qquad -\infty < t < \infty$$

is also a covariance function. What is the corresponding spectral density?

2.27. Let

$$p(\lambda) = 1 - 2|\lambda|, \qquad -1/2 \leq \lambda \leq 1/2,$$

denote the spectral density of a stationary discrete-parameter process $\{x_n\}$,

$n \in I$. Find the corresponding covariance function. Let

$$y_n = x_{np},$$

where p is an even positive integer. Calculate the spectral density of the process $\{y_n\}$.

Hint:

$$R(n) = 2 \int_0^{1/2} (1 - 2\lambda) \cos 2\pi n\lambda \, d\lambda$$

$$= \frac{1}{2} \qquad \text{for } n = 0.$$

Integrate by parts for $n \neq 0$ to obtain

$$R(n) = 4 \int_0^{1/2} \frac{\sin 2\pi n\lambda}{2\pi n} \, d\lambda = \frac{1 - \cos \pi n}{\pi^2 n^2}$$

$$= 0 \qquad \text{for } n \text{ even}$$

$$= 2/\pi^2 n^2 \qquad \text{for } n \text{ odd}.$$

2.28. Show that

$$R(t) = (\cos 2\pi t)^n, \qquad -\infty < t < \infty,$$

is a covariance function for every integer $n \geq 1$. Determine the corresponding spectral density for

(a) $n = 2$
(b) $n = 3$.

Hint: Use

$$(\cos x)^n = \frac{1}{2^n} \sum_0^n \frac{n!}{p!(n-p)!} \cos (n - 2p)x.$$

2.29. Determine whether the covariance function corresponding to the Von Karman turbulence spectral density (Example 2.4) is twice differentiable at the origin.

2.30. Consider the general case of Problem 2.22. Thus let $\{x_n\}$ be a stationary stochastic process, $n \in I$, with spectral density $p(\lambda)$, $-1/2 \leq \lambda \leq 1/2$. Define

$$z_n = x_{qn},$$

where q is a nonzero integer. Using (2.39) or otherwise, show that the spectral density of the process $\{z_n\}$ can be expressed as

$$\left(\frac{1}{q}\right)\sum_{n=-\infty}^{\infty}\tilde{p}\left(\frac{\lambda+n}{q}\right), \qquad -1/2 \le \lambda \le 1/2,$$

where $\tilde{p}(\cdot)$ is defined as

$$\tilde{p}(\lambda) = p(\lambda), \qquad -1/2 \le \lambda \le 1/2$$
$$= 0, \qquad |\lambda| > 1/2.$$

2.31. Let $\{x_n\}$ be a 1×1 stationary Gaussian process with mean zero and stationary covariance function $R(m)$. Find the mean and covariance function of the process

$$y_n = \cos x_n.$$

Answer:

$$E[y_n] = e^{-R(0)/2}$$

$$E[y_{n+m} \ y_n] = e^{-R(0)} \cosh R(m).$$

Calculate the spectral density of the process $\{y_n\}$, when $R(m) = \rho^{|m|}$, $0 < \rho < 1$.

2.32. Let $x(t)$, $-\infty < t < \infty$, be a stationary process with spectral density $P(\cdot)$. Let v be a 1×1 random variable with density function denoted $p(\cdot)$. Show that the process

$$x(vt), \qquad -\infty < t < \infty$$

is also a stationary random process with spectral density given by

$$\int_0^\infty P\left(\frac{f}{v}\right)\left(\frac{p(v)+p(-v)}{v}\right)dv, \qquad -\infty < f < \infty.$$

2.33. Let $R(t)$, $-\infty < t < \infty$, denote a 1×1 stationary covariance function.

(a) Show that

$$\begin{vmatrix} R(t) & R(v+t) \\ R(v-t) & R(t) \end{vmatrix}, \qquad -\infty < t < \infty$$

defines a 2×2 stationary covariance function for any (real) v. Calculate the corresponding spectral density.

Answer:

$$\begin{vmatrix} 1 & e^{-2\pi i f v} \\ e^{2\pi i f v} & 1 \end{vmatrix} p(f).$$

(b) Show that (assuming the necessary differentiability)

$$\frac{R'(t)}{t} \geq R''(0)$$

and

$$R''(t) \geq R''(0), \qquad \text{for } 0 \leq t < \infty.$$

2.34. Show that if a 1×1 (real-valued) stationary covariance function $R(t)$, $-\infty < t < \infty$, satisfies the condition

$$R(0) = R(1),$$

then it must be of the form

$$R(t) = \sum_0^\infty a_n^2 \cos 2\pi n t, \qquad \sum_0^\infty a_n^2 < \infty.$$

Repeat the problem for the condition

$$R''(1) = R''(0).$$

Hint: Let $P(\cdot)$ denote the spectral density.

$$R(0) - R(1) = 2 \int_0^\infty (1 - \cos 2\pi f) P(f) \, df$$

$$= 0$$

implies that $P(f) = 0$ except where

$$1 - \cos 2\pi f = 0.$$

Hence $P(\cdot)$ must reduce to a δ-function at $f = n$, nonnegative integer.

2.35. Let

$$v_k = s_k + N_k, \qquad k \in \mathbf{I},$$

where $\{s_k\}$, $\{N_k\}$ are independent zero-mean stationary processes with spectral density $P_{ss}(\cdot)$ and $P_{NN}(\cdot)$, respectively. Find the spectral density of the process

$$\begin{vmatrix} s_k \\ v_k \end{vmatrix}, \qquad k \in \mathbf{I},$$

Answer:

$$\begin{vmatrix} P_{ss}(\lambda) & P_{ss}(\lambda) \\ P_{ss}(\lambda) & P_{ss}(\lambda) + P_{NN}(\lambda) \end{vmatrix}, \qquad -1/2 \leq \lambda \leq 1/2.$$

NOTES AND COMMENTS

The authoritative reference for the rigorous mathematical treatment is still reference 1; otherwise the material is mostly standard fare in engineering [3, 4].

REFERENCES

Mathematical Treatises

1. J. L. Doob. *Stochastic Processes.* John Wiley and Sons, 1953.
2. R. L. Royden. *Real Analysis.* Macmillan, 1988.

Recent Publications

3. R. E. Mortensen. *Random Systems and Signals.* John Wiley and Sons, 1986.
4. A. Papoulis. *Probability, Random Variables and Stochastic Processes.* McGraw-Hill, 1984.

3

RESPONSE OF LINEAR SYSTEMS
TO RANDOM INPUTS:
DISCRETE-TIME MODELS

In this chapter and the next, we consider the problem of system response to random inputs—specifically we show how to express the second-order properties of the output process of a linear system in terms of the second-order properties of the input process. Important on its own in studying the response of systems to random disturbances, it turns out to be also crucial in simulating random processes, as we shall see in Chapter 7. To minimize mathematical technicalities we shall begin in this chapter with discrete-time models.

3.1 "WEIGHTING PATTERN" SYSTEM MODELS

Some familiarity with the rudiments of how we describe linear systems will be helpful at this point. Our analysis depends on how the system response or "output" is specified in terms of the "input." We shall only be concerned with "physically realizable" systems. The simplest description is in terms of the system weighting pattern:

$$v_n = \sum_0^n W_{n,j} u_j, \qquad n \geq 0, \tag{3.1.1}$$

where we have taken $n = 0$ as the "start time" prior to which the system was quiescent, so that it is responding only to the input $\{u_k\}$, $k \geq 0$. At each time instant n, the system output is a weighted average of "all the available" input(s) up to time n ("physical realizability"), and

$$\{W_{ij}\}, \qquad 0 \leq j \leq i < \infty$$

is the "weighted pattern" specified for the system. Each W_{ij} is a rectangular matrix: say $p \times m$, with inputs u_k being $m \times 1$ and the output v_k being $p \times 1$.

There is no problem in allowing the input $\{u_k\}$ to be a random process. The output is then readily seen to be a random process in our definition. Using (3.1.1) we can deduce, in principle, the joint distribution of any finite number of the $\{v_n\}$. The second-order properties of the output process can be explicitly calculated in terms of those of the input process (and no more!) as follows

$$E[v_n] = \sum_0^n W_{n,k} E[u_k]$$

$$E[v_n \ v_m^*] = E\left[\left(\sum_0^n W_{n,k} u_k\right)\left(\sum_0^n W_{m,j} u_j\right)^*\right] = \sum_{k=0}^n \sum_{j=0}^m W_{n,k} E[u_k \ u_j^*] W_{m,j}^*,$$

where

$$E[u_k], \qquad E[u_k \ u_j^*]$$

are given as part of the description of the input process. What then is the big deal, one may ask? The answer is that we are primarily interested in the "steady-state" response of "time-invariant" systems to **stationary** input processes. In our weighting pattern description, time invariance means that

$$W_{n,k} = W_{n-k}, \qquad n \geq k,$$

where of course by physical realizability we have

$$W_j = 0, \qquad j < 0,$$

so that we can rewrite (3.1.1) as

$$v_n = \sum_0^n W_{n-k} u_k, \qquad n \geq 0. \tag{3.1.2}$$

3.1.1 Systems with Finite Memory

To simplify matters we shall first consider systems with "finite memory," where

$$W_k = 0, \qquad k > M,$$
$$W_M \neq 0.$$

The system "memory" here is M. In that case we can rewrite (3.1.2) as

$$v_n = \sum_0^n W_j u_{n-j}, \qquad n \le M \tag{3.1.3}$$

$$= \sum_0^M W_j u_{n-j}, \qquad n \ge M. \tag{3.1.4}$$

Suppose now that the input $\{u_k\}$ is a (second-order) stationary process. The immediate question is whether the output process is (second-order) stationary. The answer is in the negative, in general as soon as $M \ge 1$. Thus let us consider the means:

$$E[v_0] \doteq W_0 E[u_0]$$

$$E[v_1] = W_1 E[u_0] + W_0 E[u_1] = (W_1 + W_0) E[u_0] = W_1 E[u_0] + E[v_0].$$

Hence

$$E[v_1] \ne E[v_0]$$

in general unless the input process has zero mean. In that case, we have still to consider the second moment. Let $\{R_u(k)\}$ denote the stationary covariance function of the input process. Then, since it has zero mean we obtain

$$E[v_0 \; v_0^*] = W_0 R_u(0) W_0^*$$

and

$$E[v_1 \; v_1^*] = W_1 R_u(0) W_1^* + W_1 R_u(1)^* W_0^* + W_0 R_u(1) W_1^* + W_0 R_u(0) W_0^*,$$

which for $M \ge 1$ is not equal to $E[v_0 \; v_0^*]$ in general. But we are not surprised at this answer since we know that in the deterministic input case there is always a "transient" part in the response and we have attained "steady state" after the transients have decayed to zero. In fact we can see that in this case for $n \ge M$, (3.1.4) holds so that

$$E[v_n] = \sum_0^M W_j E[u_{n-j}] = \left(\sum_0^M W_j \right) E[u_k] = \text{constant}.$$

Next using the notation

$$\tilde{u}_k = u_k - E[u_k]$$

$$\tilde{v}_k = v_k - E[v_k]$$

we can calculate the covariance function of the output process:

$$R_v(m, n) = E[\tilde{v}_m \quad \tilde{v}_n^*] = E\left[\left(\sum_0^M W_j \tilde{u}_{m-j}\right)\left(\sum_0^M W_k \tilde{u}_{n-k}^k\right)^*\right], \qquad n, m > M$$

$$= \sum_0^M \sum_0^M W_j R_u(m - n + k - j) W_k^*, \qquad (3.1.5)$$

where

$$R_u(m) = E[\tilde{u}_{m+p} \quad \tilde{u}_p^*].$$

Thus for $m, n \geq M$

$$R_v(m, n) = R_v(m - n), \qquad (3.1.6)$$

where

$$R_v(p) = \sum_0^M \sum_0^M W_j R_u(p + k - j) W_k^*. \qquad (3.1.7)$$

Thus the output process is stationary if we consider only $n > M$. We have only to wait for M samples to attain "steady state"—after which the output process becomes stationary, even if M may be "large". We can also express this as

$$E[v_m \quad v_n^*] = R_v(m - n), \qquad n, m \geq M. \qquad (3.1.8)$$

We can also calculate the spectral density of the output process in the "steady state." Let $p_u(\lambda)$ denote the spectral density of the input. Let

$$\psi(\lambda) = \sum_0^M W_k e^{-2\pi i k \lambda}, \qquad -1/2 \leq \lambda \leq 1/2, \qquad (3.1.9)$$

which is recognized as the system "transfer function." Note that with W_k being real, we obtain

$$\overline{\psi(\lambda)} = \psi(-\lambda).$$

Then substituting

$$R_u(p + k - j) = \int_{-1/2}^{1/2} e^{2\pi i \lambda(p+k-j)} p_u(\lambda) \, d\lambda$$

in (3.1.7) we have

$$R_v(p) = \int_{-1/2}^{1/2} \sum_0^M \sum_0^M W_j e^{2\pi i \lambda(p+k-j)} p_u(\lambda) W_k^* \, d\lambda$$

$$= \int_{-1/2}^{1/2} e^{2\pi i \lambda p} \left(\sum_0^M W_j e^{-2\pi i \lambda j}\right) p_u(\lambda) \left(\sum_0^M W_k^* e^{2\pi i \lambda k}\right) d\lambda$$

$$= \int_{-1/2}^{1/2} e^{2\pi i \lambda p} \psi(\lambda) p_u(\lambda) \psi(\lambda)^* \, d\lambda$$

or the spectral density of the output

$$p_v(\lambda) = \psi(\lambda) p_u(\lambda) \psi(\lambda)^*. \qquad (3.1.10)$$

In the one-dimensional case this becomes

$$= p_u(\lambda) |\psi(\lambda)|^2. \qquad (3.1.11)$$

From (3.1.11) we see that the steady-state properties of the output process depend only on the magnitude of the transfer function and not on its phase. This is dramatically illustrated by a "time delay" system:

$$W_k = W, \qquad k = N \geq 0$$

$$= 0, \qquad k \neq N.$$

This defines a finite-memory system with

$$\psi(\lambda) = W e^{-2\pi i N \lambda} \qquad (3.1.12)$$

and substituting in (3.1.10) yields

$$p_v(\lambda) = W p_u(\lambda) W^*$$

$$= p_u(\lambda) |W|^2 \qquad \text{in the one-dimensional case}$$

and is independent of the delay.

3.1.2 Physically Nonrealizable Systems

Let us now extend our consideration to linear time-invariant systems with finite memory but no longer restricted to be "physically realizable"—allowing W_k to be nonzero for k negative. Consider then a weighting pattern $\{W_k\}$, subject to

$$W_k = 0, \qquad |k| > M$$

and not necessarily zero for $k < 0$. Then (3.1.2) generalizes to

$$v_n = \sum_{k=-M}^{n} W_k u_{n-k}, \qquad n \leq M \qquad (3.1.13)$$

$$= \sum_{-M}^{M} W_k u_{n-k}, \qquad n > M. \qquad (3.1.14)$$

The response at the instant n now anticipates the future of the input up to $n + M$, which of course no physical system can do. On the other hand, for

$n > 2M$ we obtain

$$v_{n-M} = \sum_{-M}^{M} W_k u_{n-M-k} = \sum_{0}^{2M} W_{k-M} u_{n-k}.$$

Defining

$$z_n = \sum_{0}^{2M} \tilde{W}_k u_{n-k}, \qquad (3.1.15)$$

where

$$\tilde{W}_k = W_{k-M}, \qquad 0 \le k \le 2M,$$

we see that $\{z_n\}$ is the output of a physically realizable system and that

$$z_n = v_{n-M}, \qquad n > 2M. \qquad (3.1.16)$$

In other words the physically nonrealizable system (3.1.14) followed by a pure time delay is physically realizable. The importance to us is that the (steady-state) spectral density of the process $\{z_n\}$ is the same as that of $\{v_n\}$. Indeed the transfer function $\phi(\cdot)$ of the system in (3.1.15) is given by

$$\phi(\lambda) = \sum_{0}^{2M} \tilde{W}_k e^{-2\pi i k\lambda} = \sum_{0}^{2M} W_{k-M} e^{-2\pi i k\lambda}$$

$$= e^{-2\pi i M\lambda} \left(\sum_{-M}^{M} W_k e^{-2\pi i k\lambda} \right)$$

$$= e^{-2\pi i M\lambda} \psi(\lambda),$$

where $\psi(\cdot)$ is the transfer function of the system corresponding to (3.1.14). Hence

$$p_v(\lambda) = \phi(\lambda) p_u(\lambda) \phi(\lambda)^* = \psi(\lambda) p_u(\lambda) \psi(\lambda)^*.$$

Thus the steady-state spectral density of the output process (in response to a white noise input) of a finite memory system that is physically nonrealizable can also be obtained as the spectral density of the output of a physically realizable system.

In going on to consider systems which cannot have a finite memory, we shall first consider "state-space" models.

3.2 STATE-SPACE MODELS

For most, if not all, of our purposes, we shall be using state-space models to describe linear systems. We shall need to recall only elementary machinery from state-space theory; no deep prior knowledge of the theory is required. But some prior exposure to the theory would be helpful in preventing "culture shock."

We now consider "state-space" models of (time-invariant) linear systems—systems with a state-space description. To characterize the system as is customary in the usual notation, let A, B, C be given matrices with the following dimensions:

$$A: n \times n$$

$$B: n \times p$$

$$C: m \times n.$$

Let $\{u_k\}$ denote a $p \times 1$ "input" (sequence). The "output" (sequence) $\{v_k\}$ which is $m \times 1$ is defined by

$$v_k = C x_k \quad \text{("state output relation")}, \tag{3.2.1}$$

where $\{x_k\}$ is the "state" sequence defined by the "state transition" law:

$$x_{k+1} = A x_k + B u_k. \tag{3.2.2}$$

Solving (3.2.2) iteratively for $n > k$ we have

$$x_n = A^{n-k} x_k + \sum_{j=k}^{n-1} A^{n-1-j} B u_j. \tag{3.2.3}$$

It is customary to take the starting "time" as 0. Then for any $n > 0$ we have

$$v_n = C A^n x_0 + \sum_{0}^{n-1} C A^{n-1-j} B u_j. \tag{3.2.4}$$

Note that the system weighting pattern $\{W_k\}$ is defined by

$$W_k = C A^k B, \quad k \geq 0,$$

and of course the memory is no longer finite.

We shall now specialize the input to be an $(m \times 1)$ white-noise sequence $\{N_k\}$, zero mean with covariance

$$E[N_k \ N_k^*] = D.$$

We have

$$x_n = A x_{n-1} + B N_{n-1}$$

$$v_n = C x_n$$

and

$$x_n = A^n x_0 + \sum_{0}^{n-1} A^k B N_{n-1-k}. \tag{3.2.5}$$

Now

$$E[v_n] = CE[x_n]$$

$$E[v_m \ v_n^*] = CE[x_m \ x_n^*]C^*$$

so that to calculate the statistics (primarily moments) of the output response $\{v_n\}$ it is enough to calculate that of the state response.

Assumption 3.2.1. We assume now that the initial state x_0 is a random variable and is Gaussian and is independent of $\{N_k\}$ for all $k \geq 0$.

As a consequence we shall show that for each $n \geq 0$

$$x_n \text{ is independent of } \{N_k\} \text{ for all } k \geq n.$$

This is immediate from

$$x_{n+1} = Ax_n + BN_n$$

by an "induction" argument. Thus suppose the statement is true for n. We shall show that it is true for $n + 1$. But from (3.2.2) we see that x_{n+1} is the sum of Ax_n and BN_n; x_n is given to be independent of $\{N_k\}$, $k \geq n$, and hence is independent of $\{N_k\}$, $k \geq n + 1$, and so is N_n. Hence x_{n+1} is independent of $\{N_k\}$, $k \geq n + 1$. To complete the "induction" argument we note that it is given to be true for $n = 0$.

Let

$$E[x_n] = m_n.$$

Then from the state equation we see that

$$E[x_n] = AE[x_{n-1}] + BE[N_{n-1}] = AE[x_{n-1}].$$

Hence

$$m_n = Am_{n-1}$$

and hence

$$m_n = A^n m_0.$$

Let us use the notation $\tilde{x}_n$ for the "centered" process

$$\tilde{x}_n = x_n - m_n.$$

Then

$$\tilde{x}_n = x_n - m_n = Ax_{n-1} + BN_{n-1} - Am_{n-1}$$

$$= A\tilde{x}_{n-1} + BN_{n-1} \qquad (3.2.6)$$

and $\tilde{x}_n$ is clearly independent of $\{N_k\}$, $k \geq n$.

From (3.2.6) we have for $k < n$:

$$\tilde{x}_n = A^{n-k}\tilde{x}_k + \sum_{j=k}^{n-1} A^{n-1-j}BN_j, \qquad (3.2.7)$$

where the first term is independent of the second term. Hence the covariance function is given by

$$R(n, k) = E[\tilde{x}_n \ \tilde{x}_k^*] = A^{n-k}E[\tilde{x}_k \ \tilde{x}_k^*] = A^{n-k}R(k, k), \qquad n \geq k. \quad (3.2.8)$$

But from (3.2.6), we have

$$R(n, n) = E[\tilde{x}_n \ \tilde{x}_n^*] = AE[\tilde{x}_{n-1} \ \tilde{x}_{n-1}^*]A^* + BE[N_{n-1} \ N_{n-1}^*]B^*,$$

since $\tilde{x}_{n-1}$ is independent of N_{n-1}.
Hence

$$R(n, n) = AR(n - 1, n - 1)A^* + BDB^*, \qquad n \geq 1, \qquad (3.2.9)$$

where

$$R(0, 0) = E[\tilde{x}(0) \ \tilde{x}(0)^*].$$

Hence we can calculate $R(n, n)$ iteratively, and $R(n, k)$ therefrom using (3.2.8) and (3.2.9). The state process $\{x_n\}$ is known as a Gaussian Markov process—see Chapter 9, Problem 9.9.

3.2.1 Steady-State (or Asymptotic) Solution

For a time-invariant system we are largely interested in the "steady-state" solution—the asymptotic or "limiting" behaviour as $n \to \infty$, which we expect to be not dependent on the "starting" or "initial" state x_0. But we need to consider what we mean by "limits" since we are dealing with random variables.

Here the notion of "system stability" plays an important role.

Definition 3.1. We call the system defined by (3.2.1), (3.2.2)† "stable" if the initial condition response

$$A^n x_0$$

goes to zero as $n \to \infty$, for every x_0.

---

† Or simply "the system" for short.

Definition 3.2. We call a matrix A "stable" if

$$A^n x \to 0 \qquad \text{as } n \to \infty \qquad (3.2.10)$$

for every x in E^n.

Suppose

$$A^n x \to 0$$

as $n \to \infty$ for every x. Then

$$[A^n x, y] \to 0$$

for arbitrary x, y. Hence every component of A^n goes to zero, or

$$\|A^n\| \qquad \text{and} \qquad \|A^n\|_0$$

goes to zero. But more is true—which we need for our purposes.

Theorem 3.2.1. Suppose the matrix A is stable: then all the eigenvalues of A are less than one in absolute value.

Proof. The proof is immediate since if e_i is an eigenvector with eigenvalue λ_i,

$$A e_i = \lambda_i e_i,$$

then by stability we have

$$\|A^k e_i\| \to 0$$

as $k \to \infty$. But

$$\|A^k e_i\| = |\lambda_i|^k$$

and hence we must have that

$$|\lambda_i| < 1.$$

Hence

$$r = \max_{i=1,\ldots,n} |\lambda_i|$$

$$< 1 \qquad (3.2.11)$$

(r is called the "spectral radius").

Theorem 3.2.2. Let r be the spectral radius defined by (3.2.11). Suppose r is less than 1. Then A is stable. In fact we can find an integer N_0 and $0 < \gamma < 1$ such that

$$\|A^n\|_0 < \gamma^n, \qquad n > N_0, \qquad (3.2.12)$$

where we can make γ as close to r as we wish for N_0 large enough.

Proof. We shall prove this for the case where the matrix is "simple"—the eigenvectors provide a basis. Each x in E^n can be expressed as

$$x = \sum_1^n a_i e_i.$$

where $\{e_i\}$ are the linearly independent eigenvectors with eigenvalues $\{\lambda_i\}$ distinct or not. (Recall that if the $\{\lambda_i\}$ are all distinct, then A is automatically simple!) Then we have

$$A^k x = \sum_1^n \lambda_i^k a_i e_i$$

and hence

$$\|A^k x\| \leq \sum_1^n |\lambda_i|^k |a_i| \leq r^k \sum_1^n |a_i| \leq r^k \sqrt{n \sum_1^n |a_i|^2}.$$

Let M denote the moment matrix

$$M = \{m_{ij} = [e_i, e_j]\},$$

which must be nonsingular. Then

$$\|x\|^2 = \sum_{i=1}^n \sum_{j=1}^n a_i m_{ij} \overline{a_j}$$

and

$$\left(\sum_1^n |a_i|^2 \right) \cdot (\text{smallest eigenvalue of } M) \leq \|x\|^2.$$

But the smallest eigenvalue of M, call it δ, must be nonzero since M is nonsingular. Hence

$$\|A^k x\| \leq r^k \sqrt{n} \, \frac{\|x\|}{\sqrt{\delta}}.$$

Thus

$$\|A^k\|_0 \leq \sqrt{n/\delta} \, r^k$$

for every positive integer k. Since $r < 1$, for any $\varepsilon > 0$, such that $(r + \varepsilon) < 1$, we can clearly choose an integer N_0 such that

$$\sqrt{n/\delta} < \left(1 + \frac{\varepsilon}{r} \right)^{N_0} = \left(\frac{r + \varepsilon}{r} \right)^{N_0}.$$

Hence taking

$$\gamma = r + \varepsilon$$

we have that

$$\|A^k\|_0 \leq (\sqrt{n/\delta}\, r^{N_0+j}) \leq (r+\varepsilon)^{N_0} r^j \leq (r+\varepsilon)^{N_0+j} = \gamma^k,$$

where $k = N_0 + j$, and $j > 0$. Obviously γ can be made as close to r as we wish by taking large enough N_0. The proof for the case where A is not simple is a little bit more complicated and will be omitted here. (See reference 5.)

Let us now explore the consequences of stability.

Theorem 3.2.3. Suppose the system is stable. Then:

(i) $\displaystyle\lim_{n \to \infty} E[x_n] = 0$.

(ii) $R(n, n)$ converges as $n \to \infty$ to a covariance matrix R_∞, which is the unique solution of the linear matrix equation (a "Liapunov" equation):

$$R_\infty = AR_\infty A^* + BB^*. \tag{3.2.13}$$

Proof. Iterating

$$R(n, n) = AR(n - 1, n - 1)A^* + BDB^* \tag{3.2.9}$$

we have

$$R(n, n) = A^n R(0, 0)A^{*n} + \sum_0^{n-1} A^k BDB^* A^{*k}. \tag{3.2.14}$$

Let us first consider the 1×1 case where all matrices A, B, C, D are 1×1. Let us use lowercase letters to denote them: a, b, c, d, respectively. Stability requires that

$$|a| < 1$$

since in this case r, the spectral radius, equals $|a|$. Also (3.2.6) becomes

$$R(n, n) = r^{2n}b^2 d + \sum_0^{n-1} r^{2k}b^2 d.$$

Hence $R(n, n)$ converges as $n \to \infty$ to

$$R_\infty = \sum_0^\infty r^{2k}b^2 d = \frac{b^2 d}{1 - r^2}.$$

Taking limits in (3.2.9) we have

$$R_\infty = r^2 R_\infty + b^2 d,$$

which has the unique solution

$$R_\infty = \frac{b^2 d}{1 - r^2}.$$

In the matrix case we invoke (3.2.12) to show convergence of $R(n, n)$. Thus in (3.1.14)

$$A^n R(0, 0) A^{*n} \to 0$$

since

$$\| A^n R(0, 0) A^{*n} \|_0 \leq \| A^n \|_0 \| A^{*n} \|_0 \| R(0, 0) \|_0 = \| A^n \|_0^2 \| R(0, 0) \|_0,$$

which by (3.2.12) is

$$\leq \gamma^{2n} \| R(0, 0) \|_0$$

for all $n > N_0$. To handle the second term in (3.2.14), we note that the infinite series

$$\sum_0^\infty A^k BDB^* A^{*k}$$

converges, since for all $n > N_0$

$$\left\| \sum_n^\infty A^k BDB^* A^{*k} \right\|_0 \leq \sum_r^\infty \gamma^{2k} \| BDB^* \|_0 = \frac{\gamma^{2n}}{1 - \gamma} \| BDB^* \|_0.$$

Hence

$$R_\infty = \sum_0^\infty A^k BDB^* A^{*k}.$$

Hence taking limits in (3.2.9), we have

$$R_\infty = A R_\infty A^* + BDB^*. \tag{3.2.15}$$

Moreover, since A is stable, the matrix equation

$$R = ARA^* + BDB^*$$

has only one solution. If it has more than one, say R_1, R_2, we have

$$R_1 = AR_1 A^* + BDB^*$$

$$R_2 = AR_2 A^* + BDB^*$$

and

$$R_1 - R_2 = A(R_1 - R_2)A^*.$$

Iteration then yields

$$(R_1 - R_2) = A^n(R_1 - R_2)A^{*n}$$

for every n and the norm of the right side

$$\|A^n(R_1 - R_2)A^{*n}\|_0 \le \|R_1 - R_2\|_0 \gamma^{2n} \qquad n > N_0$$

$$\to 0 \qquad\qquad \text{as } n \to \infty$$

or

$$R_1 - R_2 = 0.$$

In particular we see that R_∞, "the steady-state covariance," does not depend on the initial covariance $R(0, 0)$.

We have thus proved

$$\lim_{n \to \infty} E[x_n \ x_n^*] = R_\infty.$$

Also

$$E[x_{n+p} \ x_n^*] = A^p R(n, n) \to A^p R_\infty \qquad \text{as } n \to \infty.$$

Thus we have the "steady-state" covariance function

$$R(p) = \lim_n R(n + p, n) = A^p R_\infty, \qquad p \ge 0$$

$$= R_\infty A^{|p|}, \qquad p < 0. \qquad (3.2.16)$$

In particular we see that the covariance function is "stationary in the steady state":

$$\lim_{n \to \infty} E[x_{m+n} \ x_{p+n}^*]$$

$$\lim_{n \to \infty} R(m + n, p + n) = R(m - p)$$

with $R(\cdot)$ given by (3.2.16).

We say that the process is asymptotically stationary, or stationary in the steady state with the stationary covariance given by (3.2.16). Let us examine what this means more precisely since the memory is no longer finite. For this purpose we go back to

$$x_n = A^n x_0 + \sum_0^{n-1} A^k B N_{n-1-k}.$$

Heuristically, the first term goes to zero for each x_0 and the second term becomes the infinite series

$$\sum_0^\infty A^k B N_{n-1-k}, \qquad (3.2.17)$$

with the white-noise sequence N_k being defined for all integers: for all $k \in \mathbf{I}$. First let us study the convergence of the infinite series (3.2.17). Fix n, and for each positive integer N let

$$z_N = \sum_0^N A^k B N_{n-1-k}. \qquad (3.2.18)$$

To simplify matters let us first consider the 1×1 case replacing A, B, C, D by a, b, c, d, as before, and now making N_k (1×1) as well. Then

$$z_N = \sum_0^N a^k b N_{n-k}, \qquad |a| < 1.$$

This is a sum of independent Gaussians, and

$$E[z_N] = 0$$

$$E[z_N^2] = \sum_0^N a^{2k} b^2 d$$

and the series

$$\sum_0^\infty a^{2k} b^2 d$$

converges, and in fact the sum becomes

$$\frac{b^2 d}{1 - a^2}.$$

Hence the sequence $\{z_N\}$ converges in the mean square (as well as with probability one) as $N \to \infty$ to a Gaussian variable which we label y_n. The probability one convergence is a standard result in probability; see any text, for example, reference 3. In other words, using explicitly the sample path notation, with ω associated with the white-noise sequence,

$$z_N(\omega) = \sum_0^N a^k b N_{n-k}(\omega)$$

converges for every ω except on an exceptional set of zero probability. Denoting the limit by $y_n(\omega)$, we have that the covariance function is given by

$$E[y_n(\omega)y_m(\omega)] = \sum_{k=0}^{\infty} \sum_{j=0}^{\infty} a^k b E[N_{n-k} \quad N_{m-j}] b a^j$$

$$= \sum_{0}^{\infty} \sum_{0}^{\infty} a^{k+j} b^2 \delta(m-n+k-j) d$$

$$= \sum_{0}^{\infty} a^{k+(k+m-n)} b^2 d = \frac{a^{m-n} b^2 d}{1-a^2}, \qquad m \geq n,$$

which agrees with (3.2.16), the process $\{y_n\}$ being stationary. Or, defining

$$y_n = \sum_{0}^{\infty} a^k b N_{n-1-k}$$

we have that the steady-state covariance function is given by

$$\lim_{n \to \infty} E[x_{n+p} \quad x_{m+n}] = E[y_p \quad y_n],$$

where $\{y_n\}$ is stationary. Moreover

$$E[|x_n - y_n|^2] = a^{2n} E[x_0^2] + E[(z_n - y_n)^2] = a^{2n} E[x_0^2] + \sum_{n}^{\infty} a^{2k} b^2 d$$

goes to zero as $n \to \infty$.

These considerations are readily extended to the matrix case. Thus, returning to (3.2.18), we see that

$$E[z_N] = 0,$$

and z_N for each N is the sum of zero-mean independent $(n \times 1)$ Gaussians. To establish that the sequence $\{z_N\}$ converges in the mean square sense, we calculate

$$E[(z_{N+p} - z_N)(z_{N+p} - z_N)^*] = E\left[\left(\sum_{N+1}^{N+p} A^k B N_{n-1-k}\right)\left(\sum_{N+1}^{N+p} A^j B N_{n-1-j}\right)^*\right]$$

$$= \sum_{N+1}^{N+p} \sum_{N+1}^{N+p} A^k B D \delta(j-k) B^* A^{j*}$$

$$= \sum_{N+1}^{N+p} A^k B D B^* A^{*k},$$

which goes to zero for all p as N goes to infinity since we have seen that the series

$$\sum_0^\infty A^k BDB^* A^{*k}$$

converges. Hence z_N converges in the mean square (and also actually pathwise, as we noted in the one-dimensional case) as $N \to \infty$. Moreover

$$\lim_{N \to \infty} E[z_N \; z_N^*] = \sum_0^\infty A^k BDB^* A^{*k}.$$

Defining now

$$y_n = \sum_0^\infty A^k B N_{n-1-k} \tag{3.2.19}$$

we see from (3.2.5) that

$$E[(x_n - y_n)(x_n - y_n)^*]$$

$$= A^n E[x_0 x_0^*] A^{*n} + E\left[\left(\sum_n^\infty A^k B N_{n-1-k}\right)\left(\sum_n^\infty A^k B N_{n-1-k}\right)^*\right]$$

$$= A^n E[x_0 x_0^*] A^{*n} + \sum_n^\infty A^k BDB^* A^{*k}$$

which, by virtue of the matrix A being stable, goes to zero as $n \to \infty$. It is readily verified that

$$E[y_n y_m^*] = \sum_0^\infty \sum_0^\infty A^k BE[N_{n-1-k} N_{m-1-j}^*] B^* A^{*j}$$

$$= \sum_0^\infty \sum_0^\infty A^k BD\delta(n - m + j - k) A^{*j}$$

$$= \sum_0^\infty A^{n-m+j} BDB^* A^{*j}, \qquad n \geq m$$

$$= A^{n-m} R_\infty \qquad\qquad n \geq m$$

$$= R_\infty A^{*m-n}, \qquad\qquad m \geq n,$$

recognized to be the same as (3.2.16). In other words, the "idealized" process defined by (3.2.19) is stationary and has the same covariance as the process $\{x_n\}$ does asymptotically—or in the steady state! From now on by the "steady-state" (state) response we shall mean (3.2.19), and correspondingly the "steady-state" output response

$$v_n = Cy_n = \sum_0^\infty CA^k B N_{n-1-k}.$$

We can rewrite (3.2.19) in state-space form as

$$v_n = Cy_n$$

$$y_{n+1} = Ay_n + BN_n, \qquad n \geq 0$$

with

$$E[y_0] = 0, \qquad E[y_0 y_0^*] \equiv R_\infty,$$

y_0 Gaussian, independent of N_k, $\qquad k \geq 0$.

Note the particular choice of the initial covariances as equal to R_∞. With this choice of the initial covariance matrix as R_∞ we can calculate that

$$E[y_n y_n^*] = A^n R_\infty A^{*n} + \sum_0^{n-1} A^k BDB^* A^{*k}$$

$$= \sum_0^{n-1} A^k BDB^* A^{*k} + A^n \left(\sum_0^\infty A^k BDB^* A^{*k} \right) A^{*k}$$

$$= \sum_0^{n-1} A^k BDB^* A^{*k} + \sum_0^\infty A^k BBD^* A^{*k} \right) A^{*n} = R_\infty$$

and that for $n \geq m$

$$E[y_n y_m^*] = A^{n-m} E[y_m y_m^*] = A^{n-m} R_\infty.$$

Hence the process is stationary for all $n \geq 0$, or "instantly" stationary.

3.2.2 Steady-State Spectral Density

Let us calculate the steady-state spectral density—the spectral density corresponding to the steady-state covariance function. Thus

$$P(\lambda) = \sum_{-\infty}^\infty R(n) e^{2\pi i n \lambda} = \sum_0^\infty A^n R_\infty e^{2\pi i n \lambda} + \sum_0^\infty R_\infty A^{*n} e^{-2\pi i n \lambda} - R_\infty$$

$$= \psi(\lambda) R_\infty - R_\infty \psi(\lambda)^* - R_\infty,$$

where

$$\psi(\lambda) = \sum_0^\infty A^n e^{2\pi i n \lambda}.$$

In the 1×1 case,

$$\psi(\lambda) = \sum_0^\infty A^n e^{2\pi i n \lambda} = \frac{1}{1 - ae^{2\pi i \lambda}}$$

since $|a| < 1$. Let us prove that in the matrix case

$$\psi(\lambda) = (I - Ae^{2\pi i\lambda})^{-1}. \qquad (3.2.20)$$

But this follows from the identity

$$(I - Ae^{2\pi i\lambda})\left(\sum_0^{n-1} (Ae^{2\pi i\lambda})^k\right) = I - (Ae^{2\pi i\lambda})^n.$$

Now the eigenvalues of $(I - Ae^{2\pi i\lambda})$ are

$$(1 - \lambda_k e^{2\pi i\lambda}), \qquad k = 1, \ldots, n,$$

where $\{\lambda_k\}$ are the eigenvalues of A. Since $|\lambda_k| < 1$, A being stable, we see that $(I - Ae^{2\pi i\lambda})$ is nonsingular. Hence

$$(I - Ae^{2\pi i\lambda})^{-1}(I - A^n e^{2\pi in\lambda}) = \sum_0^{n-1} A^k e^{2\pi ik\lambda}.$$

Thus letting $n \to \infty$, we have that

$$(I - Ae^{2\pi i\lambda})^{-1} = \sum_0^\infty A^k e^{2\pi ik\lambda} = \psi(\lambda).$$

Hence

$$\begin{aligned}
P(\lambda) &= \psi(\lambda)[R_\infty \psi(\lambda)^{*-1} + \psi(\lambda)^{-1}R_\infty - \psi(\lambda)^{-1}R_\infty\psi(\lambda)^{*-1}]\psi(\lambda)^* \\
&= \psi(\lambda)[R_\infty(I - A^*e^{-2\pi i\lambda}) + (I - Ae^{2\pi i\lambda})R_\infty \\
&\quad - (I - Ae^{2\pi i\lambda})R_\infty(I - A^*e^{-2\pi i\lambda})]\psi(\lambda)^* \\
&= \psi(\lambda)[R_\infty - AR_\infty A^*]\psi(\lambda)^*.
\end{aligned}$$

But by the Liapunov equation we have

$$R_\infty - AR_\infty A^* = BDB^*.$$

Hence

$$P(\lambda) = \psi(\lambda)BDB^*\psi(\lambda)^*. \qquad (3.2.21)$$

We note that the rank of $P(\lambda)$ is constant, being equal to that of BDB^*. The steady-state spectral density of the output process is

$$CP(\lambda)C^*. \qquad (3.2.22)$$

An important feature of the spectral density given by (3.2.21) is that it is a rational function of $e^{2\pi i\lambda}$—every term in the matrix is a ratio of polynomials in $e^{2\pi i\lambda}$. This follows readily from the fact that every term in the matrix

$$(I - Ae^{2\pi i\lambda})^{-1}$$

is a ratio of polynomials in $e^{2\pi i\lambda}$. As for $\psi(\lambda)^*$ we can express it as

$$\psi(\lambda)^* = (I - A^*e^{-2\pi i\lambda})^{-1} = e^{2\pi i\lambda}(e^{2\pi i\lambda} - A^*)^{-1}$$

and thus again all ratios of polynomials in $e^{2\pi i\lambda}$. The denominator of all the terms in $P(\lambda)$ is given by

$$|f(e^{2\pi i\lambda})|^2,$$

where

$$f(z) = \det (I - Az).$$

We now turn our analysis of response of linear systems around and observe that a Gaussian process with spectral density specified by (3.2.22) can be generated in the steady state by specifying

$$v_n = Cx_n, \qquad n \geq 0$$
$$x_{n+1} = Ax_n + BN_n, \tag{3.2.23}$$

where $\{N_n\}$ is white Gaussian and x_0 Gaussian, independent of N_n, $n \geq 0$. We have thus a signal "generation model" for the process. We shall call it a Kalman signal generation model, after R. E. Kalman who proposed it first in the 1960s (see reference 1). It is fair to say that it has essentially become a new paradigm for random processes in engineering applications, particularly in simulation.

Remark. If (3.2.23) is looked upon only as a representation for the signal $\{v_n\}$, then the matrices (C, A, B) need not be unique. See Example 3.2.2—there can be more than one such set. However, if in addition we demand that the state space be "observable" (see reference 5), then the dimension of the state space is fixed and one state space can be obtained from the other by a nonsingular transformation. These considerations are not so important for us as to warrant further discussion except in Section 3.4, dealing with ARMA models.

Let us consider an illustrative example where we can actually calculate the covariance function analytically.

Example 3.2.1. In (3.2.23) let

$$A = \begin{vmatrix} 0 & 1 \\ a & 0 \end{vmatrix}, \qquad \text{where } 0 < a < 1$$

$$B = \begin{vmatrix} 0 \\ 1 \end{vmatrix}$$

$$C = |1 \quad 0|.$$

Here the white-noise sequence is one-dimensional so that we use d in place of

D. Note that

$$A^{2n} = (A^2)^n = a^{2n}I$$

$$A^{2n+1} = A^{2n}A = a^{2n}\begin{vmatrix} 0 & 1 \\ a & 0 \end{vmatrix}.$$

Hence substituting in (3.2.14) we have

$$R(2n, 2n) = a^{4n}R(0, 0) + d\sum_{0}^{n-1} a^{2k}\begin{vmatrix} 0 & 0 \\ 0 & 1 \end{vmatrix} + \sum_{0}^{n-1} a^{2j}d\begin{vmatrix} 1 & 1 \\ a & 0 \end{vmatrix}\begin{vmatrix} 0 & 0 \\ 0 & 1 \end{vmatrix}\begin{vmatrix} 0 & a \\ 1 & 0 \end{vmatrix}$$

$$= a^{4n}R(0, 0) + d\left(\frac{1 - a^{2n}}{1 - a^2}\right)I.$$

Hence for n even we obtain

$$R(n, n) = a^{2n}R(0, 0) + d\left(\frac{1 - a^n}{1 - a^2}\right)$$

$$R(n + 1, n + 1) = AR(n, n) + BDB^*$$

$$= a^{2n}AR(0, 0)A^* + \left(\frac{1 + a^{2n}}{1 - a^2}\right)d\begin{vmatrix} 0 & 1 \\ a & 0 \end{vmatrix}\begin{vmatrix} 0 & a \\ 1 & 0 \end{vmatrix} + d\begin{vmatrix} 0 & 0 \\ 0 & 1 \end{vmatrix}$$

$$= a^{2n}AR(0, 0)A^* + \left(\frac{1 - a^{2n}}{1 - a^2}\right)d\begin{vmatrix} 1 & 0 \\ 0 & a^2 \end{vmatrix} + \frac{d}{1 - a^2}\begin{vmatrix} 0 & 0 \\ 0 & 1 \end{vmatrix}.$$

Note that as $n \to \infty$, both yield the limit

$$R_\infty = \frac{d}{1 - a^2}I. \qquad (3.2.24)$$

It is of course a lot "quicker" to obtain this by solving the Liapunov equation. Since R_∞ the solution must be symmetric, let

$$R_\infty = \begin{vmatrix} r_{11} & r_{12} \\ r_{12} & r_{22} \end{vmatrix}.$$

There are three unknowns and the Liapunov equation yields three equations in these unknowns from

$$\begin{vmatrix} r_{11} & r_{12} \\ r_{12} & r_{22} \end{vmatrix} = \begin{vmatrix} 0 & 1 \\ r & 0 \end{vmatrix}\begin{vmatrix} r_{11} & r_{12} \\ r_{12} & r_{22} \end{vmatrix}\begin{vmatrix} 0 & a \\ 1 & 0 \end{vmatrix} + \begin{vmatrix} 0 & 0 \\ 0 & d \end{vmatrix}$$

$$= \begin{vmatrix} r_{22} & ar_{12} \\ ar_{12} & a^2r_{22} + d \end{vmatrix}.$$

After equating terms we have four equations, but one is of course redundant. The other three are

$$r_{11} = r_{22}$$

$$r_{12} = r_{21}$$

$$r_{22} = a^2 r_{11} + d.$$

Hence

$$r_{12} = 0 \qquad r_{22} = r_{11} = a^2 r_{11} + d$$

or

$$r_{12} = 0 \qquad r_{11} = \frac{d}{1 - a^2},$$

checking with (3.2.24). Let us next calculate the spectral density matrix. Using (3.2.20) we have

$$\psi(\lambda) = (I - A e^{2\pi i \lambda})^{-1} = \frac{1}{1 - a e^{4\pi i \lambda}} \begin{vmatrix} 1 & e^{2\pi i \lambda} \\ a e^{2\pi i \lambda} & 1 \end{vmatrix}.$$

Hence

$$P(\lambda) = \frac{1}{|1 - a e^{4\pi i \lambda}|^2} \begin{vmatrix} 1 & e^{2\pi i \lambda} \\ a e^{2\pi i \lambda} & 1 \end{vmatrix} \begin{vmatrix} 0 & 0 \\ 0 & d \end{vmatrix} \begin{vmatrix} 1 & a e^{-2\pi i \lambda} \\ e^{-2\pi i \lambda} & 1 \end{vmatrix}$$

$$= \frac{d}{|1 - a e^{4\pi i \lambda}|^2} \begin{vmatrix} 1 & e^{2\pi i \lambda} \\ e^{-2\pi i \lambda} & 1 \end{vmatrix}. \tag{3.2.25}$$

Note that the rank of $P(\lambda)$ is one. If

$$C = |1 \quad 0| \qquad \text{or} \qquad |0 \quad 1|$$

the spectral density of the output process

$$= \frac{d}{|1 - a e^{4\pi i \lambda}|^2}$$

$$= \frac{d}{1 + a^2 - 2a \cos 4\pi \lambda}.$$

Example 3.2.2. Our next example illustrates the possible nonuniqueness of the representation (3.2.23). Thus in (3.2.23) let A, B be as in Example 3.2.1, but let

$$C = |-\sqrt{a} \quad 1|.$$

For any
$$x = \begin{vmatrix} x_1 \\ x_2 \end{vmatrix} \quad \text{in } \mathbf{E}^2$$
we obtain
$$CAx = -\sqrt{a}\,x_2 + ax_1 = -\sqrt{a}[-\sqrt{a}\,x_1 + x_2]$$
$$= -\sqrt{a}\,Cx.$$

In particular, if we take x such that
$$x_2 - \sqrt{a}\,x_1 = 0$$
we have that
$$CA^k x = 0$$

for every $k \geq 0$—or the state is *not* observable. On the other hand,
$$v_{n+1} = Cx_{n+1} = C(Ax_n + BN_n)$$
$$= -\sqrt{a}\,v_n + CBN_n.$$

But this is a state-space representation for $\{v_n\}$:
$$y_{n+1} = -\sqrt{a}\,y_n + CBN_n$$
$$v_n = 1 \cdot y_n.$$

The advantage is that the new state-space dimension is less, and now it is observable and further reduction in dimension is not possible, in particular. Of course, the formula (3.2.22) for the spectral density of the process $\{v_n\}$ is valid whichever representation is used.

3.3 STEADY-STATE NOISE RESPONSE: GENERAL CASES

In this section we consider response of more general classes of linear systems to more general classes of stationary processes, our primary interest being in the steady-state response.

3.3.1 State-Space Systems Models and Kalman Signal Generation Models

We first consider the case where the linear system is given in terms of a state-space model and the input process described by a Kalman signal generation model rather than white noise. Thus let the process $\{u_n\}$ given by

$$u_n = Cx_n, \quad n \geq 0$$
$$x_{n+1} = Ax_n + BN_n \tag{3.3.1}$$

define the input to the system which is described by the state-space model

$$v_n = C_1 Y_n$$
$$Y_{n+1} = A_1 Y_n + B_1 u_n \tag{3.3.2}$$

where: $\{v_n\}$ is the output; A, and A_1, are square matrices which are assumed to be stable; and the numerical values of the dimensions of A_1, B_1, C_1 are immaterial to us. To handle this case we introduce the enhanced (or "compound") state vector

$$Z_n = \begin{vmatrix} x_n \\ Y_n \end{vmatrix}$$

and rewrite (3.3.2) as

$$Z_{n+1} = \mathscr{A} Z_n + \mathscr{B} \mathscr{N}_n$$
$$v_n = \mathscr{C} Z_n, \tag{3.3.3}$$

where

$$\mathscr{A} = \begin{vmatrix} A & 0 \\ B_1 C & A_1 \end{vmatrix}$$

$$\mathscr{N}_n = \begin{vmatrix} N_n \\ 0 \end{vmatrix}$$

$$\mathscr{B} = \begin{vmatrix} B & 0 \\ 0 & 0 \end{vmatrix}$$

$$\mathscr{C} = \begin{vmatrix} 0 & C_1 \end{vmatrix}.$$

We note that $\{\mathscr{N}_n\}$ is white noise with covariance

$$\mathscr{D} = \begin{vmatrix} D & 0 \\ 0 & 0 \end{vmatrix}.$$

Now $\mathscr{A}$ is stable as soon as A and A_1. In fact if λ is an eigenvalue of $\mathscr{A}$,

$$\begin{vmatrix} A & 0 \\ B_1 C & A_1 \end{vmatrix} \begin{vmatrix} x \\ y \end{vmatrix} = \begin{vmatrix} \lambda x \\ \lambda y \end{vmatrix},$$

we have that
$$Ax = \lambda x$$
$$B_1 Cx + A_1 y = \lambda y.$$

If $x = 0$, λ is an eigenvalue of A_1 and hence $|\lambda| < 1$. If $x \neq 0$, λ is an eigenvalue of A and hence $|\lambda| < 1$. (See also Problem 3.3.1.)

Thus (3.3.3) is a Kalman signal generation model. We assume that $n = 0$ is the start of initial time and assume (Assumption 3.2.1) that the initial state Z_0 is zero-mean (for simplicity) and Gaussian independent of $\mathcal{N}_k$, $k \geq 0$. This will in turn as before yield by induction that Z_n is independent of $\mathcal{N}_k$, $k \geq n$. We can then calculate the steady-state covariance function and spectral density of the process $\{Z_n\}$ from the general formulas of Section 3.2.

Let us calculate the spectral density using (3.2.21), which yields

$$P_Z(\lambda) = (I - \mathscr{A}e^{2\pi i\lambda})^{-1}\mathscr{B}\mathscr{D}\mathscr{B}^*(I - \mathscr{A}e^{-2\pi i\lambda})^{-1}. \tag{3.3.4}$$

Now

$$(I - \mathscr{A}e^{2\pi i\lambda})^{-1} = \begin{vmatrix} \psi(\lambda) & 0 \\ \psi_1(\lambda)B_1 C\psi(\lambda) & \psi_1(0) \end{vmatrix},$$

where

$$\psi(\lambda) = (I - Ae^{2\pi i\lambda})^{-1}$$
$$\psi_1(\lambda) = (I - A_1 e^{2\pi i\lambda})^{-1}$$

and

$$\mathscr{B}\mathscr{D}\mathscr{B}^* = \begin{vmatrix} BDB^* & 0 \\ 0 & 0 \end{vmatrix};$$

hence

$$p_Z(\lambda) = \begin{vmatrix} \psi(\lambda)BDB^*\psi(\lambda)^* & \psi(\lambda)BDB^*\psi(\lambda)^*C^*B_1^*\psi_1(\lambda)^* \\ \psi_1(\lambda)B_1 C\psi(\lambda)BDB^*\psi(\lambda)^* & \psi_1(\lambda)B_1 C\psi(\lambda)BDB^*\psi(\lambda)^*C^*B_1^*\psi_1(\lambda)^* \end{vmatrix}.$$
$$\tag{3.3.5}$$

In particular therefore the steady-state spectral density of the output process $\{v_n\}$ is given by

$$C_1(\psi_1(\lambda)B_1 C)(\psi(\lambda)BDB^*\psi(\lambda)^*)(\psi_1(\lambda)B_1 C)^*C_1^* \tag{3.3.6}$$
$$= (C_1\psi_1(\lambda)B_1)p_u(\lambda)(C_1\psi_1(\lambda)B_1)^*, \tag{3.3.7}$$

where

$$p_u(\lambda) = C\psi(\lambda)BDB^*\psi(\lambda)^*C^* = \text{spectral density of input process } \{u_k\},$$

and we recognize

$$C_1\psi_1(\lambda)B_1$$

as the "input-output" transfer function of the system (3.3.2).

3.3.2   Steady-State Response: Weighting Pattern Models

Let us return now to Section 3.1.1, where we began with "weighting pattern" models (3.1.2). In Section 3.1.2 we considered steady-state response, but only for systems with finite memory. Let us examine now the general case—where the memory is *not* constrained to be finite. Thus let $\{W_k\}$, $k \geq 0$, again denote the weighting pattern with W_k being, to be specific, of dimension $p \times m$ and let

$$v_n = \sum_0^n W_j u_{n-j}, \tag{3.3.8}$$

where $\{u_n\}$ is again an $m \times 1$ stationary process with mean μ and stationary covariance function $R_u(\cdot)$. We assume that the process $\{u_n\}$ is defined for $n \in \mathbf{I}$. We calculate (as in Section 3.1.2) that

$$E[v_n] = \sum_0^n W_k E[u_{n-k}] = \left(\sum_0^n W_k \right) \mu,$$

where

$$\mu = E[u_k].$$

We see that

$$\lim_n E[v_n] = \left(\sum_0^n W_k \right) \mu,$$

provided that the series in parentheses converges, for which we will need to place suitable conditions on the sequence $\{W_k\}$. Here we shall assume that

$$\sum_0^\infty \| W_k \|_0 < \infty \tag{3.3.9}$$

or, as we have seen, equivalently that

$$\sum_{k=0}^\infty |w_{ij}(k)| < \infty \qquad \text{for each } i, j,$$

where

$$W_k = \{w_{ij}(k)\}.$$

Note that this condition is satisfied by

$$W_k = CA^k B, \qquad k \geq 0,$$

for A stable.

To study the steady-state response, let us first show that the series

$$\sum_0^\infty W_k u_{n-k}$$

converges in the mean square for each n. For this is enough to show that

$$E\left\|\sum_N^\infty W_k u_{n-k}\right\|^2 \to 0 \quad \text{as } N \to \infty.$$

Now

$$E\left\|\sum_N^{N+p} W_k u_{n-k}\right\|^2 = \sum_N^{N+p}\sum_N^{N+p} E[W_k u_{n-k}, W_j u_{n-j}].$$

But

$$|E[W_k u_{n-k}, W_j u_{n-j}]| \le \sqrt{E[\|W_k u_{n-k}\|^2]}\sqrt{E[\|W_j u_{n-j}\|^2]}$$

and

$$\|W_k u_{n-k}\|^2 \le (\|W_k\|_0 \|u_{n-k}\|)^2$$

and

$$E[\|u_{n-k}\|^2] = \operatorname{Tr} E[u_{n-k}\ u_{n-k}^*] = \operatorname{Tr} R_u(0) + \|\mu\|^2.$$

Hence it follows that

$$E\left[\left\|\sum_N^{N+p} W_k u_{n-k}\right\|^2\right] \le \sum_N^{N+p}\sum_N^{N+p} \|W_k\|_0(\operatorname{Tr} R_u(0) + \|\mu\|^2)\|W_j\|_0$$

$$= (\operatorname{Tr} R_u(0) + \|\mu\|^2)\left(\sum_N^{N+p} \|W_k\|_0\right)^2,$$

which goes to zero as N goes to infinity, since (3.3.9):

$$\sum_N^\infty \|W_k\|_0 \to 0 \quad \text{as } N \to \infty.$$

Hence we have shown that the series converges to a random variable in the mean square.

We now define for each n:

$$v_{n,s} = \sum_0^\infty W_k u_{n-k}. \tag{3.3.10}$$

Then

$$E[v_{n,s}] = \left(\sum_0^\infty W_k\right)\mu. \tag{3.3.11}$$

Defining

$$\tilde{v}_{n,s} = v_{n,s} - E[v_{n,s}],$$

we can now calculate the covariance function:

$$E[\tilde{v}_{m,s}\tilde{v}_{n,s}^*] = E\left[\left(\sum_0^\infty W_k\tilde{u}_{m-k}\right)\left(\sum_0^\infty W_j\tilde{u}_{n-j}\right)^*\right], \qquad \tilde{u}_n = u_n - \mu,$$

$$= \sum_0^\infty\sum_0^\infty W_k R_u(m-n+j-k)W_j^*.$$

Hence the process $\{v_{n,s}\}$ is stationary with stationary covariance matrix $R_v(\cdot)$ given by

$$R_v(p) = \sum_0^\infty\sum_0^\infty W_k R_u(p+j-k)W_j^*. \qquad (3.3.12)$$

As in the case of the state-space model we show now that

$$E[(v_n - v_{n,s})(v_n - v_{n,s})^*]$$

goes to zero as $n \to \infty$. We have

$$E[(v_n - v_{n,s})(v_n - v_{n,s})^*] = E\left[\left(\sum_{n+1}^\infty W_k u_{n-k}\right)\left(\sum_{n+1}^\infty W_j u_{n-j}\right)^*\right]$$

$$= \sum_{n+1}^\infty\sum_{n+1}^\infty W_k\mu\mu^*W_j^* + \sum_{n+1}^\infty\sum_{n+1}^\infty W_k R_u(j-k)W_j^*.$$

As $n \to \infty$, the first term goes to zero because

$$\sum_1^\infty W_k$$

is convergent. As for the second term, we have

$$\|W_k R_u(j-k)W_j^*\|_0 \le \|W_k\|_0\|R_u(j-k)\|_0\|W_j^*\|_0$$

and

$$\|R_u(j-k)\|_0 \le \|R_u(0)\|_0$$

$$\|W_j^*\|_0 = \|W_j\|_0.$$

Hence

$$\left\|\sum_{n+1}^\infty\sum_{n+1}^\infty W_k R_u(j-k)W_j^*\right\|_0 \le \sum_{n+1}^\infty\sum_{n+1}^\infty \|W_k\|_0\|R_u(0)\|_0\|W_j\|_0$$

$$= \left(\sum_{n+1}^\infty \|W_k\|_0\right)^2 \|R_u(0)\|_0$$

and goes to zero as $n \to \infty$ because of (3.3.9). Thus for $n \geq m$ we obtain

$$\lim_{N \to \infty} E[v_{n+N} v^*_{m+N}] = E[v_{n,s} v^*_{m,s}] = R_v(n - m).$$

In other words, the asymptotic or steady-state properties of $\{v_n\}$ are the same as those of $\{v_{n,s}\}$. For this reason we shall define (3.3.10) as the "steady-state response" of the system.

We can use (3.3.12) to calculate the steady-state spectral density. Thus denoting the spectral density of the process $\{u_k\}$ by $p_u(\lambda)$, we have

$$R_u(n) = \int_{-1/2}^{1/2} e^{2\pi i n \lambda} p_u(\lambda) \, d\lambda,$$

and substituting this in (3.3.12) we have

$$R_v(p) = \sum_0^\infty \sum_0^\infty W_k \int_{-1/2}^{1/2} e^{2\pi i \lambda (p+j-k)} p_u(\lambda) W_j^* \, d\lambda$$

$$= \int_{-1/2}^{1/2} e^{2\pi i \lambda p} \psi(\lambda) p_u(\lambda) \psi(\lambda)^* \, d\lambda,$$

where

$$\psi(\lambda) = \sum_0^\infty W_k e^{-2\pi i k \lambda} \qquad (3.3.13)$$

is the transfer function of the system. Hence the spectral density of the output process

$$= \psi(\lambda) p_u(\lambda) \psi(\lambda)^*$$

$$= \begin{pmatrix} \text{system} \\ \text{transfer} \\ \text{function} \end{pmatrix} \begin{pmatrix} \text{spectral} \\ \text{density of} \\ \text{input process} \end{pmatrix} \begin{pmatrix} \text{system} \\ \text{transfer} \\ \text{function} \end{pmatrix}^*. \qquad (3.3.14)$$

Note that all the formulas for output spectral density we have obtained before agree with this general formula.

Remark. The condition (3.3.9) (which is in fact necessary if the mean of $\{u_k\}$ is not equal to zero) can be weakened to

$$\sum_0^\infty \| W_k \|^2 < \infty \qquad (3.3.9a)$$

if $p_u(\lambda)$ is continuous in $-1/2 \leq \lambda \leq 1/2$ *and* the process $\{u_k\}$ has zero mean, because in that case

$$E \left\| \sum_N^{N+P} W_k u_{n-k} \right\|^2 = \mathrm{Tr} \sum_N^{N+P} \sum_N^{N+P} W_k R_u(j-k) W_j^*$$

$$= \mathrm{Tr} \int_{-1/2}^{1/2} \left(\sum_N^{N+P} W_k e^{-2\pi i \lambda k} \right) p_u(\lambda) \left(\sum_N^{N+P} W_k e^{-2\pi i \lambda k} \right)^* d\lambda$$

$$\leq \int_{-1/2}^{1/2} \| p_u(\lambda) \| \left\| \sum_N^{N+P} W_k e^{-2\pi i \lambda k} \right\|^2 d\lambda$$

$$\leq (\max \| p_u(\lambda) \|, 0 \leq \lambda \leq 1/2) \int_{-1/2}^{1/2} \left\| \sum_N^{N+P} W_k e^{-2\pi i \lambda k} \right\|^2 d\lambda.$$

Then we know from Fourier series theory (see the Review Chapter) that

$$\int_{-1/2}^{1/2} \left\| \sum_N^{N+P} W_k e^{-2\pi i \lambda k} \right\|^2 d\lambda = \sum_N^{N+P} \| W_k \|^2.$$

Hence

$$E \left[\left\| \sum_N^{N+P} W_k u_{n-k} \right\|^2 \right] \leq (\max \| p_u(\lambda) \|, 0 \leq \lambda \leq 1/2) \sum_N^{N+P} \| W_k \|^2,$$

or

$$\sum_0^N W_k u_{n-k}$$

converges in the mean square as $N \to \infty$.

3.3.3 Weighting Pattern Signal Generation Model

Let us specialize the input $\{u_k\}$ in Section 3.3.2 to be white noise $\{N_n\}$ with spectral density matrix D, and define

$$v_n = \sum_0^n W_k N_{n-k}, \qquad n \geq 0. \tag{3.3.15}$$

In this case we note that

$$E[v_n] = 0$$

and defining

$$Z_N = \sum_0^N W_k N_{n-k}$$

we see that

$$E[(Z_{N+p} - Z_N)(Z_{N+p} - Z_N)^*] = \sum_{k=N+1}^{N+p} \sum_{j=N+1}^{N+p} W_k D \delta(j-k) W_j^*$$

$$= \sum_{N+1}^{N+p} W_k D W_k^*$$

and

$$\left\| \sum_{N+1}^{N+p} W_k D W_k^* \right\|_0 \leq \sum_{N+1}^{N+p} \| W_k \|_0^2 \| D \|_0.$$

Hence Z_N converges in the mean square sense if

$$\sum_0^\infty \| W_k \|^2 < \infty. \tag{3.3.16}$$

We could have seen this also from the fact (cf. Remark on page 137) that the spectral density of white noise is a constant. In particular we note that the process defined by (3.3.15) is asymptotically stationary under condition (3.3.16) with covariance function given by

$$R_v(p) = \sum_0^\infty \sum_0^\infty W_k D \delta(p+j-k) W_j^* = \sum_0^\infty W_k D W_{k-p}^*$$

$$= \sum_0^\infty W_{j+p} D W_j^* \tag{3.3.17}$$

with the understanding that

$$W_k = 0, \qquad k < 0, \tag{3.3.18}$$

The corresponding steady-state spectral density

$$P_v(\lambda) = \psi(\lambda) D \psi(\lambda)^* \tag{3.3.19}$$

follows from

$$R_v(p) = \sum_0^\infty \sum_0^\infty W_k \int_{-1/2}^{1/2} e^{2\pi i \lambda n(p+j-k)} D W_j^* \, d\lambda$$

$$= \int_{-1/2}^{1/2} e^{2\pi i \lambda p} \psi(\lambda) D \psi(\lambda)^* \, d\lambda,$$

where

$$\psi(\lambda) = \sum_0^\infty W_k e^{-2\pi i k \lambda}, \qquad -1/2 < \lambda < 1/2, \tag{3.3.20}$$

which, because of the weaker condition (3.3.16), must be interpreted as the mean square limit of the sequence $\psi_N(\cdot)$ defined by

$$\psi_N(\lambda) = \sum_0^N W_k e^{-2\pi i k \lambda}, \qquad -1/2 < \lambda < 1/2,$$

which converges in the mean square sense as functions of λ, $-1/2 \leq \lambda \leq 1/2$, because

$$\int_{-1/2}^{1/2} \|\psi_{N+p}(\lambda) - \psi_N(\lambda)\|^2 \, d\lambda = \sum_{N+1}^{N+p} \|W_k\|^2$$

and

$$\sum_0^\infty \|W_k\|^2 < \infty, \tag{3.3.21}$$

since (cf. Section 0.1.3)

$$\|W_k\|^2 \leq (\text{const.})\|W_k\|_0^2,$$

and (3.3.21) is of course equivalent to (3.3.16). A consequence of the relaxation of (3.3.9) to (3.3.16) is that $\psi(\lambda)$ need not be continuous for every λ, $-1/2 \leq \lambda \leq 1/2$. See Example 3.3.1 below.

We can consider (3.3.15) as a "signal generation model" for a process with steady-state spectral density $p(\lambda)$, where $p(\lambda)$ is given by (3.3.19) where $\psi(\lambda)$ satisfies (3.3.20). We shall refer to (3.3.15) as a weighting pattern signal generation model with the weighting pattern satisfying (3.3.16).

Given an arbitrary matrix spectral density function $p(\lambda)$, $-1/2 \leq \lambda \leq 1/2$, the question as to when it can be "factorized" in the form (3.3.19), with $\psi(\lambda)$ given by (3.3.20), is therefore of interest. We note that in the one-dimensional case, a necessary and sufficient condition is that

$$\int_{-1/2}^{1/2} |\log p(\lambda)| \, d\lambda < \infty. \tag{3.3.22}$$

See reference 1 of Chapter 2 for more on this. See also Chapter 9, Section 8. For an example where the condition is violated, take

$$p(\lambda) = \exp\left(\frac{-1}{2|\lambda|}\right), \qquad -1/2 \leq \lambda \leq 1/2.$$

Finally we note that the weighting pattern signal generation model (3.3.15) is more general than the Kalman signal generation model (3.2.23) in that the system transfer function $\psi(\lambda)$ is not restricted to be a rational function of λ as in the latter. Of course the former can be approximated by the latter.

Example 3.3.1. Example of a weighting pattern model, chosen for illustrative purposes only:

$$W_k = 0 \qquad k = 0$$

$$= \frac{1}{k}, \qquad k \geq 1.$$

Note that $\{W_k\}$ does not satisfy (3.3.9) but does satisfy (3.3.16).

$$v_n = \sum_1^n \frac{N_{n-k}}{k}, \qquad n \geq 1$$

$$\psi(\lambda) = \sum_1^\infty \frac{e^{2\pi i k \lambda}}{k}, \qquad \lambda \neq 0, \; -1/2 \leq \lambda \leq 1/2,$$

which is clearly not defined at $\lambda = 0$. On the other hand, by defining the function of a complex variable z in the unit circle,

$$f(z) = \sum_1^\infty \frac{z^k}{k}, \qquad |z| < 1,$$

we see that

$$f(z) = \log(1 - z),$$

which is defined for $|z| \leq 1$, except at $z = 1$. Hence the spectral density is given by

$$|\psi(\lambda)|^2 = p(\lambda) = |\log(1 - e^{2\pi i \lambda})|^2, \qquad \lambda \neq 0.$$

This function is of course not rational in $e^{2\pi i \lambda}$. The corresponding covariance function is

$$R(p) = \sum_1^\infty \frac{1}{k(k + |p|)}.$$

A plot of $10 \log_{10} p(\lambda)$ is given in Figure 3.1. Note the singularity at $\lambda = 0$.

Example 3.3.2. Here is a simple example where we know that there is a weighting pattern generation model but cannot evaluate the weighting sequence explicitly. Thus let

$$p(\lambda) = 1 - 2|\lambda|, \qquad -1/2 < \lambda < 1/2. \tag{3.3.23}$$

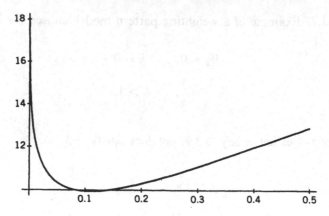

Figure 3.1. $10 \log_{10} p(\lambda)$; $p(\lambda) = |\log (1 - e^{2\pi i\lambda})|^2$.

Then

$$\int_{-1/2}^{1/2} |\log p(\lambda)| \, d\lambda = 2 \int_0^{1/2} |\log (1 - 2\lambda)| \, d\lambda$$

$$= -\int_0^1 \log x \, dx = 1 < \infty.$$

Hence we know that

$$p(\lambda) = |\psi(\lambda)|^2,$$

where

$$\psi(\lambda) = \sum_0^\infty W_k e^{-2\pi i k\lambda}$$

$$\sum_0^\infty W_k^2 < \infty.$$

Finding $\{W_k\}$, however, is another matter. There is no "explicit" solution known. We can follow the general theory in reference 3 and obtain numerical approximation, but it is too involved to go into here. See below for an alternate technique.

Physically Nonrealizable Nonfinite-Memory Weighting Patterns   As in the case of finite memory systems, we may now consider weighting patterns which are not physically realizable, where W_k does not vanish for $k < 0$. Thus we have the steady-state version; in place of (3.3.10) we have

$$v_n = \sum_{-\infty}^\infty W_k u_{n-k} \qquad (3.3.24)$$

with

$$\sum_{-\infty}^{\infty} \|W_k\|_0 < \infty \qquad (3.3.25)$$

weakened to

$$\sum_{-\infty}^{\infty} \|W_k\|^2 < \infty \qquad (3.3.25a)$$

if the process $\{u_k\}$ has zero mean, and $p_u(\lambda)$ is continuous in $-1/2 \le \lambda \le 1/2$. Since the memory is not finite we can no longer represent this as the delayed output of a physically realizable system—unless we allow for "infinite delay." However, we can create a sequence of approximations. Thus choose integer $M > 0$ and let

$$v_{n|M} = \sum_{-M}^{\infty} W_k u_{n-k}, \qquad M > 0.$$

Then, fixing M, we obtain

$$v_{n-M|M} = v_n',$$

where

$$v_n' = \sum_0^{\infty} W_k' u_{n-k}$$

$$W_k' = W_{k-M}.$$

The (steady-state) spectral density of the process $\{v_n'\}$ given by (3.3.14)

$$= \psi_M(\lambda) p_u(\lambda) \psi_M(\lambda)^*, \qquad (3.3.26)$$

where

$$\psi_M(\lambda) = \sum_0^{\infty} W_k' e^{-2\pi i k \lambda} = \sum_0^{\infty} W_{k-M} e^{-2\pi i k \lambda}$$

$$= e^{-2\pi i M \lambda} \sum_{-M}^{\infty} W_k e^{-2\pi i k \lambda}$$

and hence (3.3.26)

$$= \left(\sum_{-M}^{\infty} W_k e^{-2\pi i k \lambda}\right) p_u(\lambda) \left(\sum_{-M}^{\infty} W_k^* e^{-2\pi i k \lambda}\right),$$

which, as $M \to \infty$ converges to

$$= \psi(\lambda)^* p_u(\lambda) \psi(\lambda),$$

where

$$\psi(\lambda) = \sum_{-\infty}^{\infty} W_k e^{-2\pi i k \lambda},$$

and thus (3.3.14) holds again.

Example 3.3.2 Revisited. Armed with this idea, let us revisit Example 3.3.2. We can now provide an alternate solution to realizing a process with spectral density given by (3.3.23). Thus we now define

$$W_k = \int_{-1/2}^{1/2} e^{2\pi i k \lambda} \sqrt{1 - 2|\lambda|} \, d\lambda = 2 \int_0^{1/2} \sqrt{1 - 2\lambda} \cos \pi k \lambda \, d\lambda.$$

Then the process defined by

$$v_n = \sum_0^{2M} W_{k-M} N_{n-k}$$

has spectral density

$$p_u(\lambda) = \left| \sum_{-M}^{M} W_k e^{2\pi i k \lambda} \right|^2,$$

which, as $M \to \infty$,

$$= \left| \sum_{-\infty}^{\infty} W_k e^{2\pi i k \lambda} \right|^2 = (\sqrt{1 - 2|\lambda|})^2 = p(\lambda).$$

Thus the spectral density can be realized by a physically realizable system as closely as we wish, bypassing the factorization procedure! In this example, $M = 10$ already yields a good enough approximation to the spectral density, but M can be prohibitively large—see problem 3.2.

This procedure can be used of course even when condition (3.3.22) is violated. See Problem 3.4.

3.3.4 "Prewhitening" Filters

Let $\{x_k\}$ be a $p \times 1$ stationary Gaussian process with mean zero and spectral density matrix $p(\cdot)$. In many applications we are interested in first processing it with a "prewhitening" filter—that is, we want to construct a (physically realizable) linear time-invariant system with transfer function $\psi(\lambda)$ such that

$$\psi(\lambda) p(\lambda) \psi(\lambda)^* = I_{p \times p}$$

or, equivalently, that the output process $\{v_n\}$ is white noise in the steady state.

To simplify matters let us specialize to the one-dimensional case, $p = 1$, so that

$$|\psi(\lambda)|^2 = \frac{1}{p(\lambda)},$$

where we shall assume

$$\int_{-1/2}^{1/2} \frac{1}{p(\lambda)} d\lambda < \infty. \tag{3.3.27}$$

To reduce complexity, we shall actually assume that

$$p(\lambda) \geq \varepsilon > 0, \qquad -1/2 \leq \lambda \leq 1/2. \tag{3.3.28}$$

Let us consider the case where the process $\{x_k\}$ is modeled as the output of a finite memory system excited by white noise so that

$$p(\lambda) = \left| \sum_0^M W_k e^{-2\pi i k \lambda} \right|^2,$$

where the $\{W_k\}$ are real. Then we must have

$$|\psi(\lambda)| = \frac{1}{\left| \sum_0^M W_k e^{-2\pi i k \lambda} \right|}.$$

If $M = 1$, we have

$$|\psi(\lambda)| = \frac{1}{|W_0 + W_1 e^{-2\pi i \lambda}|}, \qquad W_0 \neq 0, W_1 \neq 0.$$

Since by (3.3.24) $p(\lambda)$ cannot be zero, for any λ, $-1/2 \leq \lambda \leq 1/2$, we see that

$$|W_0| \neq |W_1|.$$

Let

$$\left| \frac{W_1}{W_0} \right| < 1.$$

Then we may take

$$\psi(\lambda) = \frac{1}{W_0(1 - \rho e^{-2\pi i \lambda})}, \qquad \rho = \frac{-W_1}{W_0}$$

$$= \frac{1}{W_0} \sum_0^\infty \rho^k e^{-2\pi i k \lambda}.$$

This is recognized as the transfer function of a system with a state-space model.

In fact we have

$$y_{n+1} = \rho y_n + x_n$$

$$v_n = \frac{1}{W_0} y_n,$$

where $\{v_n\}$ is yielding white noise in the steady state. If

$$\frac{W_1}{W_0} > 1$$

we define

$$\psi(\lambda) = \frac{1}{W_1\left(1 + \dfrac{W_0}{W_1} e^{-2\pi i \lambda}\right)}$$

since

$$\frac{1}{\left|W_1\left(1 + \dfrac{W_0}{W_1} e^{-2\pi i \lambda}\right)\right|} = \left|\frac{1}{W_1 + W_0 e^{-2\pi i \lambda}}\right| = \frac{1}{|W_1 + W_0 e^{2\pi i \lambda}|}$$

$$= \left|\frac{1}{W_0 + W_1 e^{-2\pi i \lambda}}\right|$$

and hence

$$\psi(\lambda) = \frac{1}{W_1} \sum_0^\infty \rho^k e^{-2\pi i k \lambda}, \qquad \rho = -\left(\frac{W_0}{W_1}\right),$$

yielding the transfer function of a system with a state-space model. Note that

$$\frac{-W_0}{W_1}$$

is the root of the polynomial in z:

$$W_0 + W_1 z = 0.$$

Thus in the more general case, we may set

$$W_M = 1$$

and note that the polynomial

$$\sum_0^M W_k z^k = \prod_{i=1}^M (z - z_i),$$

where $\{z_i\}$ are the roots. And we have seen how to construct the prewhitening transfer function if all the roots are real. If not real, the roots must come in conjugate pairs since the $\{W_k\}$ are real. Hence let us consider a pair of conjugate roots

$$(z - \bar{z}_i)(z - \bar{z}_i).$$

Suppose

$$|z_i| > 1.$$

Then

$$\frac{1}{(z - z_i)}\frac{1}{(z - z_i)} = \frac{1}{|z_i|^2}\left(\sum_0^\infty \frac{z^k}{z_i^k}\right)\left(\sum_0^\infty \frac{z^k}{\bar{z}_i^k}\right)$$

$$= \frac{1}{|z_i|^2}\sum_0^\infty W_k z^k,$$

where

$$W_k = \sum_0^n \frac{1}{z_i^{(n-k)}}\frac{1}{\bar{z}_i^k}$$

and is real since

$$\bar{W}_k = \sum_0^n \frac{1}{\bar{z}_i^{(n-k)}}\frac{1}{z_i^k} = \sum_0^n \frac{1}{\bar{z}_i^k}\frac{1}{z_i^{(n-k)}} = W_k.$$

Hence we may set

$$\psi(\lambda) = \frac{1}{|z_i|^2}\sum_0^\infty W_k e^{-2\pi i\lambda k}.$$

Since

$$\psi(\lambda) = \frac{1}{|z_i|^2}\frac{1}{(e^{-2\pi i\lambda} - z_i)(e^{-2\pi i\lambda} - \bar{z}_i)}$$

we see that the corresponding system can be represented with a state-space model of dimension two.

The case

$$|z_i| < 1$$

can be handled following our treatment for

$$\left|\frac{W_0}{W_1}\right| < 1.$$

By (3.3.24) there cannot be any root z_i such that

$$|z_i| = 1.$$

Thus we can express the prewhitening filter transfer function as the product of physically realizable transfer functions, each of which corresponds to a system with a state-space model.

If the process $\{x_k\}$ is such that it can be modeled by a Kalman signal generation model, then we can essentially reverse our steps. Indeed, we know that

$$p(\lambda) = |\phi(\lambda)|^2,$$

where

$$\phi(\lambda) = C(I - Ae^{-2\pi i\lambda})^{-1}B,$$

where A is $n \times n$, C is $1 \times n$, B is $n \times 1$, and A is stable so that the eigenvalues λ_i are such that

$$|\lambda_i| < 1$$

and hence $\phi(\lambda)$ is of the form

$$\frac{N(\lambda)}{D(\lambda)}.$$

where

$$D(\lambda) = \prod_1^n (1 - \lambda_k e^{-2\pi i\lambda})$$

and $N(\lambda)$ is a polynomial in $(e^{-2\pi i\lambda})$ of degree less than n. Hence the transfer function of the prewhitening filter $\psi(\lambda)$ must be such that

$$|\psi(\lambda)| = |D(\lambda)| \frac{1}{|N(\lambda)|},$$

where

$$|N(\lambda)| > 0, \quad \text{in } -1/2 \le \lambda \le 1/2$$

by virtue of (3.3.24). Clearly $D(\lambda)$ is the transfer function of a physical realizable finite memory system. Under condition (3.3.26), we may, as we have just shown, realize

$$\frac{1}{|N(\lambda)|}$$

as a physically realizable system with a state-space model.

In the general $(p = 1)$ case, condition (3.3.24) implies (3.3.23) and we are assured that we can find a physically realizable prewhitening filter. Otherwise we can always approximate it by constructing a sequence of finite-memory systems that we would use for simulating a process with spectral density

$$\frac{1}{p(\lambda)}, \quad -1/2 \le \lambda \le 1/2$$

as in Chapter 7. This approximation of course carries over to the multidimensional case as well; condition (3.3.24) becomes

$$\gamma(\lambda) > 0, \quad -1/2 \le \lambda \le 1/2,$$

where $\gamma(\lambda)$ is the smallest eigenvalue of $P(\lambda)$.

3.4 DIFFERENCE EQUATION AND ARMA MODELS

The signal generation model is given by

$$\begin{align} v_n &= Cx_n, \quad n \ge 0 \\ x_{n+1} &= Ax_n + BN_n, \end{align} \tag{3.4.1}$$

where $\{N_k\}$ being white Gaussian can be expressed as a "difference equation" characterization of the process $\{v_k\}$. For this purpose recall the Cayley–Hamilton theorem for matrices (compare Review Chapter).

$$A^n + \sum_{k=0}^{n-1} a_k A^k = 0,$$

where the coefficients $\{a_k\}$ are those of

$$\det(\lambda I - A) = \lambda^n + \sum_{k=0}^{n-1} a_k \lambda^k.$$

Hence for any x in $\mathbf{E}^n$, defining $a_n = 1$, we have

$$C\left(\sum_0^n a_k A^k\right) x = 0$$

or

$$\sum_0^n a_k CA^k x = 0.$$

In other words, the $n+1$ matrices $C, CA, \ldots, CA^n$ are linearly dependent. It may happen that this is true for $m < n$. We say that the system is "observable" (a system-theoretical concept: see reference 5) if

$$C, CA, \ldots, CA^{n-1}$$

are linearly independent. We shall go ahead and use m with the understanding that $m \leq n$. Since n is already used, we shall use j as the index in (3.4.1). Thus we assume there are constants $\{c_k\}$, $k \leq m \leq n$, such that

$$\sum_0^m c_k CA^k = 0,$$

where we may take $c_m = 1$. Then

$$\sum_0^m c_k CA^k x_j = 0 \tag{3.4.2}$$

for every j. (We do not use $\{a_k\}$ since for $m < n$, the coefficients may be different!) From (3.2.7) we can calculate that

$$A^k x_j = x_{j+k} - \sum_{i=0}^{k-1} A^{k-1-i} BN_{j+i}, \qquad k \geq 1.$$

Hence

$$c_0 v_j + \sum_1^m c_k \left(C x_{j+k} - \sum_{i=0}^{k-1} CA^{k-1-i} BN_{j+i} \right) = 0$$

or

$$\sum_0^m c_k v_{j+k} = \sum_{k=0}^m c_k \sum_{i=0}^{k-1} CA^{k-1-i} BN_{j+i}$$

$$= c_1 CBN_j + c_2 (CABN_j + CBN_{j+1}) + \cdots$$

$$+ c_m (CA^{m-1} BN_j + CBN_{j+m-1})$$

$$= (c_1 CB + c_2 CAB + \cdots + c_m CA^{m-1} B) N_j$$

$$+ (c_2 CB + c_3 CAB + \cdots + c_m CA^{m-2} B) N_{j+1}$$

$$+ c_m CBN_{j+m-1}$$

$$= \sum_0^{m-1} b_k N_{j+k},$$

where

$$b_k = c_{k+1} CB + c_{k+2} CAB + \cdots + c_m CA^{m-1} B.$$

Hence finally we have

$$v_{j+m} + \sum_0^{m-1} c_k v_{j+k} = \sum_0^{m-1} b_k N_{j+k}. \tag{3.4.3}$$

We have thus a "difference equation" characterization of the signal process

$\{v_n\}$. This is known as an ARMA model in the time series literature (see, e.g., reference 4). Usually (3.4.3) is generalized slightly by adding a term corresponding to N_{j+m} on the right side:

$$v_{j+m} + \sum_0^{m-1} c_k v_{j+k} = \sum_0^m b_k N_{j+k}. \tag{3.4.4}$$

This is an AR—"autoregressive"—model if all but one b_k are zero. It is an MA—"moving average"—model if all the $\{c_k\}$ are zero. Thus (3.1.1) is an MA model if the $\{u_i\}$ is white noise.

Starting with (3.4.4) we can also produce a state-space model. This is of course a system-theoretic problem and of secondary importance to us. Hence in order not to complicate purely notational problems we shall here consider the case where both the input process $\{N_k\}$ and the output process are 1×1, the so-called "SISO"—"single-input single-output"—system. In particular, the coefficients b_k are all 1×1. Thus starting with (3.4.4) let

$$x_j = \begin{vmatrix} v_j \\ \vdots \\ v_{j+m-1} \end{vmatrix}, \qquad x_j \in \mathbf{E}^m.$$

Then we can express x_{j+1} as

$$x_{j+1} = Fx_j + G\left(\sum_0^m b_k N_{j+k}\right), \tag{3.4.5}$$

where F is the $m \times m$ matrix (so-called "companion" matrix)

$$F = \begin{vmatrix} 0 & 1 & 0 & \cdots & 0 \\ 0 & 0 & 1 & \cdots & 0 \\ \vdots & \vdots & \vdots & \ddots & \vdots \\ 0 & 0 & 0 & \cdots & 1 \\ -c_0 & -c_1 & -c_2 & \cdots & -c_{m-1} \end{vmatrix} \tag{3.4.6}$$

and G is the $m \times 1$ matrix:

$$G = \begin{vmatrix} 0 \\ 0 \\ \vdots \\ 1 \end{vmatrix}. \tag{3.4.7}$$

Let

$$H = |1 \quad 0 \quad \cdots \quad 0|, \qquad 1 \times m.$$

Then it is readily verified that

$$HF^kG = 0, \qquad k = 0, \ldots, m-2,$$
$$HF^{m-1}G = 1. \qquad\qquad (3.4.8)$$

And of course

$$v_j = Hx_j.$$

If in (3.4.5) all b_k except one are zero we already have a state-space model in (3.4.5), (3.4.9). More generally, define

$$y_j = x_j - \sum_1^m b_k \sum_0^{k-1} F^{k-1-i}GN_{j+i}. \qquad (3.4.9)$$

Then by (3.4.8)

$$Hy_j = Hx_j - b_m N_j = v_j - b_m N_j. \qquad (3.4.10)$$

Next,

$$y_{j+1} = x_{j+1} - \sum_1^m b_k \sum_0^{k-1} F^{k-1-i}GN_{j+1+i}$$

$$= x_{j+1} - \sum_1^m b_k \sum_1^k F^{k-i}GN_{j+1}$$

so that

$$Fy_j = Fx_j - \sum_1^m b_k \sum_0^{k-1} F^{k-i}GN_{j+i}.$$

Hence

$$y_{j+1} - Fy_j = x_{j+1} - Fx_j - \sum_1^m b_k(GN_{j+k} - F^kGN_j)$$

$$= \left(\sum_0^m b_k F^k G\right) N_j,$$

using (3.4.5). Hence writing $\tilde{G}$ for the matrix

$$\sum_0^m b_k F^k G,$$

we have

$$y_{j+1} = Fy_j + \tilde{G}N_j$$
$$v_j = Hy_j + b_m N_j, \qquad (3.4.11)$$

which is then the state space corresponding to the ARMA model (3.4.4). Note the slight generalization of (3.4.1) in the appearance of the added noise term in (3.4.11) in the output. Since y_j and N_j are independent and $\{N_j\}$ is white noise, this presents no problem for us in terms of calculating the spectral density or covariance function of the process $\{v_j\}$. For most of our work, $b_m = 0$.

Suppose we start with (3.4.1), produce the ARMA model (3.4.3) and go back to the state-space model (3.4.11) (with $b_m = 0$). Do we get the same representation back? The answer is no, in general. Let us consider the SISO case. First of all, the state space in (3.4.11) is observable, since

$$HF = |0 \quad 1 \quad 0 \quad \cdots \quad 0 \quad 0|$$
$$\vdots$$
$$HF^{m-1} = |0 \quad 0 \quad 0 \quad \cdots \quad 0 \quad 1|$$

so that HF^k, $k = 0, \ldots, m - 1$, are linearly independent. Hence the system (3.4.1) must also be observable, and in particular the state-space dimension must be the same so that $m = n$. Moreover, we define the transformation on E^n into E^n by

$$y = Tx$$

and

$$Tx = \begin{vmatrix} Cx \\ CAx \\ \vdots \\ CA^{n-1}x \end{vmatrix}$$

or in matrix form, since CA^k is $1 \times n$:

$$T = \begin{vmatrix} C \\ CA \\ \vdots \\ CA^{n-1} \end{vmatrix}_{n \times n}.$$

We can readily verify that

$$y_n = Tx_n$$
$$F = TAT^{-1}$$
$$G = TB$$
$$H = CT^{-1}$$
$$(I - Fe^{2\pi i\lambda})^{-1}G = T(I - Ae^{2\pi i\lambda})^{-1}B.$$

Hence the spectral density $P_y(\cdot)$ of the process $\{y_n\}$

$$= TP_x(f)T^*,$$

where $P_x(\cdot)$ is the spectral density of the process $\{x_n\}$. The spectral density of the process $\{v_n\}$, according to (3.4.11),

$$= HTP_x(f)T^*H^*$$
$$= CP_x(f)C^*,$$

which is the same as according to (3.4.1).

3.5 TIME-VARYING SYSTEMS

We may extend the Kalman signal generation model (3.4.1) to include "time-varying" systems where the system matrices depend on the index n. Thus

$$v_k = C_k x_k \tag{3.5.1}$$

$$x_{k+1} = A_k x_k + B_k N_k, \tag{3.5.2}$$

where

$$A_k: \quad n \times n$$
$$B_k: \quad n \times p$$
$$C_k: \quad m \times n$$

and N_k is white Gaussian. Under the assumption (3.2.1), we again have that x_n is independent of $\{N_k\}$, $k \geq n$. This implies that $\{x_k\}$ is a Gaussian Markov process. See Chapter 9, Problem 9.9.

PROBLEMS

3.1. Let $\{v_j\}$ be defined as in (3.1.3), (3.1.4), where $\{u_k\}$ is white noise. Find the steady-state covariance of the output $\{v_j\}$. Show that the steady-state covariance function $R_v(m)$ vanishes for $m > M$.

3.2. Show that for $0 < |\rho| \leq 1$,

$$p(\lambda) = 1 + \rho \cos 2\pi\lambda, \qquad -1/2 \leq \lambda \leq 1/2$$

can be represented as the spectral density of the response to white noise of a finite memory system with memory 2. Is such a system unique?
Hint: Find a, b such that

$$1 + \rho \cos 2\pi\lambda = |a + be^{-2\pi i \lambda}|^2.$$

Next for $\rho = 1$, calculate

$$W_k = 2 \int_0^{1/2} (\sqrt{1 + \cos 2\pi\lambda}) \cos 2\pi k\lambda \, d\lambda$$

$$= 2 \int_0^{1/2} \sqrt{2} \cos \pi\lambda \cos 2\pi k\lambda \, d\lambda$$

and show that

$$W_k = \frac{(2\sqrt{2}) \cos k\pi}{\pi(a - 4k^2)}$$

and hence in this approximation:

$$1 + \cos 2\pi\lambda \approx \left(W_0 + 2 \sum_1^M W_k \cos 2\pi k\lambda \right)^2.$$

M has to be at least 10, or system memory $(2M + 1) = 21$, much higher than 2!

3.3. Let

$$p(\lambda) = 6 - 8 \cos 2\pi\lambda + 2 \cos 4\pi\lambda$$

denote the spectral density corresponding to the covariance sequence in the second part of Problem 2.8. Show that, unlike in Problem 3.2,

$$W_k = 2 \int_0^{1/2} \sqrt{p(\lambda)} \cos 2\pi k\lambda \, d\lambda = 0, \qquad \text{for } |k| > 1$$

and construct the finite memory system weighting pattern whose steady-state response to white noise will yield this spectral density.

Answer:

$$\psi(\lambda) = (2 - e^{2\pi i\lambda} - e^{-2\pi i\lambda}).$$

3.4. Let

$$p(\lambda) = e^{-1/|\lambda|}, \qquad -1/2 \le \lambda \le 1/2.$$

Show how to approximate this as closely as required as the spectral density of the output of a finite memory system with white noise input.

3.5. Using (3.3.14) or otherwise, show that the steady-state response of a time-invariant linear system to an input process with a pure δ-function spectral density has also a pure δ-function spectral density. Specialize to Example (1.1.9) of Chapter 1, using as input process (1.1.16) or more generally the discretized version of (1.1.6) for given Δ.

3.6. Let A be $m \times m$ square matrices and let A_1 be $p \times p$ square matrices. Let $\mathscr{A}$ denote the $(m + p) \times (m + p)$ matrix:

$$\begin{vmatrix} A & 0 \\ J & A_1 \end{vmatrix},$$

where J is any $p \times m$ matrix. Show that if $\mathscr{A}$ is stable, then both A and A_1 must be stable.

3.7. Show by example that the Liapunov equation

$$R = ARA^* + FF^*$$

(a) need not have a solution if A is not stable [Hint: Take $A = I$.] and
(b) can have many solutions if A is not stable and that
(c) if it has any solution, it has a self-adjoint solution.

3.8. For the general case of

$$A: \quad 2 \times 2$$
$$B: \quad 2 \times 1$$
$$D: \quad 1 \times 1$$

obtain the solution for the Liapunov equation (3.2.13) assuming A stable. Where does the stability of A figure in your solution?

Hint: Using

$$A = \begin{vmatrix} a_{11} & a_{12} \\ a_{21} & a_{22} \end{vmatrix}$$

show that

$$\det \begin{vmatrix} a_{11}^2 - 1 & 2a_{11}a_{12} & a_{12}^2 \\ a_{11}a_{12} & a_{11}a_{22} + a_{12}a_{21} - 1 & a_{12}a_{22} \\ a_{21}^2 & 2a_{21}a_{22} & a_{22}^2 - 1 \end{vmatrix} \neq 0$$

if A is stable!

3.9. If the signal model is given by

$$v_n = Cx_n$$
$$x_n = Ax_{n-1} + FN_{n-1},$$

where $\{N_n\}$ is white Gaussian with identity covariance matrix, then for

$$C = |1, \quad 0|$$

$$F = \begin{vmatrix} 0 \\ 1 \end{vmatrix}$$

$$A = \begin{vmatrix} 0.75 & 0.25 \\ 0.25 & 0.25 \end{vmatrix}$$

calculate:

(a) the steady-state covariance function of the state process $\{x_n\}$; of $\{v_n\}$.

(b) the spectral density matrix of the state process $\{x_n\}$; of $\{v_n\}$.

3.10. Let $R(k)$, $k \in I$, denote a 1×1 stationary covariance function, and let M denote the covariance matrix

$$M = \{R(i - j)\}, \quad 1 \leq i, j \leq N.$$

Such a matrix is called a "Toeplitz" matrix. Show the smallest eigenvalue

$$\geq \min p(\lambda), \quad -1/2 \leq \lambda \leq 1/2$$

and the largest eigenvalue

$$\leq \max p(\lambda), \quad -1/2 \leq \lambda \leq 1/2,$$

where $p(\cdot)$ is the spectral density. Show that as N increases, the largest eigenvalue is nondecreasing and the smallest eigenvalue is nonincreasing.

Hint: Use

$$[Ma, a] = \int_{-1/2}^{1/2} \left| \sum_{0}^{N-1} a_k e^{2\pi i k \lambda} \right|^2 p(\lambda) \, d\lambda$$

$$a = \text{column: } [a_0, \ldots, a_{N-1}]$$

$$\int_{-1/2}^{1/2} \left| \sum_{0}^{N-1} a_k e^{2\pi i k \lambda} \right|^2 d\lambda = \|a\|^2.$$

3.11. Let $\{y_n\}$ denote the response of a finite memory system to a stationary input $\{u_n\}$:

$$y_n = \sum_{0}^{L} w_k u_{n-k}, \quad n \in I.$$

Suppose $y_n = 0$ for every $n > L$. Is this response possible if both the system and the process are nonzero?

3.12. Calculate the steady-state covariance function of the process $\{x_n\}$ in Problem 3.9, using the series solution:

$$R = \sum_0^\infty A^k F F^* A^{*k}.$$

Hint: Use the Cayley–Hamilton theorem to reduce A^k for $k > 2$.

3.13. Given the signal generation model

$$v_n = Cx_n$$

$$x_{n+1} = Ax_n + Bz_n$$

$$z_{n+1} = \rho z_n + N_n,$$

where $\{N_n\}$ is white Gaussian, 1×1 with unit covariance, A, B, C as in Problem 3.9, and $\rho = 0.9$, find the steady-state covariance and spectral density of $\{v_n\}$.

3.14. Given that that spectral density defined by (3.2.21) is nonsingular, show that D must be nonsingular and that B^* is 1:1—that is, if

$$B^*u = 0$$

for some u, then u must be zero, or, equivalently, BB^* is nonsingular.

3.15. Deduce an ARMA model for $\{v_n\}$ in Problem 3.9. Evaluate explicitly the state equivalence transformation matrix T.

3.16. For the model (3.4.1)

$$A = \begin{vmatrix} 0 & 1 & 0 & \cdots & 0 \\ 0 & 0 & 1 & \cdots & 0 \\ \vdots & \vdots & \vdots & \ddots & \vdots \\ 0 & 0 & 0 & \cdots & 1 \\ -a_0 & -a_1 & -a_2 & \cdots & -a_{n-1} \end{vmatrix}$$

$$B = \begin{vmatrix} 0 \\ 0 \\ \vdots \\ 1 \end{vmatrix}; \quad C = |1 \quad 0 \quad 0 \quad \cdots \quad 0|.$$

Show that the spectral density of the process $\{v_n\}$ is given by

$$p_v(\lambda) = \frac{1}{|\psi(z)|^2}, \qquad z = e^{2\pi i\lambda},$$

where

$$\psi(z) = z^n + \sum_0^{n-1} a_k z^k.$$

What is the corresponding ARMA model for this case?

3.17. Show that the process

$$v_n = Cx_n$$

$$x_{n+1} = Ax_n + FN_n,$$

where A is stable, $\{N_k\}$ white noise, x_0 independent of N_k, $k \geq 0$, is stationary if

$$E[x_0] = 0$$

$$E[x_0 x_0^*] = R,$$

where

$$R = ARA^* + FF^*.$$

3.18. Let M be a (nilpotent) 3×3 matrix:

$$M = \begin{vmatrix} 0 & 1 & 0 \\ 0 & 0 & 1 \\ 0 & 0 & 0 \end{vmatrix}.$$

Note that $M^3 = 0$. Let $\Delta > 0$ and calculate

$$A = e^{m\Delta}.$$

Note that A is not stable. Let

$$C = |1 \quad 0 \quad 0|.$$

Show that

$$\text{Tr}\left(\sum_0^N A^{*k} C^* C A^k\right) \to \infty \qquad \text{as } N \to \infty.$$

3.19. Show directly (without going through the "physically realizable" limiting procedure) that (3.3.14) holds for weighting patterns $\{W_k\}$ not restricted to be physically realizable; that is, allow

$$W_k \neq 0, \qquad k < 0$$

and define

$$v_n = \sum_{-\infty}^{\infty} W_k u_{n-k}$$

with

$$\psi(\lambda) = \sum_{-\infty}^{\infty} W_k e^{-2\pi i k \lambda}, \qquad |\lambda| \leq 1/2$$

$$\sum_{-\infty}^{\infty} \|W_k\|_0 < \infty,$$

which is relaxed to

$$\sum_{-\infty}^{\infty} \|W_k\|_0^2 < \infty$$

when $\{u_n\}$ has mean zero.

3.20. Define the (physically realizable) "differencing operator" $\delta(\cdot)$ on the class of sequences $\{x_n\}$, $n \in I$, by

$$\delta(x_n) = x_n - x_{n-1}$$

and for each positive integer m define

$$\delta^m(\cdot) = \delta(\delta^{m-1}(\cdot)).$$

Show that if $\{x_n\}$ is a stationary stochastic process, so is the process

$$\delta^k(x_n)$$

and it has spectral density given by

$$|1 - e^{-2\pi i \lambda}|^{2k} p(\lambda), \qquad |\lambda| \leq 1/2,$$

where $p(\cdot)$ is the spectral density of the process $\{x_n\}$. Specializing to 1×1 processes, let

$$x_n = p_m(n) + z_n,$$

where $p_m(\cdot)$ is a polynomial of degree m and $\{z_n\}$ is a stationary process with zero mean. Show that

$$y_n = \delta^{m+1}(x_n)$$

is a stationary process with zero mean and calculate its spectral density.

Note that the process $\{y_n\}$ is defined knowing only the degree of the polynomial.

Hint:

$$\delta^{m+1}(p_m(n)) \equiv 0$$

and

$$\delta^{m+1}(x_n) = \delta^{m+1}(z_n).$$

3.21. Let $\{N_k\}$ denote one-dimensional white noise with unit variance. Let

$$v_k = N_{k+2} - 6N_{k+1} + 8N_k.$$

Find the transfer function of a physically realizable weighting pattern system model $\psi(\lambda)$ such that $|\psi(\lambda)|^2$ is the spectral density of the process. Suppose $\{v_k\}$ is the input to a linear time-invariant system. Find the transfer function of this system such that the steady-state output is white noise. Show that this system can be described by a state-space model.

NOTES AND COMMENTS

The idea of combining random processes and state-space theory originated with R. E. Kalman [1]. For ARMA models in time series analysis see reference 4. Because of the essential equivalence to state-space models the latter is more common in control engineering.

REFERENCES

Classic Paper

1. R. E. Kalman. "A New Approach to Linear Filtering and Prediction Problems," *Transactions of the ASME: Journal of Basic Engineering*, vol. 82, series D, no. 1 (1960), pp. 35–45.

Mathematical Treatises

2. J. L. Doob. *Stochastic Processes*. John Wiley and Sons, 1953.

3. K. L. Chung. *A Course in Probability Theory*. Academic Press, 1974.

Recent Publications

4. P. J. Brockwell and R. A. Davis. *Time Series Theory and Methods*. Springer-Verlag, 1993.

5. A. V. Balakrishnan. *State Space Theory of Systems*. Optimization Software Publications, 1988.

4

RESPONSE OF LINEAR SYSTEMS TO RANDOM INPUTS: CONTINUOUS-TIME MODELS

In this chapter we extend our study of the response of linear systems to continuous-time models: of systems as well as signals.

In analyzing electrical networks or mechanical structures, we need to deal with continuous-time models: "differential" equations based on "continuous-time" concepts such as (a) acceleration and velocity in structural dynamics or (b) currents and voltages in electric circuits. The inputs are also modeled as functions of time, defined for every instant of time in any time interval, not just a discrete subset. While the nature of the results remains essentially the same as in the discrete-time case treated in Chapter 3, continuous-time models require more sophisticated mathematics and will involve many "technicalities" which we shall need to point out.

4.1 DEFINITION OF INTEGRALS WITH STOCHASTIC INTEGRANDS

The first of these involves the definition of an integral when the integrand is a random process. The traditional input-output description of a continuous-time linear system is in terms of a "weighting function'": The output $v(\cdot)$ is defined for each t, $0 \leq t$, by

$$v(t) = \int_0^t W(t, s)u(s)\, ds, \qquad 0 \leq t, \tag{4.1.1}$$

where the "system input" is $u(\cdot)$ (say $(m \times 1)$ to be specific) defined on the interval $0 \leq t < \infty$ and the $(q \times m)$ "weighting function" is a function of two variables $0 \leq s, t < \infty$ and

$$W(t, s) = 0, \qquad s > t.$$

accounting for "physical realizability" just as in the discrete-time models. Avoiding technicalities, we may assume that the weighting function is continuous in both variables. In the deterministic case, if the input is also continuous in $0 \leq t < \infty$, we can define the integral as a limit of "partial sums"

$$\sum_{k=0}^{N} W(t, s_k) u(s_k)(s_{k+1} - s_k), \tag{4.1.2}$$

taking the subdivisions smaller and smaller.

We can imitate the same procedure when the input is random, with the "partial sum" approximation (4.1.2) as the key. If the sample paths of the input are continuous, then we can define (4.4.1) as an ordinary ("Riemann") integral on each sample path. Thus we have

$$v(t, \omega) = \int_0^t W(t, s) u(s, \omega)\, ds \tag{4.1.3}$$

and of course for each ω, $v(t, \omega)$ will be continuous in t as well, for $0 \leq t < \infty$. This requires that the assumptions we make regarding the input process assume that the sample paths are continuous—a "technicality." We shall return to this point below.

On the other hand, we can interpret the integral in a different way—by showing that the partial sums which are random variables converge in the mean square sense. For this purpose we shall only need to require that the mean and covariance functions be continuous. Let us state this as a theorem.

Theorem 4.1.1. Suppose the input mean and covariance functions are continuous; that is,

$$E[u(t)] = m(t)$$

is continuous in $0 \leq t < \infty$, and

$$R_u(t_2, t_1) = E[(u(t_1) - m(t_2))\ (u(t_2) - m(t_1))^*]$$

is continuous $0 \leq t_1, t_2 < \infty$. Suppose further that $W(t, s)$ is continuous $0 \leq s \leq t < \infty$. Then the partial sums (4.1.2) converge to a random variable $v(t)$ for each t, defining a random process in $0 \leq t < \infty$, such that

$$E[v(t)] = \int_0^t W(t, s) m(s)\, ds \tag{4.1.4}$$

and whose covariance function $R_v(\cdot, \cdot)$ is given by

$$R_v(t_2, t_1) = E[(v(t_2)v(t_1)^*) - E[v(t_2)]E[v(t_1)^*]$$

$$= \int_0^{t_2} \int_0^{t_1} W(t_2, s) R_u(s, \sigma) W(t_1, \sigma)^* \, ds \, d\sigma. \qquad (4.1.5)$$

Proof. The basis mathematical tool we shall need is the Hölder inequality for sums of random variables (vectors):

$$E\left\| \sum_1^N x_i \right\|^2 \le \left(\sum_1^N (E[\|x_i\|^2]^{1/2}) \right)^2, \qquad (4.1.6)$$

which we already used in Chapter 3.

Let us fix t and use the notation

$$x(s) = W(t, s)u(s), \qquad 0 \le s < \infty$$

so that the partial sums (4.1.2) are

$$\sum_1^N x(s_k)(s_{k+1} - s_k), \qquad s_k < s_{k+1}. \qquad (4.1.7)$$

The mean of the sum is

$$\sum_1^N E[x(s_k)](s_{k+1} - s_k) = \sum_1^N W(t, s_k)m(s_k)(s_{k+1} - s_k),$$

which clearly converges, as $\max |s_{k+1} - s_k| \to 0$, to the integral

$$\int_0^t W(t, s)m(s) \, ds = E[v(t)],$$

which is (4.1.4). We may therefore continue, setting the mean of $x(s)$ to be zero. To show mean square convergence we use the same technique as in the deterministic case by considering "refinements of partitions." Thus, given any subdivision, a refinement is obtained by increasing the number of subdivision points $\{s_k\}$ and at the same time reducing the "mesh size"

$$\max |s_{k+1} - s_k|.$$

The difference between two such partial sums (4.1.7), one being a refinement of the other, can be expressed

$$\sum_1^N (x(s_i) - x(\tilde{s}_i))(s_{i+1} - s_i)$$

where $\tilde{s}_i$ is any point in the interval (s_i, s_{i+1}), $s_i \leq \tilde{s}_i \leq s_{i+1}$. The mean square of the difference

$$E\left[\left\| \sum_1^N (x(s_i) - x(\tilde{s}_i))(s_{i+1} - s_i) \right\|^2 \right] \tag{4.1.8}$$

by the Hölder inequality

$$\leq \left(\sum_1^N (E[\| x(s_i) - x(\tilde{s}_i) \|^2])^{1/2}(s_{i+1} - s_i) \right)^2. \tag{4.1.9}$$

But

$$E[\| x(s_i) - x(\tilde{s}_i) \|^2] = \text{Tr}\,[R(s_i, s_i) - R(s_i, \tilde{s}_i) - R(\tilde{s}_i, s_i) + R(\tilde{s}_i, \tilde{s}_i)], \quad (4.1.10)$$

where

$$R(s_i, s_j) = E[x(s_i)x(s_j)^*],$$

which in our case

$$= W(t, s_i)R_u(s_i, s_j)W(t, s_j)^*$$

and hence is continuous in $0 \leq s_i, s_j \leq t$; and uniformly continuous, since $t < \infty$. Hence given $\varepsilon > 0$, we can find δ such that for all $s_i, s_i^*, 0 \leq s_i, s_i^* \leq t$,

$$|\text{Tr}\,(R(s_i, s_i) - R(s_i, s_i^*))| < \varepsilon$$

and

$$|\text{Tr}\,(R(s_i^*, s_i) - R(s_i^*, s_i^*))| < \varepsilon$$

for

$$|s_i - s_i^*| < \delta.$$

Hence the mean square error (4.1.8) for mesh size $< \delta$ is

$$\leq \varepsilon \left(\sum_1^N (s_{i+1} - s_i) \right)^2 = \varepsilon t^2.$$

Hence given any two partial sums (4.1.2) denoted S_1, S_2 with mesh size less than δ, let S_3 denote the sum corresponding to a refinement which is obtained by using subdivision points of both. Then

$$E\| S_1 - S_3 \|^2 \leq \varepsilon t^2$$

$$E\| S_2 - S_3 \|^2 \leq \varepsilon t^2.$$

But

$$\|S_1 - S_2\| = \|S_1 - S_3 + S_3 - S_2\|^2$$
$$\leq (\sqrt{\|S_1 - S_3\|^2} + \sqrt{\|S_3 - S_2\|^2})^2.$$

Hence

$$E[\|S_1 - S_2\|^2] \leq 4\varepsilon t^2.$$

Thus the partial sums converge in the mean square sense as the mesh size goes to zero. Thus defined, to calculate

$$E[v(t_1)v(t_2)^*]$$

we use the approximating partial sums corresponding to t_1, t_2:

$$E\left[\left(\sum_1^{N_2} W(t_2, s_k)u(s_k)(s_{k+1} - s_k)\right)\left(\sum_1^{N_1} W(t_1, \tilde{s}_j)u(\tilde{s}_j)(\tilde{s}_{j+1} - \tilde{s}_j)\right)^*\right]$$

$$= \sum_{k=1}^{N_2} \sum_{j=1}^{N_1} W(t_2, s_k)(R_u(s_k, \tilde{s}_j) + m(s_k)m(\tilde{s}_j)^*) W(t_1, \tilde{s}_j)^*,$$

which, as the mesh size goes to zero in each subdivision, converges to

$$\int_0^{t_2} \int_0^{t_1} W(t_2, s)(R_u(s, \sigma) + m(s)m(\sigma)^*) W(t_1, \sigma)^* \, d\sigma \, ds.$$

This is (4.1.5). We should also note the estimate:

$$E[\|v(t)\|^2] \leq E\left[\left\|\int_0^t W(t, s)u(s) \, ds\right\|^2\right]$$

$$= \int_0^t \int_0^t E[W(t, s)u(s), W(t, \sigma)u(\sigma)] \, ds \, d\sigma$$

$$\leq \int_0^t \int_0^t |E[W(t, s)u(s), W(t, \sigma)u(\sigma)]| \, ds \, d\sigma, \quad (4.1.11)$$

and by the Schwarz inequality we obtain

$$|E[W(t, s)u(s), W(t, \sigma)u(\sigma)]| \leq (E[\|W(t, s)u(s)\|^2])^{1/2}(E[\|W(t, \sigma)u(\sigma)\|^2])^{1/2}$$

$$E[\|W(t, s)u(s)\|^2] \leq \|W(t, s)\|_0^2 E[\|u(s)\|^2].$$

Hence by substituting in (4.1.11) we have

$$E[\|v(t)\|^2] \le \left(\int_0^t \|W(t, s)\|_0 (E\|u(s)\|^2)^{1/2} \, ds \right)^2 \tag{4.1.12}$$

$$\le \int_0^t \|W(t, s)\|_0^2 \, ds \cdot \int_0^t E\|u(s)\|^2 \, ds, \tag{4.1.13}$$

yielding a useful estimate.

Let us note in particular that for any $q \times m$ continuous function $h(t)$, $0 \le t \le T$, and $q \times 1$ random process $x(\cdot)$, Theorem 4.1.1 establishes that the integral

$$\int_0^T h(t) x(t) \, dt \tag{4.1.14}$$

is defined in the mean square sense, assuming that the covariance function

$$R(t, s) = E[x(t) x(s)^*]$$

is continuous in the rectangle $0 \le s \le T, 0 \le t \le T$. Moreover, if $g(t), 0 \le t \le T$, is another $q \times m$ continuous function, we obtain

$$E\left[\left(\int_0^T h(t) x(t) \, dt \right) \left(\int_0^T g(t) x(t) \, dt \right)^* \right] = \int_0^T \int_0^T h(t) R(t, s) g(s)^* \, ds \, dt. \tag{4.1.15}$$

If $h(t)$ is also $m \times 1$, we obtain

$$E\left(\int_0^T [h(t), x(t)] \, dt \right)^2 = \int_0^T \int_0^T E[[h(t), x(t)][h(s), x(s)]] \, ds \, dt$$

$$= \int_0^T \int_0^T \mathrm{Tr} \, h(t)^* R(t, s) h(s) \, ds \, dt$$

$$= \int_0^T \int_0^T [h(t), R(t, s) h(s)] \, ds \, dt \tag{4.1.16}$$

and from (4.1.13), this is

$$\le \int_0^T \|h(t)\|^2 \, dt \cdot \int_0^T \mathrm{Tr} \, R(t, t) \, dt. \tag{4.1.17}$$

We note in conclusion that if the integral is definable sample-path-wise, then both definitions will coincide—the difference is nonzero with probability zero.

4.2   STEADY-STATE RESPONSE: WEIGHTING FUNCTION MODELS

Let us now assume that the input $u(\cdot)$ is stationary. Let $R_u(\cdot)$ denote the stationary covariance function and $P_u(\cdot)$ the spectral density. As in the case of discrete-time models we want to study the steady-state response of time-invariant systems. Thus the weighting function is given by

$$W(t, s) = W(t - s), \qquad t \geq s, \tag{4.2.1}$$

where, analogous to the stability condition we now assume that

$$\int_0^\infty \| W(s) \|_0 \, ds < \infty. \tag{4.2.2}$$

We note that this is equivalent to the condition that

$$\int_0^\infty |a_{ij}(s)| \, ds < \infty \qquad \text{for every } i, j,$$

where

$$\{a_{ij}(s)\} = W(s).$$

First we note that the process $v(t)$, $0 \leq t$, defined now by

$$v(t) = \int_0^t W(t - s)u(s) \, ds = \int_0^t W(\sigma)u(t - \sigma) \, d\sigma,$$

is "asymptotically" stationary, or "steady-state" stationary. Thus the steady-state mean is given by

$$\lim_{t \to \infty} E[v(t)] = \left(\int_0^\infty W(\sigma) \, d\sigma \right) m,$$

where

$$m = E[u(t)].$$

Next, using (4.1.5), for any $L \geq 0$, and denoting the stationary covariance function of the process $u(\cdot)$ by $R_u(\cdot)$, we have

$$E[v(t_2 + L)v(t_1 + L)^*] = \left(\int_0^{t_2 + L} W(\sigma) \, d\sigma \right) mm^* \left(\int_0^{t_1 + L} W(\sigma) \, d\sigma \right)^*$$

$$+ \int_0^{t_2 + L} \int_0^{t_1 + L} W(t_2 + L - s)R_u(s - \sigma)W^*(t_1 + L - \sigma) \, ds \, d\sigma.$$

As $L \to \infty$, the first term goes to

$$\left(\left(\int_0^\infty W(\sigma)\,d\sigma\right)m\right)\left(\int_0^\infty W(\sigma)\,d\sigma\,m\right)^* = \lim_{L\to\infty} E[v(t_2 + L)]E[v(t_1 + L)]^*$$

while the second term by a change of variable can be expressed

$$\int_0^{t_2+L}\int_0^{t_1+L} W(s)R_u(t_2 - t_1 + \sigma - s)W(\sigma)^*\,ds\,d\sigma,$$

which, as $L \to \infty$, becomes

$$\int_0^\infty\int_0^\infty W(s)R_u(t_2 - t_1 + \sigma - s)W(\sigma)^*\,ds\,d\sigma,$$

where the integral converges since

$$\int_0^\infty\int_0^\infty \|W(s)\|_0\|R_u(t_2 - t_1 + \sigma - s)\|_0\|W(\sigma)^*\|_0\,ds\,d\sigma$$

$$\leq \|R_u(0)\|_0\left(\int_0^\infty \|W(s)\|_0\,ds\right)^2.$$

Our next step is to show that the "steady-state" response entirely analogous to the discrete-time version in Chapter 3, can be defined by

$$v(t) = \int_0^\infty W(s)u(t - s)\,ds, \qquad 0 \leq t. \qquad (4.2.3)$$

Thus we need to define the limit as $L \to \infty$ of

$$\int_0^L W(s)u(t - s)\,ds.$$

We take this limit in the mean square sense. Now, for finite $L_2 \geq L_1$, following (4.1.12) we obtain

$$E\left[\left\|\int_{L_1}^{L_2} W(s)u(t - s)\,ds\right\|^2\right] \leq \left(\int_{L_1}^{L_2} \|W(s)\|_0(E[\|u(t - s)\|^2])^{1/2}\,ds\right)^2. \qquad (4.2.4)$$

But

$$E[\|u(t - s)\|^2] = \|m\|^2 + \mathrm{Tr}\,R_u(0)$$

and hence (4.2.4) is

$$\leq (\|m\|^2 + \text{Tr } R_u(0)) \left(\int_{L_1}^{L_2} \|W(s)\|_0 \, ds \right)^2$$

$$\leq (\|m\|^2 + \text{Tr } R_u(0)) \left(\int_{L_1}^{\infty} \|W(s)\|_0 \, ds \right)^2 \qquad (4.2.5)$$

and goes to zero as L_1 goes to infinity by (4.2.2). Hence the limit may be taken in the mean square sense. In particular we see that

$$E[v(t)] = \int_0^{\infty} W(s) E[u(t - s)] \, ds = \left(\int_0^{\infty} W(s) \, ds \right) m$$

and that the covariance is given by

$$E[\tilde{v}(t_1)\tilde{v}(t_2)^*] = E\left[\left(\int_0^{\infty} W(s)\tilde{u}(t_1 - s) \, ds \right)\left(\int_0^{\infty} W(\sigma)\tilde{u}(t_2 - \sigma) \, d\sigma \right)^* \right]$$

$$= \int_0^{\infty} \int_0^{\infty} W(s) R_u(t_1 - t_2 + \sigma - s) W(\sigma)^* \, d\sigma \, ds$$

$$= R_v(t_1 - t_2),$$

where

$$R_v(t) = \int_0^{\infty} \int_0^{\infty} W(s) R_u(t + \sigma - s) W(\sigma)^* \, d\sigma \, ds. \qquad (4.2.6)$$

Thus the steady-state process is stationary with covariance function given by (4.2.6). The corresponding spectral density is obtained as in the discrete-time analysis, by substituting

$$R_u(t) = \int_{-\infty}^{\infty} e^{2\pi i f t} P_u(f) \, df$$

into (4.2.6). Thus we have

$$R_v(t) = \int_{-\infty}^{\infty} e^{2\pi i f t} \left(\int_0^{\infty} W(s) e^{-2\pi i f s} \, ds \right) P_u(f) \left(\int_0^{\infty} W(s) e^{-2\pi i f s} \, ds \right)^* df.$$

Or, the spectral density of the steady-state output process $P_v(\cdot)$ is given by

$$P_v(f) = \psi(f) P_u(f) \psi(f)^*, \qquad (4.2.7)$$

where $\psi(\cdot)$ is the system transfer function:

$$\psi(f) = \int_0^\infty e^{-2\pi i f t} W(t)\, dt. \tag{4.2.8}$$

Note the complete similarity of (4.2.7) to the discrete-time result (3.3.16).

Example 4.2.1. Consistent with the constraint (4.2.2), we may "extend" the weighting-pattern functions to include δ-functions. Our interest is mainly in the time-invariant case. For example, let

$$W(s) = \left(\frac{\delta(s) - \delta(s - \Delta)}{\Delta} \right) I, \qquad 0 \le s < \infty,$$

where Δ is positive. Then

$$v(t) = \int_0^t W(t - s) u(s)\, ds$$

yields

$$v(t) = \frac{u(t) - u(t - \Delta)}{\Delta}.$$

The corresponding system transfer function (cf. 4.2.8) is given by

$$\psi(f) = \left(\frac{1 - e^{-2\pi i f \Delta}}{\Delta} \right) I.$$

We see that the steady-state spectral density of $v(\cdot)$, using (4.2.7), is given by

$$P_v(f) = \left| \frac{1 - e^{-2\pi i f \Delta}}{\Delta} \right|^2 P_u(f).$$

Example 4.2.2. Let Δ be a fixed positive number and define

$$z_\Delta(t) = \frac{1}{\Delta} \int_{t-\Delta}^t x(s)\, ds.$$

We can rewrite this as

$$z_\Delta(t) = \int_0^\Delta W(s) x(t - s)\, ds,$$

where

$$W(s) = \frac{1}{\Delta} I, \qquad 0 \le s \le \Delta$$

$$= 0, \qquad \text{otherwise.}$$

We can see that $z_\Delta(\cdot)$ is a stationary process with spectral density using (4.2.7) given by

$$\left|\frac{e^{2\pi i f \Delta} - 1}{2\pi i f \Delta}\right|^2 P(f).$$

As $\Delta \to 0$, this converges to $P(f)$, the spectral density of $P(\cdot)$. In fact,

$$z_\Delta(t) - x(t)$$

is a stationary process with spectral density

$$\left|1 - \left(\frac{e^{2\pi i f \Delta} - 1}{2\pi i f \Delta}\right)\right|^2 P(f)$$

and goes to zero as $\Delta \to 0$, and in particular we have

$$E(\|z_\Delta(t) - x(t)\|^2) \to 0 \qquad \text{as } \Delta \to 0.$$

4.3   WHITE NOISE AND SIGNAL GENERATION MODELS

The state-space "signal-generation model" for discrete-time Gaussian processes involved a white-noise input. In extending this to continuous time we have first to examine the concept of time-continuous white noise.

4.3.1   White Noise: Continuous Parameter

The direct extension of white noise from discrete time to continuous time might be Example 1.2.1, where we constructed a zero-mean Gaussian process $x(t)$, $-\infty < t < \infty$, such that

$$\begin{aligned} E[x(t)x(s)] &= 0, \qquad t \neq s \\ &= 1, \qquad t = s. \end{aligned} \tag{4.3.1}$$

One suspects immediately that something is wrong with this process because the corresponding stationary covariance function

$$\begin{aligned} R(t) &= 1, \qquad t = 0 \\ &= 0, \qquad t \neq 0 \end{aligned}$$

is not continuous at the origin, neither left nor right. If we make the attempt to calculate formally the spectral density by taking the Fourier transform

$$\int_{-\infty}^{\infty} e^{2\pi i f t} R(t)\, dt,$$

we get zero, which contradicts the fact that $R(0) = 1$. So something is definitely wrong with this process and whatever it is, it is not the extension we are seeking.

In fact, we have to proceed differently and go back to physics where the notion of "white light" is of light whose spectrum contains all frequencies at equal intensity. First we consider "band-limited white noise." This is a zero-mean stationary Gaussian process $x(t)$, $-\infty < t < \infty$, with spectral density defined by (recall Problem 2.2)

$$P(f) = D, \qquad -W < f < W$$
$$= 0, \qquad |f| > W, \tag{4.3.2}$$

where D is nonnegative (nonnegative definite in the matrix case); the corresponding covariance function is given by

$$R(t) = (2WD)\frac{\sin 2\pi W t}{2\pi W t}. \tag{4.3.3}$$

We see that $R(\cdot)$ is (more than) twice differentiable at the origin. Indeed we can readily calculate the derivative from

$$R(t) = \int_{-W}^{W} e^{2\pi i f t} D\, df, \tag{4.3.4}$$

yielding

$$R''(0) = -\int_{-W}^{W} f^2 D\, df = -\frac{2W^3 D}{3},$$

and hence we see from Theorem 4.7.1 and (4.7.1) (see p. 202) that the sample paths of band-limited white noise are continuous. We shall also see this independently in Chapter 6.

Let us consider what happens as we allow the bandwidth $(2W)$ to increase without bound. Let $h(\cdot)$ be any real-valued continuous function on $[-L, L]$, $0 < L < \infty$, and defined to be zero for $|t| > L$. Then the integral is

$$\int_{-\infty}^{\infty} h(t) R(t)\, dt = \int_{-\infty}^{\infty} h(t)\frac{\sin 2\pi W t}{2\pi W t}(2W)\, dt\, D. \tag{4.3.5}$$

Let

$$\psi_h(f) = \int_{-L}^{L} e^{2\pi i f t} h(t)\, dt.$$

Then, using the Fourier convolution theorem [cf. Eq. (4.7) on page 26], (4.3.5) can be expressed as

$$\int_{-W}^{W} \psi_h(f) \, df \, D.$$

As $W \to \infty$, this becomes

$$\int_{-\infty}^{\infty} \psi_h(f) \, df = h(0)D$$

by the inverse Fourier transform formula applied to $\psi_h(\cdot)$. Hence

$$\lim_{W \to \infty} \int_{-\infty}^{\infty} h(t)R(t) \, dt = h(0)D.$$

$R(t)$ goes to a δ-function at the origin in the limit as $W \to \infty$. (Note that as we can see from (4.3.3), $R(t)$ does not converge for any t as $W \to \infty$. Of course

$$R(0) = 2WD$$

goes to infinity.) Formally the Fourier transform of the δ-function

$$\int_{-\infty}^{\infty} e^{-2\pi i f t} \delta(t)D \, dt = D, \qquad -\infty < f < \infty. \tag{4.3.6}$$

Thus the δ-function is indeed a "correlation function" in that its Fourier transform is nonnegative. But the latter is a constant for all frequencies, $-\infty < f < \infty$, and hence if there were a process with this spectral density, the covariance function at the origin would be infinite and the variance of the Gaussian random variable for any instant t would be infinite. In other words, it is *not* a random variable for any t and certainly does not define a random process in our definition.

Nevertheless, we can endow an "operational" meaning to such a process, and only in this sense we can define the concept of white noise in continuous time. Thus we define the response of a linear system to white noise as the limit of the response to band-limited white noise as the bandwidth expands to infinity.

Let $N(t)$, $-\infty < t < \infty$, denote an $m \times 1$ "white noise" with spectral density D. Let $h(\cdot)$ be a $q \times m$ continuous function on $0 \le t \le L$, $L < \infty$. Then the integral

$$\int_0^L h(t)N(t) \, dt \tag{4.3.7}$$

is defined as follows: Begin with the integral

$$\int_0^L h(t)x_W(t)\, dt,$$

where $x_W(\cdot)$ is $m \times 1$ band-limited white Gaussian with spectral density defined by (4.3.2). This is Gaussian with mean zero and covariance

$$E\left[\left(\int_0^L h(t)x_W(t)\, dt\right)\left(\int_0^L h(s)x_W(s)\, ds\right)^*\right] = \int_0^L \int_0^L h(t)R(t-s)h(s)^*\, ds\, dt.$$

$$(4.3.8)$$

Now let the bandwidth increase to infinity, in "arbitrary" fashion. To calculate the limit, use (4.3.4) and substitute into (4.3.8), yielding

$$\int_{-W}^{W} \int_0^L \int_0^L e^{2\pi i f(t-s)} h(t)Dh(s)^*\, ds\, dt = \int_{-W}^{W} \psi_h(f)D\psi_h(f)^*\, df, \quad (4.3.9)$$

where

$$\psi_h(f) = \int_0^L e^{2\pi i f t} h(t)\, dt.$$

As $W \to \infty$, (4.3.9) becomes

$$= \int_{-\infty}^{\infty} \psi_h(f)D\psi_h(f)^*\, ds$$

and by Parseval's theorem [see Eq. (4.14) on page 28], this is

$$= \int_0^L h(t)Dh(t)^*\, dt. \quad (4.3.10)$$

We may incorporate this directly in (4.3.8) by saying that the covariance function of white noise is given by

$$E[N(s)N(t)^*] = \delta(t-s)D, \quad (4.3.11)$$

so that we may treat white noise as a Gaussian process with a δ-function for its covariance function. In particular, if $h(\cdot)$ is an $m \times 1$ continuous function, we have that

$$E\left(\int_0^L [h(t), N(t)]\, dt\right)^2 = \int_0^L [Dh(t), h(t)]\, dt.$$

Let us note that instead of a band-limited white noise we may begin with any process and introduce a limiting procedure which yields the spectral density (4.3.6). Thus let $x_n(t)$, $-\infty < t < \infty$, be a zero-mean Gaussian process with covariance function $R_n(\cdot)$ and spectral density $P_n(f)$ such that

$$P_n(f) \to D \qquad \text{as } n \to \infty, \text{ for every } f, \ -\infty < f < \infty.$$

Then with $h(\cdot)$ as before, we form

$$\int_0^L h(t)x_n(t)\,dt$$

and note that the covariance is

$$\int_0^L \int_0^L h(t)R_n(t-s)h(s)^*\,ds\,dt = \int_{-\infty}^{\infty} \psi_h(f)P_n(f)\psi_h^*(f)\,df$$

$$\to \int_{-\infty}^{\infty} \psi_h(f)D\psi_h^*(f)\,df = \int_0^L h(t)Dh(t)^*\,dt.$$

As a canonical way of generating the bandwidth expansion sequence, let $x(t)$, $-\infty < t < \infty$, be a stationary process with spectral density $P(\cdot)$ such that

$$P(0) \neq 0.$$

Define

$$x_n(t) = \sqrt{n}\,x(nt).$$

Then

$$E[x_n(t)x_n(s)^*] = nR(nt - ns).$$

Hence the corresponding spectral density is given by

$$P_n(f) = n\int_{-\infty}^{\infty} e^{2\pi i f t}R(nt)\,dt = P(f/n).$$

Hence as $n \to \infty$,

$$P_n(f) \to P(0) \qquad \text{for all } f, \ -\infty < f < \infty.$$

This arbitrariness in how the "bandwidth expands" is an essential point of the definition. This also makes it difficult to define operations on white noise which are not linear. (Indeed, specializing to one dimension, $m = 1$,

$$\int_0^L N(t)^2\,dt$$

cannot be given an unambiguous definition by a bandwidth expansion limiting process.)

An important property of white noise we shall need to note is that nonoverlapping segments are independent. That is to say, given

$$-\infty < a < b < c < \infty$$

the integrals

$$\int_a^b h(s)N(s)\,ds, \qquad \int_b^c g(s)N(s)\,ds$$

are independent, for arbitrary $q \times m$ continuous functions $h(\cdot), g(\cdot)$. The covariance is given by

$$E\left[\left(\int_a^b h(s)N(s)\,ds\right)\left(\int_b^c g(\sigma)N(\sigma)\,d\sigma\right)^*\right]$$

$$= \int_c^b \int_b^c h(s)E[N(s)N(\sigma)^*]g(\sigma)^*\,ds\,d\sigma$$

$$= \int_a^b \int_b^c h(s)D\delta(s-\sigma)g(\sigma)^*\,ds\,d\sigma = 0,$$

since the integrand is zero.

4.3.2 Steady-State Response to White Noise Inputs

Armed with this definition, we can now consider steady-state response to white noise input, defined by

$$v(t) = \int_0^\infty W(s)N(t-s)\,ds \tag{4.3.12}$$

where $W(\cdot)$ satisfies

$$\int_0^\infty \|W(s)\|_0^2\,ds < \infty \tag{4.3.13}$$

or, equivalently,

$$\int_0^\infty \|W(s)\|^2\,ds < \infty.$$

Thus we begin by

$$\int_0^L W(s)N(t-s)\,ds, \qquad L < \infty$$

and show that (4.3.12) is the limit in the mean square sense as $L \to \infty$. Thus let $L_1 < L_2$. Then

$$E\left[\left(\int_{L_2}^{L_2} W(s)N(t-s)\,ds\right)\left(\int_{L_1}^{L_2} W(s)N(t-s)\,ds\right)^*\right] = \int_{L_1}^{L_2} W(s)DW(s)^*\,ds,$$

which goes to zero as $L_1 \to \infty$, by virtue of (4.3.13) since

$$\|W(s)DW(s)^*\|_0 \le \|W(s)\|_0^2 \|D\|_0.$$

Thus defined, using (4.3.12) we have

$$E[v(t_1)v(t_2)^*] = \int_0^\infty \int_0^\infty W(s)E[N(t_1-s)N(t_2-\sigma)^*]W(\sigma)^*\,d\sigma\,ds,$$

and by the white noise formalisms we obtain

$$E[N(t_1-s)N(t_2-\sigma)^*] = \delta(t_2-t_1+s-\sigma)D.$$

Substituting this into (4.3.13) yields

$$E[v(t_1)v(t_2)^*] = \int_0^\infty \int_0^\infty W(s)\delta(t_2-t_1+s-\sigma)DW(\sigma)^*\,d\sigma\,ds$$

$$= \int_0^\infty W(s)DW(s+t_2-t_1)^*\,ds = R(t_1-t_2), \quad (4.3.14)$$

where, in the integral, it is understood that

$$W(s) = 0, \qquad s < 0.$$

We see from (4.3.14) that the output process is stationary with covariance function

$$R(t) = \int_0^\infty W(s+t)DW(s)^*\,ds, \qquad -\infty < t < \infty. \qquad (4.3.15)$$

The corresponding spectral density function is readily obtained from the Fourier convolution theorem [cf. Eq. (4.16) on page 28]

$$R(t) = \int_{-\infty}^\infty e^{2\pi i f t}\psi(f)D\psi(f)^*\,df,$$

where $\psi(\cdot)$ is the system transfer function:

$$\psi(f) = \int_0^\infty e^{-2\pi i f t} W(t)\, dt, \qquad (4.3.16)$$

where the integral is to be interpreted as a mean square limit if condition (4.2.2) is *not* satisfied. This distinction is of course unimportant in "practical" problems.

Formally by using

$$R(t) = \int_0^\infty \int_0^\infty W(s)\delta(t + s - \sigma)DW(\sigma)^*\, d\sigma\, ds$$

and substituting the "formal" relation (inverse transform corresponding to (4.3.6))

$$\delta(t)D = \int_{-\infty}^\infty e^{2\pi i f t} D\, df$$

we get

$$R(t) = \int_{-\infty}^\infty e^{2\pi i f t}\psi(f)D\psi(f)^*\, df$$

again. Either way, the spectral density is

$$P(f) = \psi(f)D\psi(f)^*. \qquad (4.3.17)$$

4.3.3 Weighting Function Signal Generation Model

As in the case of discrete-time models in Chapter 3, we can turn (4.3.16) around to yield a signal generation model. Thus, given an arbitrary spectral density function $P(\cdot)$, if we can "factorize" in the form (4.3.17) where $\psi(\cdot)$ is given by (4.3.16), then of course we can use the generation model

$$v(t) = \int_0^t W(\sigma)N(t - \sigma)\, d\sigma \qquad (4.3.18)$$

to simulate a process which asymptotically has the given spectral density. It is known, however, that given any spectral density it is not necessarily factorizable in the form (4.3.17), where the weighting function $W(\cdot)$ is "physically realizable":

$$W(s) = \int_{-\infty}^\infty e^{2\pi i f s}\psi(f)\, df = 0 \qquad \text{for } s < 0.$$

In fact, in general this will be the case (*not* physically realizable) if for example we take the obvious factorization:

$$\psi(f) = \sqrt{P(f)}.$$

Indeed in the one-dimensional case, a necessary and sufficient condition is that (Paley-Wiener; see reference 1 in Chapter 2).

$$\int_{-\infty}^{\infty} \frac{|\log P(f)|}{1 + f^2} df < \infty. \tag{4.3.19}$$

This condition is of course satisfied by spectral densities arising from state-space models where in particular the spectral density is rational. An example of a nonrational spectral density where it is satisfied is provided by the Von Karman turbulence spectrum, given in Example 2.4 in Chapter 2, where in fact the factorization is also indicated.

A standard example where this condition is not satisfied is

$$P(f) = \exp(-f^2), \qquad -\infty < f < \infty.$$

If we take

$$\psi(f) = \sqrt{P(f)} = \exp\left(\frac{-f^2}{2}\right)$$

we see that

$$W(t) = \int_{-\infty}^{\infty} e^{2\pi i f t} \psi(f) \, df = \frac{1}{\sqrt{2\pi}} \exp\left(\frac{-t^2}{2}\right) \neq 0 \qquad \text{for } t < 0.$$

The multidimensional extension of this condition will take us too far from our scope, and in any case it is not of great significance in engineering application. We shall refer to (4.3.18) as the "weighting function signal generation" model.

4.4 STATE-SPACE MODELS

Next we shall study state-space models of linear systems, now that we have defined integrals and white noise. As in the discrete-time case, some familiarity with state-space theory would be helpful.

Let $N(t)$, $-\infty < t < \infty$, denote white noise with spectral density D. Let the system with input $N(\cdot)$ and output $v(\cdot)$ be specified by the state-space model

$$v(t) = Cx(t)$$

$$\frac{dx(t)}{dt} = Ax(t) + BN(t) \tag{4.4.1}$$

for $t \geq 0$, with initial condition $x(0)$ "given." Here we may specify the dimensions as in Section 3.2:

$$A: \quad n \times n$$

$$B: \quad n \times p$$

$$C: \quad m \times n$$

$$N(\cdot): \quad p \times 1.$$

For our purposes it is convenient to work with the equivalent "integral" version (obtained by "solving" the differential equation):

$$x(t) = e^{At}x(0) + \int_0^t e^{A(t-s)}BN(s)\, ds \qquad (4.4.2)$$

$$v(t) = Cx(t). \qquad (4.4.3)$$

In particular, we can consider the case where the input is a white-noise process with spectral density D, since we know how to define the integral in (4.4.2). Thus let $N(\cdot)$ be a $p \times 1$ white Gaussian with zero mean and

$$E[N(t)N(s)^*] = \delta(t - s)D.$$

Then let us invoke the analog of Assumption 3.2.1.

Assumption 4.4.1.

$$x(0)$$

is Gaussian with mean m (which we will take to be zero from now on) and covariance Λ, and it is "independent of the white-noise process $N(t)$, $t \geq 0$." By the statement in quotes we mean that

$$x(0) \text{ is independent of } \int_0^t h(s)N(s)\, ds$$

for any $h(\cdot)$ and any t, $0 < t < \infty$. Now

$$x(t) = e^{At}x(0) + \int_0^t e^{A(t-s)}BN(s)\, ds \qquad (4.4.4)$$

and $x(0)$ is independent of

$$\int_t^{t+L} h(s)N(s)\, ds$$

and so is the second term in (4.4.4) because of the property that the white-noise process is independent on non-overlapping intervals. Hence $x(t)$ is independent of $N(s)$, $s \geq t$.

Let $R(t, s)$ denote the covariance function:

$$R(t, s) = E[x(t)x(s)^*].$$

Then because the two terms in (4.4.4) are independent and the covariance of the second term is

$$= \int_0^t \int_0^t e^{A(t-s)} BD\delta(s - \sigma) B^* e^{A^*(t-\sigma)} \, ds \, d\sigma$$

$$= \int_0^t e^{A(t-s)} BDB^* e^{A^*(t-s)} \, ds \tag{4.4.5}$$

$$= \int_0^t e^{A\sigma} BDB^* e^{A^*\sigma} \, d\sigma, \tag{4.4.6}$$

we have

$$R(t, t) = e^{At} \Lambda e^{A^*t} + \int_0^t e^{A\sigma} BDB^* e^{A^*\sigma} \, d\sigma. \tag{4.4.7}$$

Now for $t > s$, we can, going from (4.4.4) alone (and not the differential equation), see that

$$x(t) = e^{A(t-s)} \left[e^{As} x(0) + \int_0^t e^{-A(t-s) + A(t-\sigma)} BN(\sigma) \, d\sigma \right]$$

$$= e^{A(t-s)} \left[e^{As} x(0) + \int_0^s e^{A(s-\sigma)} BN(\sigma) \, d\sigma \right] + \left[\int_s^t e^{A(s-\sigma)} BN(\sigma) \, d\sigma \right]$$

$$= e^{A(t-s)} x(s) + \int_s^t e^{A(t-\sigma)} BN(\sigma) \, d\sigma. \tag{4.4.8}$$

Hence

$$E[x(t)x(s)^*] = E[e^{A(t-s)} x(s) x(s)^*],$$

the second term in (4.4.8) being independent of $x(s)$, and hence

$$R(t, s) = e^{A(t-s)} R(s, s), \qquad t \geq s$$
$$= R(s, s) e^{A^*(s-t)}, \qquad t \leq s. \tag{4.4.9}$$

We have thus calculated the covariance function of the state process $x(\cdot)$, and the covariance function of the output process

$$= CR(t, s)C^*. \tag{4.4.10}$$

Finally we note that $R(t, t)$ can be characterized as the solution of a (matrix) differential equation. Thus

$$\frac{d}{dt}[e^{At}\Lambda e^{A^*t}] = Ae^{At}\Lambda e^{A^*t} + e^{At}Ae^{A^*t}A^*$$

and the derivative of the second term in (4.4.7) taken in the form (4.4.5) yields

$$\frac{d}{dt}\left[\int_0^t e^{A(t-s)}BDB^* e^{A^*(t-s)}\, ds\right] = A\int_0^t e^{A(t-s)}BDB^* e^{A^*(t-s)}\, ds$$

$$+ \int_0^t e^{A(t-s)}BDB^* e^{A^*(t-s)}\, ds\, A^* + BDB^*.$$

Hence

$$\frac{d}{dt}R(t, t) = AR(t, t) + R(t, t)A^* + BDB^*, \qquad t > 0 \qquad (4.4.11)$$

with the initial condition

$$R(0, 0) = \Lambda.$$

This is then the continuous-time analog of (3.2.9). The state process $x(\cdot)$ is called a Markov process and is of course Gaussian.

4.4.1   Steady-State Solution

As in the discrete-time case, for a time-invariant system our main interest is in the steady-state behavior of the response. In particular, we expect the steady-state properties not to depend on the initial conditions. We begin again with the definition of "stability."

Definition 4.1. The system with state-space description (4.4.1) is said to be stable (or "asymptotically" stable) if the initial condition response is given by

$$e^{At}x(0) \to 0 \qquad \text{as } t \to \infty \qquad (4.4.12)$$

for every choice of $x(0)$.

Theorem 4.4.1. (Compare with Theorems 3.2.1, 3.2.2.)

The system with the state-space description (4.4.1) is stable if and only if the eigenvalues of A all have strictly negative real parts. Moreover, taking

$$\omega_0 = \max_i \text{Re } \lambda_i,$$

where $\{\lambda_i\}$ are the eigenvalues of A, we can find $0 < \gamma$ and $0 < t_0$ such that

$$\|e^{At}\|_0 \leq e^{-\gamma t}, \qquad t > t_0,$$

where $-\gamma$ can be as close to ω_0 as possible by choosing t_0 large enough.

Proof. As in the proof of Theorem 3.2.2, A being simple, the linearly independent eigenvectors $\{e_i\}$ (with eigenvalues $\{\lambda_i\}$) of A form a basis for $\mathbf{R}^n$. Hence each x in $\mathbf{R}^n$ can be expressed as

$$x = \sum_1^n a_i e_i$$

so that

$$e^{At}x = \sum_1^n e^{\lambda_i t} a_i e_i.$$

Hence

$$\|e^{At}x\| \leq e^{\omega_0 t} \sum_1^n |a_i| \leq e^{\omega_0 t} \sqrt{n \sum_1^n |a_i|^2}.$$

As in the proof of Theorem 3.2.2, we obtain

$$\sum_1^n |a_i|^2 \leq \frac{\|x\|^2}{\delta}, \qquad \delta > 0.$$

Hence

$$\|e^{At}x\| \leq \frac{e^{\omega_0 t}(\sqrt{n})\|x\|}{\sqrt{d}}.$$

Given $\varepsilon > 0$, we can find t_0 such that

$$\frac{\sqrt{n}}{\sqrt{\delta}} \leq e^{\varepsilon t}, \qquad t \geq t_0.$$

Hence

$$\|e^{At}x\| \leq e^{(\omega_0 + \varepsilon)t}\|x\|, \quad t \geq t_0$$
$$= e^{-(|\omega_0| - \varepsilon)t}\|x\|,$$

and we can clearly make ε smaller by allowing t_0 to be larger, so that

$$\gamma = |\omega_0| - \varepsilon > 0.$$

Hence

$$\|e^{At}x\| \leq e^{-\gamma t}\|x\|, \qquad t > t_0,$$

or

$$\|e^{At}\|_0 \le e^{-\gamma t}, \qquad t > t_0,$$

and hence also

$$\|e^{At}\| \le \sqrt{n}\, e^{-\gamma t}, \qquad t > t_0.$$

For a proof of the result without the restriction that A be simple, see reference 7.

As a consequence of stability we have:

Theorem 4.4.2. Suppose the system with the state-space description (4.4.1) is stable. Then

(i)

$$E[x(t)] \to 0 \qquad \text{as } t \to \infty. \tag{4.4.13}$$

(ii) $R(t, t)$ converges as $t \to \infty$ to a matrix R, which is the unique solution of the Liapunov equation:

$$0 = AR + RA^* + BDB^*. \tag{4.4.14}$$

Proof. Statement (i) follows from (4.4.4) by taking expectation on both sides and noting that the second term has zero mean so that

$$E[x(t)] = e^{At} E[x(0)] \to 0$$

as $t \to \infty$, where $E[x(0)]$ is zero or not.

As for (ii), we see from (4.4.7) that as $t \to \infty$, the first term therein goes to zero, since

$$\|e^{At} \Lambda e^{A^*t}\| \le \|e^{At}\|_0 \|\Lambda\|_0 \|e^{A^*t}\|_0$$

and

$$\|e^{A^*t}\|_0 = \|e^{At}\|_0.$$

The second term converges, since for $0 < L_1 < L_2$ we have

$$\left\| \int_{L_1}^{L_2} e^{A\sigma} BDB^* e^{A^*\sigma}\, d\sigma \right\|_0 \le \int_{L_1}^{L_2} \|e^{A\sigma} BDB^* e^{A^*\sigma}\|_0\, d\sigma$$

$$\le \int_{L_1}^{L_2} \|e^{A\sigma}\|_0 \|BDB^*\|_0 \|e^{A\sigma}\|_0\, d\sigma$$

$$\le \|BDB^*\|_0 \int_{L_1}^{\infty} e^{-2\gamma\sigma}\, d\sigma \qquad \text{for } L_1 > t_0$$

$$\to 0 \qquad \text{as } L_1 \to \infty.$$

Denote the limit by R. Then

$$R = \int_0^\infty e^{A\sigma}BDB^*e^{A^*\sigma}\,d\sigma. \tag{4.4.15}$$

Now using (4.4.7) we obtain

$$\frac{d}{dt}R(t,t) = Ae^{At}\Lambda e^{A^*t} + e^{At}\Lambda e^{A^*t}A^* + e^{At}BDB^*e^{A^*t}$$

$$\to 0 \quad \text{as } t \to \infty.$$

Hence from (4.4.11), taking limits on both sides as $t \to \infty$, we have

$$0 = AR + RA^* + BDB^*,$$

which is (4.4.14).

Let us pause now to discuss briefly the nature of solutions of the Liapunov equation (4.4.14).

Theorem 4.4.3. The Liapunov equation

$$0 = AR + RA^* + BDB^* \tag{4.4.14}$$

has a unique solution if all eigenvalues of A have negative real parts. Moreover, if it has a unique positive definite solution, then all eigenvalues of A must have negative real parts.

Proof. Under the stability condition on A—that all eigenvalues have negative real parts—we see that R given by (4.4.15) is a solution. Note that R is self-adjoint and nonnegative definite. Suppose there is another solution; denote it R_1, so that

$$AR_1 + R_1A^* + BDB^* = 0.$$

Subtracting this equation from (4.4.14) we have

$$A(R - R_1) + (R - R_1)A^* = 0.$$
Let
$$z(t) = e^{At}(R - R_1)e^{A^*t}.$$
Then
$$\dot{z}(t) = e^{At}[A(R - R_1) + (R - R_1)A^*]e^{A^*t} = 0.$$
Hence
$$z(t) = z(0) \quad \text{for all } t.$$

But

$$e^{At}, e^{A^*t} \to 0 \qquad \text{as } t \to \infty.$$

Hence

$$z(t) \to 0 \qquad \text{as } t \to \infty.$$

Hence

$$z(0) = 0$$

or

$$R = R_1$$

or the solution is unique and is self-adjoint and nonnegative definite.

Suppose next (4.4.14) has a positive definite solution; denote it R. Let

$$\lambda = \sigma + i\omega$$

be an eigenvalues of A^*, so that

$$A^*z = \lambda z, \qquad \|z\| = 1.$$

We shall show that σ cannot be nonnegative. From (4.4.14) we have

$$0 = [ARz, z] + [RA^*z, z] + [BDB^*z, z]$$

$$= 2\sigma[Rz, z] + [BDB^*z, z].$$

If $\sigma > 0$, since $[Rz, z]$ is ≥ 0, we must have that

$$[Rz, z] = 0 \qquad \text{or} \qquad Rz = 0,$$

which is impossible. If $\sigma = 0$, we must have

$$0 = [BDB^*z, z]$$

or

$$BDB^*z = 0.$$

But by (4.4.14)

$$ARz + (i\omega)Rz = 0$$

or

$$ARz = (-i\omega)Rz.$$

Denote the projection matrix P by

$$Px = \frac{[x, Rz]Rz}{[Rz, Rz]}.$$

Then P is nonzero, unless z is zero, R being nonsingular. Now

$$APx = [x, Rx]ARz = [x, Rz](-i\omega)Rz$$

$$PA^*x = (i\omega) = (i\omega)[x, Rz]Rz$$

and hence

$$AP + PA^* = 0.$$

Thus

$$A(R + P) + (R + P)A^* + BDB^* = 0.$$

Hence $(R + P)$ is another positive definite solution, violating the uniqueness. Hence z must be zero, or A^* and hence A cannot have eigenvalues with nonnegative real parts.

Invoking the notion of controllability (see reference 7), we can state a useful corollary:

Corollary. If $(A, B\sqrt{D})$ is "controllable" and the Liapunov equation (4.4.14) has a nonnegative definite solution, then A cannot have eigenvalues with nonnegative real parts and the solution must be nonsingular, and it is the only solution.

Proof. The controllability condition is equivalent to

$$BDB^*A^{*^k}x = 0, \qquad \text{for every integer } k \geq 0$$

and implies that $x = 0$. Let

$$A^*z = (\sigma + i\omega)z, \qquad \|z\| = 1.$$

Let R denote any nonnegative definite solution of (4.4.14). Then, as before, if $\sigma \geq 0$ we obtain

$$2\sigma[Rz, z] + [BDB^*z, z] = 0.$$

Since $[Rz, z]$ is ≥ 0, we must have that

$$[BDB^*z, z] = 0 \qquad \text{or} \qquad BDB^*z = 0.$$

Hence

$$0 = (\sigma + i\omega)^k BDB^*z = BDB^*A^{*^k}z \qquad \text{for every } k \geq 0.$$

Hence

$$z = 0$$

or A^* (and hence A) cannot have eigenvalues with nonnegative real parts. Hence the solution R is unique. It must be nonsingular. Suppose

$$Rx = 0.$$

Then

$$[ARx, x] + A^*[RA^*x, x] = 0$$

and hence

$$BDB^*x = 0.$$

Again from

$$ARx + RA^*x + BDB^*x = 0$$

we see that

$$RA^*x = 0.$$

Hence

$$(BDB^*)A^*x = 0.$$

By induction we see that

$$(BDB^*)A^{*k}x = 0 \qquad \text{for every } k \geq 0.$$

By controllability we have

$$x = 0$$

or R is nonsingular, concluding the proof of the corollary.

Getting back to the state process, we see that it is asymptotically stable in the sense that for $t > s$ we have

$$\lim_{L \to \infty} E[x(t + L)x(s + L)^*] = e^{A(t-s)} \lim_{L \to \infty} R(s + L, s + L)$$

$$= e^{A(t-s)}R, \qquad t \geq s$$

$$= Re^{A^*(s-t)}, \qquad s \geq t$$

(4.4.16)

and of course so is

$$v(t) = Cx(t)$$

with

$$\lim_{L \to \infty} E[v(t + L)v(s + L)^*] = Ce^{A(t-s)}RC, \qquad t \geq s$$

$$= CRe^{A^*(s-t)}C^*, \qquad s \geq t.$$

Moreover, we can obtain a representation for the steady-state process from (4.4.4) rewriting it as

$$x(t) = e^{At}x(0) + \int_0^t e^{As}BN(t - s) \, ds.$$

As $t \to \infty$, the first term goes to zero and hence defines

$$x_a(t) = \int_0^\infty e^{As}BN(t - s) \, ds$$

because by setting

$$e^{As}B = W(s), \quad s \geq 0$$

we see that $W(\cdot)$ satisfies (4.2.2) by virtue of (4.4.13). Hence (4.4.17) defines a stationary stochastic process with steady-state covariance function, as calculated by (4.3.14)

$$
\begin{aligned}
R(t) &= \int_0^\infty e^{As}BDB^*e^{A^*(s+t)} \, ds \\
&= Re^{A^*|t|}, \qquad t < 0 \\
&= e^{At}R, \qquad t > 0,
\end{aligned}
$$
(4.4.17)

which checks with (4.4.16). In particular, the corresponding steady-state spectral density is, using (4.3.15) (or otherwise),

$$
\begin{aligned}
P(f) &= \psi(f)BDB^*\psi(f)^* \\
&= (2\pi if I - A)^{-1}BDB^*(2\pi if - A)^{*-1}
\end{aligned}
$$
(4.4.18)

since

$$\int_0^\infty e^{At}Be^{-2\pi ift} \, dt = (2\pi if I - A)^{-1}B.$$

The steady-state spectral density of the output process correspondingly

$$= CP(f)C^*.$$
(4.4.19)

We shall call (4.4.1) where the system is assumed to be stable, the Kalman or state-space signal generation model.

A distinguishing feature of the spectral density (4.4.18) is that it is a rational function of the frequency f: each term in (4.4.18) is a ratio of polynomials in f. This follows from the fact that every term in

$$(2\pi if I - A)^{-1}$$

is a ratio of polynomials in f, the denominator polynomial being

$$\det |2\pi i f I - A|,$$

and the same is true for every term in

$$((2\pi i f I - A)^*)^{-1} = (-2\pi i f I - A^*)^{-1}.$$

Now let

$$\psi(f) = C(2\pi i f I - A)^{-1}B.$$

Then

$$CP(f)C^* = \psi(f)D\psi(f)^*,$$

which is a "factorization" of $P(\cdot)$, as noted in Section 4.3, in which

$$\psi(f) = \int_0^\infty e^{-2\pi i f t} C e^{At} B \, dt$$

or

$$W(t) = C e^{At} B, \qquad t \geq 0$$
$$= 0, \qquad t < 0$$

is the weighting function of the system. Thus we have the weighting function model (explained in the previous section) for the process:

$$v(t) = \int_0^t C e^{As} B N(t - s) \, ds. \tag{4.4.20}$$

Comparing this with (4.4.3) we see that the initial condition term therein,

$$C e^{At} x(0),$$

has been omitted. But because the system is stable, the steady-state properties of (4.4.20) are the same as those of (4.4.3). Thus the Kalman signal generation model is a special case where advantage has been taken of the fact that the spectral density is rational. This is thus less general, but it nevertheless can be of value as an approximation: The spectral density corresponding to a weighting function signal generation model where the weighting function is physically realizable can be approximated by rational functions. The details of the approximation will take us too far beyond our scope here as an introductory course in random processes.

A final note concerning the model (4.4.1): In many problems (see Problem 4.1) the state space comes with the problem. But, as in the discrete-time state-space model (3.2.21), as a representation for a process the matrices (C, A, B) in (4.4.4) are not unique—the output response $v(\cdot)$ with input $N(\cdot)$

can be represented by different state spaces. In particular, we may always assume, if we wish, that the state space is observable—same definition as before. However, as in the discrete-time case, this is more a system-theoretic problem (see reference 5) and of less concern to us. In particular, of course (4.4.19) is always valid and remains the same.

4.5   EXAMPLES

Example 4.5.1. Consider the general second-order differential equation model with white noise as input:

$$\frac{d^2v}{dt^2} + 2b\frac{dv}{dt} + cv(t) = N(t), \qquad t > 0, \tag{4.5.1}$$

where b and c are given (real) constants and $N(\cdot)$ is one-dimensional white noise with spectral density D. It is convenient to rewrite this in state-space form. Setting, as usual,

$$x(t) = \begin{vmatrix} v(t) \\ \dot{v}(t) \end{vmatrix}$$

we have

$$v(t) = Cx(t)$$

$$\frac{dx}{dt} = Ax(t) + BN(t), \tag{4.5.2}$$

where

$$C = |1 \quad 0|$$

$$B = \begin{vmatrix} 0 \\ 1 \end{vmatrix}$$

$$A = \begin{vmatrix} 0 & 1 \\ -c & -2b \end{vmatrix}.$$

Since we are interested primarily in the steady-state properties, let us make sure first that the system is stable. The eigenvalues of A are the roots of

$$\lambda(\lambda + 2b) + c = 0$$

$$\lambda^2 + 2b\lambda + c = 0$$

and hence given by

$$-b \pm \sqrt{b^2 - c}.$$

For stability therefore we need (and assume from now on)

$$b > 0, \qquad c > 0. \tag{4.5.3}$$

Next

$$(\lambda - A)^{-1} = \left(\frac{1}{\lambda^2 + 2b\lambda + c} \right) \begin{vmatrix} \lambda + 2b & 1 \\ -c & \lambda \end{vmatrix}$$

$$BDB^* = \begin{vmatrix} 0 & 0 \\ 0 & D \end{vmatrix}.$$

Hence the steady-state spectral density of $x(\cdot)$, using (4.4.18), is given by

$$P(f) = \frac{1}{|s^2 + 2bs + c|^2} \begin{vmatrix} D & \bar{s}D \\ sD & |s|^2 D \end{vmatrix}, \qquad s = 2\pi if. \tag{4.5.4}$$

The steady-state spectral density of the output process $v(\cdot)$ is

$$= CP(f)C^* = \frac{D}{|-4\pi^2 f^2 + 4b\pi if + c|^2}. \tag{4.5.5}$$

Note that both (4.5.4) and (4.5.5) are rational functions of f. To find the steady-state covariance

$$R = \lim_{t \to \infty} E[x(t)x(t)^*]$$

we use the Liapunov equation (4.4.14), which in our case becomes

$$0 = A \begin{vmatrix} r_{11} & r_{12} \\ r_{12} & r_{22} \end{vmatrix} + \begin{vmatrix} r_{11} & r_{12} \\ r_{12} & r_{22} \end{vmatrix} A^* + \begin{vmatrix} 0 & 0 \\ 0 & D \end{vmatrix}, \tag{4.5.6}$$

where

$$R = \begin{vmatrix} r_{11} & r_{12} \\ r_{12} & r_{22} \end{vmatrix}$$

exploiting symmetry. This yields

$$0 = \begin{vmatrix} 2r_{12} & r_{22} - cr_{11} - 2br_{12} \\ r_{22} - cr_{11} - 2br_{12} & D - 2cr_{12} - 4br_{22} \end{vmatrix}.$$

Hence

$$r_{12} = 0$$

$$r_{22} = cr_{11}$$

$$4br_{22} = D$$

or

$$R = \begin{vmatrix} \dfrac{D}{4bc} & 0 \\ 0 & \dfrac{D}{4b} \end{vmatrix}.$$ (4.5.7)

Let us use (4.5.7) to illustrate Theorem 4.4.3. Thus (4.5.7) is a unique solution of (4.5.6) with A defined by

$$A = \begin{vmatrix} 0 & 1 \\ -c & -2b \end{vmatrix}$$

whether the system is stable or not. For (4.5.7) to be a covariance we must have

$$bc > 0$$
$$b > 0.$$

Hence $b > 0$ and $c > 0$, but these are the conditions also for A to have all eigenvalues with negative real parts.

To calculate the steady-state covariance function

$$\lim_{L \to \infty} E[x(t + L)x(s + L)^*] = e^{A(t-s)}R, \qquad t \geq s$$

we need to calculate

$$e^{At}$$

which (we omit the details of calculation)

$$= e^{-bt} \begin{vmatrix} \cosh \omega_s t + \dfrac{b \sinh \omega_s t}{\omega_s} & \dfrac{\sinh \omega_s t}{\omega_s} \\ \dfrac{-c \sinh \omega_s t}{\omega_s} & \cosh \omega_s t - \dfrac{b \sinh \omega_s t}{\omega_s} \end{vmatrix},$$

where

$$\omega_s = \sqrt{b^2 - c}.$$

In this example we began with formally setting

$$x(t) = \begin{vmatrix} v(t) \\ \dot{v}(t) \end{vmatrix}$$ (4.5.8)

and immediately went to the integral forms, as in (4.4.4). Expressing $x(t)$ in terms of its components

$$x(t) = \begin{vmatrix} x_1(t) \\ x_2(t) \end{vmatrix}$$

we see that indeed

$$x_1(t) = Cx(t) = v(t).$$

But since $x_1(\cdot)$ is now a random process and the derivative is a limit, we have to specify first in what sense the derivative is going to be taken. Since

$$\dot{x}(t) = Ax(t) + BN(t)$$

has no meaning for any t, because the white noise is not defined as a process for any t, $x(t)$ itself cannot have a pointwise derivative in any sense. On the other hand,

$$C\dot{x}(t) = CAx(t)$$

since

$$CB = 0$$

and it does not matter what the value of $N(t)$ is. Hence we shall show that

$$v(t) = Cx(t)$$

does have a derivative in the mean square sense which is

$$CAx(t) = x_2(t).$$

Thus we shall show that

$$\lim_{\Delta \to \infty} E\left[\left\| \frac{Cx(t + \Delta) - Cx(t)}{\Delta} - CAx(t) \right\|^2 \right] = 0.$$

But

$$\frac{Cx(t + \Delta) - Cx(t)}{\Delta} - CAx(t)$$

$$= C\left(\left(\frac{e^{A\Delta} - I}{\Delta} \right) - A \right) x(t) + \frac{1}{\Delta} \int_t^{t+\Delta} Ce^{A(t+\Delta-s)}BN(s)\, ds.$$

The two terms are independent. From

$$\frac{d}{dt} e^{At} \bigg|_{t=0} = A$$

it follows that the first term goes to zero in the mean square sense. The second term can be written

$$= \frac{1}{\Delta} \int_0^\Delta Ce^{A(\Delta - \sigma)} BN(t + \sigma) \, d\sigma,$$

which, since $CB = 0$, can be expressed as

$$= \frac{1}{\Delta} \int_0^\Delta C(e^{A(\Delta - \sigma)} - I) BN(t + \sigma) \, d\sigma.$$

But the covariance matrix of this

$$= \frac{1}{\Delta^2} \int_0^\Delta C(e^{A(\Delta - \sigma)} - I) BDB^* (e^{A^*(\Delta - \sigma)} - I) C^* \, d\sigma$$

$$= \frac{1}{\Delta^2} \int_0^\Delta C(e^{A\sigma} - I) BDB^* (e^{A^*\sigma} - I) C^* \, d\sigma \qquad (4.5.9)$$

and

$$C(e^{A\sigma} - I) B = C\left(\sigma A + \frac{\sigma^2}{2} A^2 + \cdots \right) B$$

$$B^* (e^{A^*\sigma} - I) C^* = B^* \left(\sigma A^* + \frac{\sigma^2}{2} A^{*2} + \cdots \right) B.$$

Hence the integrand in (4.5.9)

$$= \frac{1}{\Delta^2} \sigma^2 (CAB) D (B^* A^* C^*) + \text{higher powers of } \sigma$$

and

$$\frac{1}{\Delta^2} \int_0^\Delta \sigma^2 \, d\sigma = \frac{\Delta}{3},$$

and hence (4.5.9) goes to zero as $\Delta \to 0$. Thus the formalism (4.5.8) is correct, interpreting the derivative in the mean square sense. Hence we can in particular "read off" from the components of the steady-state covariance matrix R that

$$\lim_{t \to \infty} E[v(t)\dot{v}(t)] = 0$$

$$\lim_{t \to \infty} E[\dot{v}(t)^2] = \frac{D}{4b} = D \int_{-\infty}^\infty \frac{4\pi^2 f^2}{|-4\pi^2 f^2 + 4b\pi i f + c|^2} \, df \qquad (4.5.10)$$

using (4.5.4). The spectral density of the process $\dot{v}(t)$ is given from (4.5.4)) by

$$\frac{4\pi^2 f^2 D}{|-4\pi^2 f^2 + 4b\pi i f + c|^2}. \qquad (4.5.11)$$

4.6 DERIVATIVES OF STOCHASTIC PROCESSES

The spectral density $CP(\cdot)C^*$ of the process $v(\cdot)$ in Example 4.5.1 given by (4.5.5) satisfies

$$\int_{-\infty}^{\infty} |f|^2 CP(f)C^* \, df < \infty.$$

We shall now show that we can define the derivative in the mean square of any stationary process $x(t)$, $-\infty < t < \infty$, whose spectral density $P(\cdot)$ satisfies

$$\int_{-\infty}^{\infty} |f|^2 \operatorname{Tr} P(f) \, df < \infty. \qquad (4.6.1)$$

Moreover, the derivative process denoted $\dot{x}(t)$, $-\infty < t < \infty$, is stationary with covariance function given by

$$-\ddot{R}(t), \qquad (4.6.2)$$

where $R(\cdot)$ is the stationary covariance of the process $x(\cdot)$.

To prove this we begin with the difference-quotient

$$y_\Delta(t) = \frac{x(t + \Delta) - x(t)}{\Delta}, \qquad -\infty < t < \infty.$$

The covariance function of this process

$$E[y_\Delta(t) y_\Delta(s)^*] = \frac{1}{\Delta^2} (R(t - s) - R(t - s - \Delta) - R(t + \Delta - s) + R(t - s))$$

$$= \int_{-\infty}^{\infty} \frac{(1 - e^{-2\pi i f \Delta} - e^{2\pi i f \Delta} + 1)}{\Delta^2} e^{2\pi i f (t - s)} P(f) \, df.$$

Showing that it is stationary, with spectral density

$$\left| \frac{1 - e^{2\pi i f \Delta}}{\Delta} \right|^2 P(f).$$

As $\Delta \to 0$, this goes to

$$4\pi^2 f^2 P(f).\tag{4.6.3}$$

This is enough to show (we omit some of the mathematical technicalities) that $y_\Delta(t)$ converges in the mean square sense to a process, denote it $y_0(t)$, with spectral density given by (4.6.3). Next we have to show that $y_0(t)$ is the (mean square) derivative of $x(t)$. In fact, we shall prove now that

$$\int_{t_1}^{t_2} y_0(s)\, ds = x(t_2) - x(t_1).\tag{4.6.4}$$

We may calculate the left-hand side as the (mean square) limit of

$$\int_{t_1}^{t_2} y_\Delta(s)\, ds$$

as $\Delta \to 0$. Now

$$\int_{t_1}^{t_2} y_\Delta(s)\, ds - (x(t_2) - x(t_1))$$

$$= \int_{t_1}^{t_2} x(s + \Delta)\, ds - \frac{1}{\Delta} \int_{t_1}^{t_2} x(s)\, ds - x(t_2) + x(t_1)$$

$$= \frac{1}{\Delta} \int_{t_1+\Delta}^{t_2+\Delta} x(s)\, ds - \frac{1}{\Delta} \int_{t_1}^{t_2} x(s)\, ds - x(t_2) + x(t_1)$$

$$= \left(\frac{1}{\Delta} \int_{t_2}^{t_2+\Delta} x(s)\, ds - x(t_2) \right) - \left(\frac{1}{\Delta} \int_{t_1}^{t_1+\Delta} x(s)\, ds - x(t_1) \right).\tag{4.6.5}$$

But

$$\frac{1}{\Delta} \int_{t}^{t+\Delta} x(s)\, ds - x(t)$$

is a stationary stochastic process (cf. Example 4.2.1) with spectral density

$$\left| \frac{e^{2\pi i f\Delta} - 1}{2\pi i f\Delta} - 1 \right|^2 P(f),$$

which goes to zero as $\Delta \to 0$. Hence each term in parentheses in (4.6.5) goes to zero, or (4.6.4) is proved. Thus the spectral density of the process $\dot{x}(t)$, $-\infty < t < \infty$, is given by (4.6.3). That the corresponding stationary covariance function is

$$-\ddot{R}(t)$$

follows from (2.41). Note also that

$$E[x(t)\dot{x}(t)^*] = \lim_{\Delta \to \infty} E[x(t)y_\Delta(t)^*]$$

$$= \lim_{\Delta \to \infty} \frac{R(-\Delta) - R(0)}{\Delta} = -R'(0) = 0.$$

We should note that the steady-state spectral density (4.4.18) of the state process $x(\cdot)$ defined by (4.4.1) is such that

$$\int_{-\infty}^{\infty} |f|^2 P(f)\, df = \infty$$

unless $BDB^* = 0$, so that we cannot define a derivative in the mean square sense! The best we can do to indicate in what sense (4.4.1) holds is to show that for any continuous function $h(t)$, $t_1 \le t \le t_2$,

$$\int_{t_1}^{t_2} \left[\frac{(x(t + \Delta) - x(t))}{\Delta}, h(t) \right] dt$$

converges, as $\Delta \to 0$, to

$$\int_{t_1}^{t_2} [(Ax(t) + BN(t)), h(t)]\, dt.$$

This follows from

$$\frac{x(t + \Delta) - x(t)}{\Delta} = \left(\frac{e^{A\Delta} - I}{\Delta} \right) x(t) + \frac{1}{\Delta} \int_0^\Delta (e^{A(\Delta - \sigma)} - I)BN(t + \sigma)\, d\sigma$$

$$+ \frac{1}{\Delta} \int_0^\Delta BN(t + \sigma)\, d\sigma$$

and only the third term needs attention. But the difference between corresponding terms

$$\int_{t_1}^{t_2} \left[\frac{1}{\Delta} \int_0^\Delta BN(t + \sigma)\, d\sigma, h(t) \right] dt - \int_{t_1}^{t_2} [BN(t), h(t)]\, dt \qquad (4.6.6)$$

can be expressed as

$$= \frac{1}{\Delta} \int_0^\Delta d\sigma \left(\int_{t_1+\sigma}^{t_2+\sigma} [BN(s), h(s-\sigma)] \, ds \right)$$

$$- \frac{1}{\Delta} \int_0^\Delta d\sigma \int_{t_1}^{t_2} [BN(s), h(s)] \, ds$$

$$= \frac{1}{\Delta} \int_0^\Delta \left(\int_{t_1+\sigma}^{t_2} [BN(s), (h(s-\sigma) - h(s))] \, ds \right) d\sigma$$

$$+ \frac{1}{\Delta} \int_0^\Delta d\sigma \left(\int_{t_2}^{t_2+\sigma} [BN(s), h(s-\sigma)] \, ds \right)$$

$$- \frac{1}{\Delta} \int_0^\Delta d\sigma \left(\int_{t_1}^{t_1+\sigma} [BN(s), h(s)] \, ds \right). \qquad (4.6.7)$$

We can now show that each term goes to zero in the mean square and hence, we know, so does the sum. Let $q_1(\cdot), q_2(\cdot), q_3(\cdot)$ denote the integrands in parentheses in (4.6.7). Then (4.6.6) becomes

$$\frac{1}{\Delta} \int_0^\Delta q_1(\sigma) \, d\sigma + \frac{1}{\Delta} \int_0^\Delta q_2(\sigma) \, d\sigma + \frac{1}{\Delta} \int_0^\Delta q_3(\sigma) \, d\sigma$$

and we apply Schwarz inequality (cf. (4.1.13)) to each. Thus

$$E\left[\left(\frac{1}{\Delta} \int_0^\Delta q_i(\sigma) \, d\sigma \right)^2 \right] \le \frac{1}{\Delta} \int_0^\Delta E[q_i(\sigma)]^2 \, d\sigma, \qquad i = 1, 2, 3.$$

Now

$$E[q_1(\sigma)^2] = \int_{t_1+\sigma}^{t_2} \|B^*(h(s-\sigma) - h(s))\|^2 \, ds$$

$$\le \|B^*\|_0^2 (t_2 - t_1) \max_{t_1 \le s \le t_2} \|h(s-\sigma) - h(s)\|^2$$

$$E[q_2(\sigma)^2] = \int_{t_1}^{t_2+\sigma} \|B^* h(s-\sigma)\|^2 \, ds \le \sigma \|B^*\|_0^2 \max_{t_1 \le s \le t_2} \|h(s)\|^2$$

$$E[q_3(\sigma)^2] = \int_{t_1}^{t_1+\sigma} \|B^* h(s)\|^2 \, ds \le \sigma \|B^*\|_0^2 \max_{t_1 \le s \le t_2} \|h(s)\|^2.$$

Because $h(\cdot)$ is continuous in the closed bounded interval† it follows that

$$\frac{1}{\Delta} \int_0^\Delta E[q_1(\sigma)^2] \, d\sigma \to 0 \qquad \text{as } \Delta \to 0$$

$$\frac{1}{\Delta} \int_0^\Delta E[q_2(\sigma)^2] \, d\sigma \le \frac{\Delta}{2} \|B_0^*\|^2 \max_{t_1 \le s \le t_2} \|h(s)\|^2 \to 0 \qquad \text{as } \Delta \to 0.$$

† By "uniform" continuity—a mathematical technicality.

Similarly for

$$\frac{1}{\Delta} \int_0^\Delta E[q_3(\sigma)^2] \, d\sigma.$$

Finally we note that stationarity is *not* essential in defining differentiability and the concept of the mean square derivative process is readily extended to nonstationary processes. Without going into details, we mention that a sufficient condition is that the covariance function $R(t, s)$ must be such that the second-mixed derivative

$$\frac{\partial^2 R(t, s)}{\partial t \, \partial s}$$

be continuous in the square $a < t < b$, $a < s < b$, where $[a, b]$ is the interval of definition of the process. And the covariance function of the derivative would be

$$\frac{\partial^2 R(t, s)}{\partial t \, \partial s}.$$

A necessary condition is that for each t:

$$\frac{R(t + \Delta, t + \Delta) - R(t + \Delta, t) - R(t, t + \Delta) + R(t, t)}{\Delta^2}$$

converge to a finite limit as $\Delta \to 0$. For an example, see Problem 4.20.

4.7 CONTINUITY OF SAMPLE PATHS

Now we are in a position to answer the question: When can we assume that the sample paths of a process are continuous?

Theorem 4.7.1. Suppose that the process $x(t)$, $-\infty < t < \infty$, is second-order stationary with zero mean. The sample paths of the process are continuous (with probability one) if the spectral density $P(\cdot)$ is such that

$$\int_{-\infty}^{\infty} |f|^2 \operatorname{Tr} P(f) \, df < \infty. \tag{4.7.1}$$

Proof. This result is due to Kolmogorov. See reference 9. The proof is beyond our scope.

Remark. Condition (4.7.1) implies that $R(\cdot)$ has a second derivative at the origin (cf. Chapter 2). We could use this as a sufficient condition in place of (4.7.1)—see reference 9. Also under this condition, we see from Section 4.6 that the process actually has a derivative in the mean square sense.

4.8 EXAMPLE: NOISE IN ELECTRIC CIRCUITS

As an illustration of our analysis of the noise response of linear systems, let us consider the problem of (thermal) noise in lumped parameter electrical networks. Let there be n meshes or loops so that we can represent the mesh charges by an $n \times 1$ vector $Q(t)$. Then Kirchhoff's laws yield the network equations

$$L \frac{d^2Q}{dt^2} + R \frac{dQ}{dt} + GQ(t) = E(t), \qquad (4.8.1)$$

where

L is the $n \times n$ inductance matrix

R is the $n \times n$ resistance matrix

G is the $n \times n$ elastance matrix

and $E(t)$ is the $n \times 1$ input voltage. The matrices L, R, and G are symmetric and nonnegative definite, and, in addition, L and G are nonsingular. Let $E(t)$ denote the voltage induced by the thermal noise due to the resistors. Each resistor generates across its terminals random noise which can be characterized as white noise with spectral density given by $2kT$ times the resistance, k being Planck's constant and T the absolute temperature in Kelvin. Thus $E(t)$ is white noise with spectral density matrix

$$2kTR.$$

By letting $x(t)$ denote the $2n \times 1$ column-vector

$$x(t) = \begin{vmatrix} Q(t) \\ \dot{Q}(t) \end{vmatrix},$$

where a superdot indicates the derivative, Kirchhoff's equations (4.8.1) can be expressed in the state-space form:

$$\dot{x}(t) = Ax(t) + BN(t), \qquad (4.8.2)$$

where

$$A = \begin{vmatrix} 0 & I \\ -L^{-1}G & -L^{-1}R \end{vmatrix}$$

$$B = \begin{vmatrix} 0 \\ L^{-1} \end{vmatrix}$$

$N(\cdot)$ white noise with spectral density matrix $2kTR$.

The steady-state covariance R_∞ of the process $x(\cdot)$ is thus given by (compare (4.4.14))

$$AR_\infty + R_\infty A^* + \begin{vmatrix} 0 & 0 \\ 0 & 2kTL^{-1}RL^{-1} \end{vmatrix} = 0. \tag{4.8.3}$$

By direct substitution we can verify that the solution is given by

$$R_\infty = kT \begin{vmatrix} G^{-1} & 0 \\ 0 & L^{-1} \end{vmatrix}, \tag{4.8.4}$$

and hence in particular the steady-state covariance matrix of the vector of mesh currents $\dot{Q}(t)$ is given by

$$kTL^{-1}$$

and the spectral density is readily seen to be

$$2kT(G + 2\pi if R - 4\pi^2 f^2 L)^{-1}R(G - 2\pi if R - 4\pi^2 f^2 L)^{-1}4\pi^2 f^2. \tag{4.8.5}$$

See Problem 4.10 for the mechanical system analog.

PROBLEMS

4.1 Aircraft Gust Response:

A standard problem in aircraft flight test data analysis is to assess the effect of atmospheric wind turbulence. Here we shall consider a "textbook" version of the problem that we can handle using the tools for analyzing linear system response to random inputs developed in this chapter. Thus we limit ourselves to linearized rigid body dynamics, longitudinal mode only so that we can formulate the equations of motion in the usual terminology (see Figure 4.1 for an explanation of the variables)

$$\dot{\alpha}_s(t) = Z_\alpha \alpha_s(t) + \dot{\theta}(t) + Z_\alpha \alpha_g(t)$$
$$\ddot{\theta}(t) = M_\alpha \alpha_s(t) + M_{\dot{\theta}}\dot{\theta} + b\theta,$$

where (as illustrated in Figure 4.1)

$\alpha_s(t)$ is the angle of attack (radians)

$\theta(t)$ is the pitch angle (radians)

$\alpha_g(t)$ is the gust input modelled as a mean zero stationary Gaussian process with spectral density

$$P_g(f) = \frac{2\sigma^2 k}{k^2 + 4\pi^2 f^2}, \qquad -\infty < f < \infty$$

(the Dryden simplification of the Kolmogorov spectrum).

We are primarily interested in the effect of turbulence on the normal acceleration:

$$a_n(t) = \frac{v}{g}(\dot{\theta}(t) - \dot{\alpha}_s(t)),$$

where v is the speed and g is the acceleration due to gravity.

The problem is to calculate the covariance and spectral density of the normal acceleration as a function of gust and system parameter with a numerical evaluation for typical values:

$$Z_\alpha = -1.46$$

$$M_\alpha = -11.31$$

$$M_{\dot{\theta}} = -1.72$$

$$v = 600$$

$$g = 32$$

$$k = \frac{v}{1000}$$

$b = -.001,$ a "small number" inserted to assure system stability.

Hint: Develop first the state-space formulation using a Kalman signal generation model for $\alpha_g(\cdot)$:

$$\frac{d}{dt}\alpha_g(t) = -k\alpha_g(t) + \sigma_g N(t)$$

and obtain

$$a_n(t) = Cx(t)$$

$$\dot{x}(t) = Ax(t) + BN(t),$$

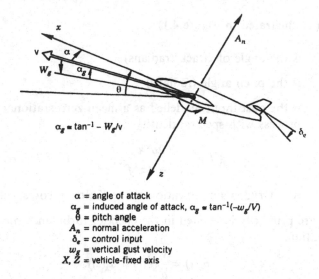

α = angle of attack
α_g = induced angle of attack, $\alpha_g = \tan^{-1}(-w_g/V)$
θ = pitch angle
A_n = normal acceleration
δ_e = control input
w_g = vertical gust velocity
X, Z = vehicle-fixed axis

Figure 4.1. Flight Data: Model variables (Problem 4.1).

where

$$
A = \begin{vmatrix} Z_\alpha & 0 & 1 & Z_\alpha \\ 0 & 0 & 1 & 0 \\ M_\alpha & b & M_{\dot\theta} & 0 \\ 0 & 0 & 0 & -k \end{vmatrix}
$$

$$
B = \begin{vmatrix} 0 \\ 0 \\ 0 \\ \sigma_g \end{vmatrix}
$$

$$
C = \begin{vmatrix} -\dfrac{v}{g}Z_\alpha & 0 & 0 & -\dfrac{v}{g}Z_\alpha \end{vmatrix}.
$$

The eigenvalues are

$$-1.59 \pm 3.36i, \qquad -0.6, \qquad -.0001.$$

Leave σ_g as a "free" parameter with answers in terms of it.

4.2. Develop a Kalman signal generation model for a zero-mean stationary 1×1 Gaussian process with spectral density

(a) $$P(f) = \frac{4k_1 k_2}{(4\pi^2 f^2 + k_1^2)(4\pi^2 f^2 + k_2^2)}$$

(b) $$P(f) = \left| \frac{2 + 2\pi i f}{1 + (2\pi i f)^3} \right|^2.$$

4.3. Use the Liapunov equation to calculate

$$\int_{-\infty}^{\infty} \frac{1}{|-4\pi^2 f^2 + 8i\pi f + 25|^2} \, df.$$

4.4. Let

$$\dot{x}(t) = \gamma A x(t) + \gamma N(t)$$
$$v(t) = C x(t),$$

where $N(\cdot)$ is white Gaussian with

$$\text{spectral density} = \text{identity matrix},$$

the system is stable, and γ is a positive number. Show that as $\gamma \to \infty$, the steady-state spectral density converges, obtaining a white-noise process in the limit. Determine the corresponding spectral density.

4.5. Let A be a square matrix. Determine conditions on A in order that

$$R(t) = e^{At}, \qquad t > 0$$
$$= e^{A^*|t|}, \qquad t < 0$$

is a stationary covariance function: determine the corresponding spectral density.

4.6. Generalize Example 4.5.1 to

$$A = \begin{vmatrix} 0 & 1 & 0 \\ 0 & 0 & 1 \\ a_0 & a_1 & a_2 \end{vmatrix}, \qquad B = \begin{vmatrix} 0 \\ 0 \\ 1 \end{vmatrix} \qquad C = |1 \quad 0 \quad 0|$$

$$\dot{x}(t) = A x(t) + B N(t)$$
$$s(t) = C x(t),$$

where $N(\cdot)$ is white Gaussian with unit spectral density. Assuming that the Liapunov equation has a covariance solution, deduce conditions on a_0, a_1, a_2 for A to be stable, and compare with the Routh criteria (see reference 8) for roots of polynomials to have negative real parts. Note that (A, B) is controllable and $D = 1$.

(a) What is the spectral density of the output process?
(b) How many mean square derivatives can be defined?
(c) What is the steady-state covariance:

$$\lim_{t \to \infty} E[s(t)^2]?$$

(d) Show that the steady-state covariance

$$R = \lim_{t \to \infty} E[x(t)x(t)^*]$$

is nonsingular.

Hint: In solving the Liapunov equation you may set

$$R = \{r_{ij}\} \quad \text{and} \quad r_{12} = r_{23} = 0.$$

4.7. Let

$$\Lambda = \int_0^\Delta e^{A\sigma} BDB^* e^{A^*\sigma} \, d\sigma, \quad D > 0.$$

Show that Λ is the unique solution of

$$A\Lambda + \Lambda A^* + BDB^* - e^{A\Delta} BDB^* e^{A^*\Delta} = 0.$$

Use this linear equation to determine Λ for

$$A = \begin{vmatrix} 0 & 1 \\ -4\pi^2 & -\pi \end{vmatrix}, \quad B = \begin{vmatrix} 0 \\ 1 \end{vmatrix}, \quad D = \text{identity}$$

for

$$\Delta = 1, \frac{1}{10}, \frac{1}{100}.$$

Verify your answer also by evaluating the integral. Compare the truth value with the approximation:

$$BDB^*\Delta$$

4.8. Show that the Liapunov equation for given A, B

$$0 = AR + RA^* + BB^*$$

(a) need not have a solution [Hint: Take $A = 0$.];
(b) can have many solutions;
(c) has a self-adjoint solution if it has a solution.

4.9. Let $x(t)$, $-\infty < t < \infty$, be an $n \times 1$ stationary stochastic process with spectral density $P(\cdot)$. Let $y(\cdot)$ denote the steady-state response of a system

$$y(t) = \int_0^\infty W(\sigma)x(t - \sigma)\,d\sigma,$$

when $W(\cdot)$ is $n \times n$ and satisfies (4.2.2), and let

$$\psi(f) = \int_0^\infty e^{-2\pi i f\sigma} W(\sigma)\,d\sigma.$$

Calculate the spectral density $P_z(\cdot)$ of

$$z(t) = x(t) + y(t).$$

For

$$W(t) = e^{At}, \qquad A \text{ stable}$$

express the spectral density of $x(\cdot)$ in terms of $P_z(\cdot)$.

4.10. (Noise response of structures) Let $x(\cdot)$ be $n \times 1$ and

$$M\ddot{x}(t) + Ax(t) + D\dot{x}(t) + BN(t) = 0,$$

where

$$M > 0, \quad A > 0, \quad D > 0$$

(self-adjoint, nonnegative definite, nonsingular) and $N(\cdot)$ is white Gaussian with unit spectral density matrix. Find the steady-state spectral density of the process $x(\cdot)$. Find the steady-state covariance matrix

$$\lim_{t \to \infty} E[x(t)x(t)^*].$$

for the case where $BB^* = \gamma D$, $\gamma > 0$ (positive real number).
   Hint: The eigenvalues of the matrix

$$\begin{vmatrix} 0 & I \\ -M^{-1}A & -M^{-1}D \end{vmatrix}$$

have negative real parts. The spectral density of $x(\cdot)$ is

$$= \psi(f)BB^*\psi(f)^*,$$

where

$$\psi(f) = (-4\pi^2 f^2 M + 2\pi i f D + A)^{-1}.$$

Use (4.8.4) for the steady-state covariance matrix.

4.11. Let $x(\cdot)$ be stationary. Let $\Delta > 0$. Show that the processes

$$y_+(t) = \frac{1}{\Delta} \int_t^{t+\Delta} x(s)\, ds, \qquad y_-(t) = \frac{1}{\Delta} \int_{t-\Delta}^t x(s)\, ds$$

have the same spectral density and yet are different as processes. Find the spectral density of the difference process

$$z(t) = y_+(t) - y_-(t).$$

What happens as $\Delta \to 0$? Show that both processes $y_+(t)$ and $y_-(t)$ have mean square derivatives, if $x(\cdot)$ does. Which process is "physically realizable"?

4.12. For the process $v(\cdot)$ in Example 4.5.1 with spectral density $P(\cdot)$ given by (4.5.5), calculate

$$\int_{-\infty}^{\infty} f^2 P(f)\, df.$$

Hint: Use (2.41).

4.13. For $R(t)$, $-\infty < t < \infty$, specified by (4.4.17), show that $R(\cdot)$ is *not* differentiable at $t = 0$ unless it is identically zero. Nevertheless,

$$CR(t)C^*$$

may well be differentiable (once or twice) at $t = 0$ for appropriate C. State a sufficient condition of C for this to hold.

Hint: Look at Problem 4.12.

4.14. (Time-scale change) Let $a > 0$. Let $N(t)$, $-\infty < t < \infty$, denote $n \times 1$ white noise with spectral density matrix D. Define

$$\tilde{N}(t) = N(at).$$

Show that $\tilde{N}(t)$, $-\infty < t < \infty$, is also white noise and find the corresponding spectral density matrix. Let the process $V(t)$ be described by the Kalman model

$$V(t) = Cx(t)$$

$$\dot{x}(t) = Ax(t) + BN(t).$$

Show that process $V(at)$ can also be described by a Kalman model and determine the same.

Answer:

$$V(at) = Cy(t)$$

$$\dot{y}(t) = aAy(t) + \sqrt{a}\, BN(t)$$

or

$$V(at) = \sqrt{a}\,Cy(t)$$

$$\dot{y}(t) = aAy(t) + BN(t).$$

4.15. (Nonlinear operation on white noise) Let $N(t)$, $-\infty < t < \infty$, denote one-dimensional white Gaussian noise with unit spectral density. Let $x(t)$, $t \geq 0$, be defined by

$$x(t) = \sin\left(2\pi f_c t + \sigma \int_0^t N(s)\,ds\right), \qquad f_c \text{ given, } \sigma > 0.$$

(Random frequency modulation) Show that the process $x(\cdot)$ is asymptotically stationary with stationary covariance function

$$R(t) = \frac{1}{2}(e^{-\sigma^2 |t|/2}) \cos 2\pi f_c t.$$

Compare with Problem 1.2.6.

4.16. Just as in the discrete-time case (Problem 3.17), show that we can generalize (4.3.17) to weighting functions $W(\cdot)$ which are not restricted to be physically realizable—that is

$$W(s) \neq 0, \qquad s < 0$$

and

$$x(t) = \int_{-\infty}^{\infty} W(s)N(t - s)\,ds,$$

where

$$\int_{-\infty}^{\infty} \|W(s)\|_0^2\,ds < \infty,$$

by defining

$$\psi(\lambda) = \int_{-\infty}^{\infty} e^{-2\pi i f t}W(t)\,dt.$$

4.17. Analog of Problem 3.18 to continuous-time case. Let $x(\cdot)$ denote a 1×1 process defined by

$$x(t) = p_m(t) + z(t),$$

where $p_m(\cdot)$ is a polynomial of degree m and $z(\cdot)$ is a stationary process with mean zero and spectral density $P(\cdot)$ where we assume that

$$\int_{-\infty}^{\infty} |f|^{2m+2}P(f)\,df < \infty.$$

Then

$$y(t) = \frac{d^{m+1}}{dt^{m+1}} x(t)$$

is a stationary process with zero mean and spectral density

$$(4\pi^2|f|^2)^{m+1} P(f), \qquad -\infty < f < \infty.$$

Note in particular that only the degree of the polynomial is needed in defining the process $y(\cdot)$.

4.18. Let $x(t), a \le t \le b$, denote an $n \times 1$ process continuous in the mean square sense. Then for any $n \times 1$ function $h(\cdot)$ defined in $a \le t \le b$ such that it is "square integrable," such as

$$\int_a^b \|h(t)\|^2 \, dt < \infty,$$

we can define

$$E[e^{i \int_a^b [x(t), h(t)] \, dt}].$$

This is called the "characteristic functional of the process" and yields a complete description of the process (including all finite-dimensional distributions) as $h(\cdot)$ ranges over the class of all "square integrable" functions. Show that

$$E[e^{i[x(t), z]}] = \lim_{\Delta \to 0} E\left[\text{Exp} \frac{i}{\Delta} \int_t^{t+\Delta} [x(s), z] \, ds \right]$$

and generalize the idea to arbitrary $t_i, i = 1, \ldots, N, a \le t_i \le b$. Calculate the characteristic functional of a Gaussian process. What is the characteristic functional of white noise?

4.19. Let $x(t), -\infty < t < \infty$, be an $n \times 1$ zero-mean, Gaussian, stationary process with stationary covariance function $R(\cdot)$ and spectral density $P(\cdot)$. Assume that $x(\cdot)$ has a mean square derivative. Let $Y(\cdot)$ denote the $2n \times 1$ process:

$$Y(t) = \left| \begin{array}{c} x(t) \\ \dot{x}(t) \end{array} \right|.$$

Calculate the stationary covariance function $R_y(\cdot)$ of the process $Y(\cdot)$.

4.20. Let $N(t)$, $-\infty < t < \infty$, denote $n \times 1$ white noise with spectral density matrix D. Define

$$W(t) = \int_0^t N(s) \, ds.$$

Show that

$$E[W(t)W(s^*)] = D \min{(t, s)}.$$

This is a Wiener process (cf. Problem 1.2.5) (for nonsingular D). Show that it does *not* have a mean square derivative.

Hint:

$$E\left\|\frac{W(t+\Delta) - W(t)}{\Delta}\right\|^2 = \frac{\mathrm{Tr}\, D}{\Delta}.$$

NOTES AND COMMENTS

The notions of continuous-time white noise and band-limited signals already occur in Shannon's basic work [1], as well as in the concurrent work of P. M. Woodward [2] in connection with radar detection. State-space theory for continuous-time processes was initiated by R. E. Kalman in his now-famous paper with R. S. Bucy [3]. For recent mathematical research on white noise theory, especially in the context of nonlinear operations, see reference 4. Problem 4.10 is generic in vibration of structures—see reference 6.

REFERENCES

Classic Treatises

1. C. E. Shannon. "A Mathematical Theory of Communication," *Bell System Technical Journal*, Vol. 27 (August 1948).

2. P. M. Woodward. *Probability and Information Theory with Applications to Radar.* McGraw-Hill, 1953.

3. R. E. Kalman and R. S. Bucy. "New Results in Linear Filtering and Prediction Theory," *Transactions of the ASME: Journal of Basic Engineering*, vol. 83, series D, no. 1 (1961), pp. 95–108.

Recent Publications

4. G. Kallianpur and R. L. Karandikar. *White Noise Theory of Prediction, Filtering and Smoothing.* Gordon and Breach, 1988.

5. R. E. Maine and K. W. Iliff. "Identification of Dynamic Systems, Theory and Formulation," NASA References Publication 1138, February 1985.

6. J. L. Humar. *Dynamics of Structures.* Prentice-Hall, 1990.

7. A. V. Balakrishnan. *State Space Theory of Systems*. Optimization Software Publications, 1988.

8. J. J. D'Azzo and C. H. Houpis. *Linear Control System Analysis and Design*. McGraw-Hill, 1988.

Mathematical Treatise

9. A. D. Wentzell. *A Course in the Theory of Stochastic Processes*. McGraw-Hill, 1981.

5

TIME AVERAGES AND THE ERGODIC PRINCIPLE

In this chapter we consider the crucial problem of determining from experiment—from observable data—the means and covariances which we have so far treated as given. Central to this problem is the "Ergodic Hypothesis" or "Ergodic Principle."

The Ergodic Principle is invoked every time any measurement is made on observed data modeled as a stationary random process. The following is a typical statement found in applications [3, p. 15]: "Usually we have a single function of time, . . . , and we must invoke the assumption of ergodicity to obtain estimates of ensemble averages. An ergodic process is one whose infinite-time average is equal to its ensemble average; and if we can make this assumption we can substitute time averages wherever we have phase averages."

5.1 DISCRETE-TIME MODELS

To examine this further, which is the object of this chapter, let us begin with an example which in fact nearly tells it all. Thus suppose we observe a discrete-time random process $\{x_n\}$ which is known to be at least second-order stationary, and we want to estimate the mean which is of course a constant. How do we do this? By "observation" we mean one sample path or sequence, however long. We can "average" in "time"; thus let

$$\zeta_N = \frac{1}{N+1} \sum_0^N x_k. \tag{5.1.1}$$

According to the Ergodic Principle, this "time average" should, for large enough

215

N, yield the "phase average"

$$E[x_k].$$

To see what is involved, let μ denote the process mean and $R(\cdot)$ the stationary covariance. First we note that

$$E[\zeta_N] = \mu.$$

Let us next calculate its covariance matrix. Since we can write

$$\zeta_N - \mu = \frac{1}{N+1} \sum_0^N (x_k - \mu)$$

we see that

$$\Lambda_N = E[(\zeta_N - \mu)(\zeta_N - \mu)^*] \qquad (5.1.2)$$

$$= \frac{1}{(N+1)^2} \sum_0^N \sum_0^N R(j-k). \qquad (5.1.3)$$

But as we did in Chapter 2, we can replace the double sum by a single sum:

$$\Lambda_N = \frac{1}{N+1} \sum_{-N}^N R(n)\left(1 - \frac{|n|}{N+1}\right). \qquad (5.1.4)$$

If the covariance Λ_N goes to zero as N goes to infinity, then the "time average" (5.1.1) converges to the phase average μ in the mean square sense, verifying the Ergodic Principle (for the mean, in the mean square sense).

Now since Λ_N is nonnegative definite we note that (cf. Problem 2.16)) Λ_N goes to zero if and only if Tr Λ_N goes to zero, or equivalently

$$\frac{1}{(N+1)^2} \sum_0^N \sum_0^N \text{Tr } R(j-k) \to 0$$

$$\frac{1}{N+1} \sum_{-N}^N (\text{Tr } R(n))\left(1 - \frac{|n|}{N+1}\right) \to 0.$$

A sufficient condition for this is that

$$\text{Tr } R(n) \to 0, \qquad n \to \infty. \qquad (5.1.5)$$

See Problem 5.1 if necessary. But this is clearly satisfied if the covariance function satisfies (2.31). In that case, using (2.34) we can express (5.1.4) as

$$\Lambda_N = \left(\frac{1}{N+1}\right) P_N(0) = \frac{1}{N+1} \int_{-1/2}^{1/2} \left(\frac{\sin^2 \pi\lambda(N+1)}{(N+1)\sin^2 \pi\lambda}\right) P(\lambda)\, d\lambda,$$

where $P_N(0)$ converges to $P(0)$, where $P(\lambda)$, spectral density of the process, is continuous in $-1/2 \le \lambda \le 1/2$.

A pertinent question of importance in applications is how large N should be taken for a given accuracy since we cannot go on forever! Assuming (2.31), we can provide an estimate, illustrating at the same time the use of the concept of "correlation time." The "correlation time" is defined by

$$\bar{N} = \sum_{-\infty}^{\infty} \frac{\text{Tr } R(n)}{\text{Tr } R(0)} \tag{5.1.6}$$

or, equivalently,

$$= \frac{\text{Tr } P(0)}{\text{Tr } R(0)}.$$

Then, the "normalized" mean square error

$$\frac{E[\|\zeta_N - \mu\|^2]}{\text{Tr } R(0)} = \left(\frac{1}{N+1}\right) \frac{\text{Tr } P_N(0)}{\text{Tr } R(0)}$$

$$\approx \left(\frac{1}{N+1}\right) \frac{\text{Tr } P(0)}{\text{Tr } R(0)} \approx \frac{\bar{N}}{N} \qquad \text{for large } N$$

gives an estimate of how large N should be in terms of $\bar{N}$, the "correlation time." Thus for

$$E[\|\zeta_N - \mu\|^2] \le \sigma^2$$

we estimate that we need

$$N > \frac{(\text{Tr } R(0))\bar{N}}{\sigma^2}$$

large enough so that $\text{Tr } P_N(0) \approx \text{Tr } P(0)$.

5.1.1 One-Dimensional Example

In the simplest one-dimensional 1×1 example where

$$R(n) = R(0)\rho^{|n|}, \qquad 0 < \rho < 1,$$

we have

$$\bar{N} = \sum_{-\infty}^{\infty} \rho^{|n|} = \frac{1+\rho}{1-\rho}.$$

We note that the larger the ρ, the larger the correlation time ($\bar{N} = 20$ for $\rho = .9$, $\bar{N} = 100$ for $\rho = .98$), increasing rapidly as ρ approaches 1.

In fact, in this particular case $R(k)$ is nonnegative so that we have

$$\sum_{-N}^{N} R(k)\left(1 - \frac{|k|}{N+1}\right) \leq \sum_{-N}^{N} R(k) \leq \sum_{-\infty}^{\infty} R(k)$$

and hence

$$\frac{E[(\zeta_N - \mu)^2]}{R(0)} \leq \frac{1}{N+1}\left(\frac{1+\rho}{1-\rho}\right), \quad \text{for all } N.$$

For a "pathwise" estimate we can invoke the Chebyshev inequality

$$\Pr[|\zeta_N - \mu| > \varepsilon] \leq \frac{E[(\zeta_N - \mu)^2]}{\varepsilon^2} \leq \frac{R(0)}{\varepsilon^2} \cdot \left(\frac{1+\rho}{1-\rho}\right) \cdot \frac{1}{N+1}.$$

Hence

$$\Pr[|\zeta_N - \mu| > \varepsilon] \leq p_\varepsilon,$$

provided that

$$N + 1 \geq \frac{1+\rho}{1-\rho} \cdot \frac{R(0)}{\varepsilon^2} \cdot \frac{1}{p_\varepsilon}.$$

Thus

$$\varepsilon = \frac{1}{10}\sqrt{R(0)}, \qquad p_\varepsilon = 10^{-3}$$

$$N + 1 \geq \left(\frac{1+\rho}{1-\rho}\right)10^5,$$

which should be considered a pessimistic estimate of how large N should be. If the process is Gaussian so that ζ_N is Gaussian, we can do better, because in that case by (0.5.9) for a $(0, \sigma)$ Gaussian ζ we have

$$\Pr[|\zeta| > 3\sigma] \approx 10^{-3}. \tag{5.1.7}$$

Hence taking

$$\varepsilon^2 = \frac{R(0)}{N+1} \cdot 9\left(\frac{1+\rho}{1-\rho}\right)$$

we have that

$$\Pr[|\zeta_N - \mu| > \varepsilon] \leq 10^{-3}$$

for

$$N + 1 = R(0)\left(\frac{1+\rho}{1-\rho}\right) \cdot \frac{9}{\varepsilon^2},$$

which is much smaller than the previous Chebyshev-inequality-based estimate for the same ε. A more detailed study is beyond our scope. For $\rho = 0$, for instance, there are better estimates based on the Central Limit Theorem (see reference 4 of Chapter 1). In any case we see that N increases rapidly as ρ approaches one, whatever kind of error is considered.

Again, if the process is Gaussian, which is usually the case for us, we can under condition (2.31) assert pathwise (probability one) convergence. See reference 1. Ultimately, however, it is a matter of "faith" that the Ergodic Principle holds, since so many intangibles are involved in a practical application.

5.1.2 Estimate of Covariance

Because of the importance in applications, we consider next estimates of covariances. Here we shall *assume* that the process $\{x_k\}$ is actually Gaussian and that the mean is known, so that we may assume it to be zero. Thus the time average estimate of the covariance $R(m)$ would be

$$\gamma_N = \left(\frac{1}{N+1}\right) \sum_0^N x_k x_{k+m}^* \tag{5.1.8}$$

for fixed integer m. Since the convergence will be term by term in this matrix, we may consider each term separately in this matrix. Hence without loss of generality, we may confine ourselves to the one-dimensional case. Thus let

$$R(m) = \{r_{ij}(m)\}$$

and in terms of its components let

$$x_k = \begin{vmatrix} x_{1k} \\ \vdots \\ x_{nk} \end{vmatrix},$$

and let us consider the time average or "sample mean" for the covariance $r_{ij}(m)$. Fixing $i, j,$ and m, let

$$\gamma_N = \left(\frac{1}{N+1}\right) \sum_0^N x_{ik} x_{j(k+m)}.$$

Let us use the notation

$$z_k = x_{ik} \cdot x_{j(k+m)} \tag{5.1.9}$$

so that

$$\gamma_N = \left(\frac{1}{N+1}\right) \sum_0^N z_k.$$

The process $\{z_k\}$ is of course stationary. Its mean is

$$E[z_k] = r_{ij}(m).$$

Hence we are using the time average estimate of the mean of the process $\{z_k\}$. To verify convergence in the mean square, therefore, we need to verify (5.1.5) for the covariance function of the process $\{z_k\}$. Because of the Gaussian assumption we can calculate the covariance

$$E[z_k z_{k+p}] = E[x_{ik} x_{j(k+m)} \quad x_{i(k+p)} x_{j(k+m+p)}],$$

which, by the rules for calculating four products of Gaussians (cf. Review Chapter),

$$= r_{ii}(p)r_{jj}(p) + r_{ij}(m)^2 + r_{ij}(m+p)r_{ij}(m-p).$$

Hence the covariance functions of the process $\{z_k\}$, call it $r(\cdot)$, is given by

$$r(p) = r_{ii}(p)r_{jj}(p) + r_{ij}(m+p)r_{ij}(m-p). \tag{5.1.10}$$

But this goes to zero as p goes to infinity if $R(p)$ does. Moreover, we can show that

$$\sum_{-\infty}^{\infty} |r(p)| < \infty \tag{5.1.11}$$

if $R(\cdot)$ satisfies (2.31). Indeed the first term on the right-hand side of (5.1.10) is given by

$$|r_{ii}(p)r_{jj}(p)| \le |r_{ii}(p)|r_{jj}(0)$$

so that

$$\sum_{-\infty}^{\infty} |r_{ii}(p)r_{jj}(p)| < \infty;$$

as for the second term, we can use the Schwarz inequality

$$\left(\sum_{p=N}^{N+k} |r_{ij}(m+p)| |r_{ij}(m-p)| \right)^2 \le \sum_{p=N}^{N+k} r_{ij}(m+p)^2 \sum_{p=N}^{N+k} r_{ij}(m-p)^2$$

and from (2.31), for N sufficiently large, we obtain

$$|r_{ij}(k)| < 1, \quad k > N$$

so that

$$\sum_{N}^{\infty} r_{ij}(k)^2 \le \sum_{N}^{\infty} |r_{ij}(k)| < \infty.$$

We can of course go on to calculate the "correlation time" of the process $\{z_k\}$—see Example 5.2.

Thus we have proved the Ergodic Principle (in the mean square sense) for covariances of stationary Gaussian processes. For proof of pathwise convergence see reference 1, p. 494.

5.1.3   More General Statistics

Let us next consider the Ergodic Principle for more general statistical (phase) averages or "statistics." For example, suppose we wish to estimate p_c as

$$p_c = \text{Probability } [x_k \in C], \tag{5.1.12}$$

where C is a closed bounded set in $\mathbf{R}^n$. To handle this we define the function $g(\cdot)$ on $\mathbf{R}^n$ such that

$$\begin{aligned} g(x) &= 1 \quad x \in C \\ &= 0 \quad \text{otherwise.} \end{aligned} \tag{5.1.13}$$

Then

$$E[g(x_k)] = \text{probability } [x_k \in C] = p_c.$$

Hence by the Ergodic Principle we would use the time average

$$\frac{1}{N+1} \sum_0^N g(x_k) \tag{5.1.14}$$

to estimate (5.1.12). One may recognize that (5.1.14) is no more than the familiar "frequency" estimate in coin-tossing. We have mean square convergence if we can prove (5.1.5) for the process $\{g(x_k)\}$ that the covariance function

$$r(p) = E[\tilde{g}(x_k)\, \tilde{g}(x_{k+p})] \to 0 \quad \text{as } p \to \infty, \tag{5.1.15}$$

where

$$\tilde{g}(x_k) = g(x_k) - E[g(x_k)].$$

Let us now prove (5.1.15).

We may without loss of generality assume that $R(0)$ is nonsingular (cf. Problem 2.13 in Chapter 2) and further that it is the identity. The covariance matrix Λ_p of

$$x_k, \qquad x_{k+p}$$

will then be

$$\Lambda_p = \begin{vmatrix} I & R(p) \\ R(p)^* & I \end{vmatrix}.$$

Since $R(p) \to 0$ as $p \to \infty$, Λ_p will be nonsingular for all p sufficiently large and we may assume that this is so in what follows. In particular, as $p \to \infty$ we have

$$\Lambda_p \to \Lambda_\infty = \begin{vmatrix} I & 0 \\ 0 & I \end{vmatrix}.$$

Let $G_p(\cdot, \cdot)$ denote the Gaussian density function with zero mean and covariance Λ_p. Then as $p \to \infty$, $G_p(x, y)$ converges for each $x, y \in \mathbf{R}^n$ to

$$G(x)G(y),$$

where $G(\cdot)$ is the Gaussian density, with zero mean and covariance matrix the $n \times n$ identity. Now

$$E[g(x_k)] = \int_{\mathbf{R}^n} g(x)G(x - \mu)\, d|x| = p_c$$

and $r(p)$ can be expressed:

$$r(p) = E[g(x_k)g(x_{k+p})] - p_c^2$$

$$= \int_{\mathbf{R}^n} \int_{\mathbf{R}^n} g(x)g(y)G_p(x - \mu, y - \mu)\, d|x|\, d|y| - p_c^2$$

$$= \int_C \int_C G_p(x - \mu, y - \mu)\, d|x|\, d|y| - p_c^2. \qquad (5.1.16)$$

The integrand in (5.1.16) converges to

$$G(x - \mu)G(y - \mu)$$

and the integral is given by

$$\int_C \int_C G(x - \mu)G(y - \mu)\, d|x|\, d|y| = \int_{\mathbf{R}^n} \int_{\mathbf{R}^n} g(x)g(y)G(x - \mu)G(y - \mu)\, d|x|\, d|y|$$

$$= E[g(x_k)]E[g(x_k)] = p_c^2. \qquad (5.1.17)$$

The integral in (5.1.15) converges to the integral (5.1.17) because the integrands converge and (the "technicality" that) the sequence

$$G_p(x - \mu, y - \mu)$$

is bounded for $x \in C$ and $y \in C$, C itself being also bounded. Indeed,

$$G_p(x - \mu, y - \mu) = \frac{1}{|\Lambda_p|^{1/2}} \cdot \frac{1}{(2\pi)^n} \exp\left(\frac{-1}{2}[\Lambda_p^{-1}Z, Z]\right),$$

where

$$Z = \begin{vmatrix} x - \mu \\ y - \mu \end{vmatrix}.$$

The determinant $|\Lambda_p|$ converges to 1 and

$$e^{(-1/2)[\Lambda_p^{-1}Z, Z]} \leq e^{(-1/2)\gamma_p^{-1}(||x - \mu||^2 + ||y - \mu||^2)}, \tag{5.1.18}$$

where γ_p is the largest eigenvalue of Λ_p and converges to 1, and hence

$$\leq e^{(-1/2)\delta(||x - \mu||^2 + ||y - \mu||^2)} \leq 1$$

for some δ, $0 < \delta$, and is bounded by 1. Hence

$$r(p) \to 0, \qquad p \to \infty,$$

as required.

Finally, we shall prove the mean square Ergodic Principle for a fairly general class of functions $g(\cdot)$ for stationary Gaussian processes $\{x_k\}$ with covariance satisfying (5.1.5).

Theorem 5.1.1. Let $\{x_k\}$ be a stationary Gaussian process with mean μ and stationary covariance function $R(\cdot)$, which satisfies (5.1.5). Let $g(\cdot)$ be a function mapping $\mathbf{R}^n$ into $\mathbf{R}^1$ such that

$$E[g(x_k)^2] < \infty. \tag{5.1.19}$$

Then the time average

$$\frac{1}{N + 1} \sum_0^N g(x_k) \tag{5.1.20}$$

converges in the mean square sense to the phase average

$$E[g(x_k)]. \tag{5.1.21}$$

Proof. We begin by noting that if the theorem holds for functions

$$g_1(\cdot), \ldots, g_m(\cdot),$$

each satisfying (5.1.19), then it holds for the sum

$$g(x) = \sum_1^m g_i(x). \qquad (5.1.22)$$

Let

$$\zeta_i = \frac{1}{N+1} \sum_0^N g_i(x_k) - E[g_i(x_k)].$$

Then

$$\frac{1}{N+1} \sum_0^N g(x_k) - E[g(x_k)] = \sum_1^m \zeta_i$$

and

$$E\left[\left(\sum_1^m \zeta_i\right)^2\right] \le \left(\sum_1^m \sqrt{E[\zeta_i^2]}\right)^2.$$

But for each i we have

$$E[\zeta_i^2] \to 0 \qquad \text{as } N \to \infty$$

by assumption, and hence the Ergodic Principle holds for (5.1.22).

We have already shown that the result holds for step functions of the form (5.1.13). But it is well known in Real Analysis (see, e.g., reference 2) that for any $g(\cdot)$ satisfying (5.1.19) we can find a sequence $f_n(\cdot)$ of "simple" functions, with each simple function $f_n(\cdot)$ being of the form (5.1.22) where each $g_i(\cdot)$ is a step function such that

$$E[|g(x_k) - f_n(x_k)|^2] \to 0 \qquad \text{as } n \to \infty.$$

Now the expression

$$\left(\frac{1}{N+1} \sum_0^N g(x_k) - E[g(x_k)]\right) \qquad (5.1.23)$$

can be expressed for any n:

$$= \left(\frac{1}{N+1} \sum_0^N (g(x_k) - f_n(x_k))\right) + E[f_n(x_k) - g(x_k)]$$

$$+ \left(\frac{1}{N+1} \sum_0^N f_n(x_k) - E[f_n(x_k)]\right)$$

$$= s_1 + s_2 + s_3,$$

where s_1 is the first term in parentheses, s_2 the second term, and s_3 the third term. We consider each term in turn.

$$E[s_1^2] = E\left|\frac{1}{N+1}\sum_0^N (g(x_k) - f_n(x_k))\right|^2$$

(by Schwarz inequality)

$$\leq \frac{1}{(N+1)^2}(N+1)\sum_0^N E[|g(x_k) - f_n(x_k)|^2]. \qquad (5.1.24)$$

If we pick n large enough so that

$$E[|g(x_k) - f_n(x_k)|^2] < \varepsilon,$$

then also

$$(E[|g(x_k) - f_n(x_k)|])^2 \leq \varepsilon$$

and (5.1.24)

$$\leq \varepsilon \quad \text{for any } N.$$

Next for the chosen n, we pick N large so that the mean square error is given by

$$E\left|\frac{1}{N+1}\sum_0^N f_n(x_k) - E[f_n(x_k)]\right|^2 < \varepsilon.$$

Hence for N large enough we can make

$$E[s_1^2] < \varepsilon$$
$$E[s_2^2] < \varepsilon$$
$$E[s_3^2] < \varepsilon$$

and

$$E[(s_1 + s_2 + s_3)^2] = \left(\sqrt{E[s_1^2]} + \sqrt{E[s_2^2]} + \sqrt{E[s_3^2]}\right)^2$$

and hence the mean square of (5.1.23) goes to zero, as required.

5.1.4 Failure of Ergodic Principle

Let us next see what happens to the Ergodic Principle when condition (5.1.5) does *not* hold. For example, let

$$x_k = A \sin (2\pi\lambda_0 k + \phi), \qquad (5.1.25)$$

where λ_0 is fixed and A, ϕ are random, A is Rayleigh, and ϕ is uniform, as in

Example 1.1.1. Thus we have a Gaussian process whose covariance function is given by

$$r(p) = E[x_k x_{k+p}] = E[A^2] \cos 2\pi\lambda_0 p \qquad (5.1.26)$$

and does not go to zero as $p \to \infty$. Let us calculate the time average to estimate the mean. We have

$$\zeta_N = \frac{1}{N+1} A \sum_0^N \sin (2\pi\lambda_0 k + \phi) \qquad (5.1.27)$$

$$E[\zeta_N^2] = \frac{1}{(N+1)^2} E\left[A^2 \sum_{k=0}^N \sum_{j=0}^N \sin (2\pi\lambda_0 k + \phi) \sin (2\pi\lambda_0 j + \phi) \right]$$

$$= \frac{1}{2} \frac{1}{(N+1)^2} E[A^2] \left[\sum_0^N \sum_0^N \cos 2\pi\lambda_0 (k-j) \right.$$

$$\left. - \sum_0^N \sum_0^N E[\cos (2\pi\lambda_0 (k+j) + 2\phi)] \right],$$

and with the second sum being zero we have

$$= \frac{E[A^2]}{2(N+1)^2} \left| \sum_0^N e^{2\pi i \lambda_0 k} \right|^2 = \frac{E[A^2]}{2(N+1)^2} \left| \left(\frac{1 - e^{2\pi i \lambda_0 (N+1)}}{1 - e^{2\pi i \lambda_0}} \right) \right|^2$$

$$\to 0, \qquad \lambda_0 \neq 0, \qquad \text{as } N \to \infty.$$

If $\lambda_0 = 0$, we have

$$E[\zeta_N^2] = \frac{E[A^2]}{2}.$$

This is consistent with our theory because for $\lambda_0 \neq 0$, even though (5.1.5) is not satisfied, we do have (5.1.3)

$$= \frac{1}{(N+1)^2} E[A^2] \sum_0^{N+1} \sum_0^{N+1} \cos 2\pi\lambda_0 (j-k)$$

$$= \frac{E[A^2]}{2(N+1)^2} \left| \left(\frac{1 - e^{2\pi i \lambda_0 (N+1)}}{1 - e^{2\pi i \lambda_0}} \right) \right|^2$$

$$\to 0, \qquad N \to \infty.$$

Hence we have mean square convergence of the time average to zero. On the other hand, for $\lambda_0 = 0$ we have

$$x_k = A \sin \phi$$

and

$$\frac{1}{N+1}\sum_0^N x_k = A \sin \phi$$

and hence the time average does not converge in the mean square sense to the phase average which is zero. Neither do we have pathwise convergence to a constant! On the other hand if λ_0 is a nonzero rational number, $\lambda_0 = m/p$, then $\zeta_N = 0$ if $N + 1 = p$.

We can generalize this example to

$$x_k = \sum_1^N A_j \sin (2\pi\lambda_j k + \phi_j), \tag{5.1.28}$$

the sampled version of Example 1.1.3, with A_j, ϕ_j as defined there. The covariance function

$$E[x_k x_{k+p}] = \sum_1^N \frac{E[A_j^2]}{2} \cos 2\pi\lambda_j p$$

and again does not satisfy (5.1.5) but (5.1.3) does go to zero if

$$\lambda_j \neq 0 \quad \text{for any } j. \tag{5.1.29}$$

Hence the Ergodic Principle holds in the mean square sense for the mean, provided (5.1.29) holds.

Note that, for $\lambda_j \neq 0$, for any j, the time average (5.1.25) is given by

$$\zeta_N = \left(\sum_0^N \frac{\cos 2\pi\lambda_j k}{N+1}\right) A \sin \phi + \left(\sum_0^N \frac{\sin 2\pi\lambda_j k}{N+1}\right) A \cos \phi$$

and hence converges to zero pathwise also as $N \to \infty$, even though (5.1.5) does *not* hold.

Next let us consider the covariance. With x_k defined by (5.1.25) let us examine what happens to the time average estimate for $E[x_k^2]$,

$$\frac{1}{N+1}\sum_0^N x_k^2.$$

We have

$$\frac{1}{N+1}\sum_0^N A^2 \sin^2(2\pi\lambda_0 k + \phi) = \frac{A^2}{2} - \frac{A^2}{N+1}\frac{1}{2}\sum_0^N \cos (4\pi\lambda_0 k + 2\phi).$$

For $\lambda_0 \neq 0$, the second term goes to zero and hence the time average converges pathwise to $A^2/2$, which is *not* the phase average. The covariance

$$E\left[\left(\frac{1}{N+1}\sum_0^N x_k^2\right)^2\right] - \frac{E[A^2]}{2}$$

can be calculated using the formula for four products of Gaussians, and of course it does not go to zero! Hence the Ergodic Principle fails—and equation (5.1.5) is *not* satisfied.

5.1.5 Spectral Density Estimates

Here is another instance where the Ergodic Principle fails to hold.

Let us attempt a sample path or time average estimate for spectral density, rather than the covariance. Let $\{x_k\}$ denote a one-dimensional stationary Gaussian process with zero mean and spectral density $p(\lambda)$, $-1/2 \le \lambda \le 1/2$. Start with the "discrete Fourier transform"

$$\zeta_N(\lambda) = \frac{1}{\sqrt{N+1}}\sum_0^N x_k e^{2\pi i k\lambda}, \qquad -1/2 \le \lambda \le 1/2. \tag{5.1.30}$$

Then

$$E[|\zeta_N(\lambda)|^2] = \frac{1}{N+1}\sum_0^N\sum_0^N E[x_k x_m]e^{2\pi i(k-m)\lambda},$$

which, as we saw in Chapter 2, converges as $N \to \infty$ to the spectral density $p(\lambda)$ for each λ. Hence we may consider

$$|\zeta_N(\lambda)|^2 = \frac{1}{N+1}\left|\sum_0^N x_k e^{2\pi i k\lambda}\right|^2 \tag{5.1.31}$$

for each λ as a "time average" estimate for the spectral density. Unfortunately the Ergodic Principle fails for this estimate, as we shall now show. In fact

$$|\zeta_N(\lambda)|^2$$

does not converge to anything. Indeed we can calculate the covariance

$$E[|\zeta_N(\lambda)|^4] - E[|\zeta_N(\lambda)|^2] \ge p(\lambda)^2. \tag{5.1.32}$$

This follows from

$$E[|\zeta_N(\lambda)|^4] = \frac{1}{(N+1)^2}\sum_0^N\sum_0^N\sum_0^N\sum_0^N E[x_m x_n x_p x_q]e^{2\pi i\lambda(m-n)-2\pi i\lambda(p-q)}, \tag{5.1.33}$$

which, using the four-product average formula for Gaussians,

$$E[x_m x_n x_p x_q] = R(m - n)R(p - q) + R(m - p)R(n - q) + R(m - q)R(n - p),$$

can be expressed as

$$= \frac{1}{(N + 1)^2} \sum_0^N \sum_0^N \sum_0^N \sum_0^N [R(m - n)e^{2\pi i\lambda(m-n)} R(p - q)e^{-2\pi i\lambda(p-q)}]$$

$$+ \frac{1}{(N + 1)^2} \sum_0^N \sum_0^N \sum_0^N \sum_0^N [R(m - p)e^{2\pi i\lambda(m-p)} R(n - q)e^{-2\pi i\lambda(n-q)}]$$

$$+ \frac{1}{(N + 1)^2} \sum_0^N \sum_0^N \sum_0^N \sum_0^N [R(m - q)e^{2\pi i\lambda(m+q)} R(n - p)e^{-2\pi i\lambda(n+p)}]. \quad (5.1.34)$$

The first term in (5.1.34)

$$= \left| \frac{1}{N + 1} \sum_0^N \sum_0^N R(m - n)e^{2\pi i\lambda(m-n)} \right|^2 = p_N(\lambda)^2$$

in the notation of Section 2.2 of Chapter 2, and as proved therein it converges to

$$p(\lambda)^2$$

as $N \to \infty$. The same holds for the second term. The third term in (5.1.34), being expressible as

$$\left| \frac{1}{N + 1} \sum_{m=0}^N \sum_{q=0}^N R(m - q)e^{2\pi i\lambda(m+q)} \right|^2, \quad (5.1.35)$$

is nonnegative. Hence it follows that the covariance is given by

$$E[|\zeta_N(\lambda)|^4] - (E[|\zeta_N(\lambda)|^2])^2 \geq 2p(\lambda)^2 - p(\lambda)^2 \geq p(\lambda)^2.$$

On the other hand we can express (5.1.35) as

$$\left| \frac{1}{N + 1} \sum_0^N \sum_0^N \int_{-1/2}^{1/2} e^{2\pi i\sigma(m-q)} e^{-2\pi i\lambda(m+q)} p(\sigma) \, d\sigma \right|^2$$

$$= \frac{1}{(N + 1)^2} \left| \int_{-1/2}^{1/2} F_N(\lambda - \sigma) F_N(\lambda + \sigma) p(\sigma) \, d\sigma \right|^2$$

$$\leq \left(\int_{-1/2}^{1/2} \frac{|F_N(\lambda - \sigma)|^2 p(\sigma)}{N + 1} \, d\sigma \right) \left(\int_{-1/2}^{1/2} \frac{|F_N(\lambda + \sigma)|^2 p(\sigma)}{N + 1} \, d\sigma \right),$$

where

$$F_N(\lambda) = \frac{1 - e^{2\pi i \lambda(N+1)}}{1 - e^{2\pi i \lambda}}$$

and

$$\frac{|F_N(\lambda)|^2}{N+1} = \frac{\sin^2 \pi \lambda (N+1)}{(N+1) \sin^2 \pi \lambda}$$

and is a δ-function sequence as we have seen in Chapter 2. Hence (5.1.35), as $N \to \infty$, becomes

$$\leq p(\lambda)^2.$$

Hence we obtain

$$p(\lambda)^2 \leq E[|\zeta_N(\lambda)|^2 - E[|\zeta_N(\lambda)|]^2]^2 \leq 2p(\lambda)^2, \qquad (5.1.36)$$

from which it follows that $|\zeta_N(\lambda)|^2$ does *not* converge in the mean square except at points where $p(\lambda) = 0$, in which case the limit is zero as well. In other words the Ergodic Principle fails at all points where $p(\lambda) \neq 0$.

Remark. It is of interest to note that even though $|\zeta_N(\lambda)|^2$ does not converge, we can show that "functionals" or "weighted averages" of the form

$$\int_{-1/2}^{1/2} H(\lambda)|\zeta_N(\lambda)|^2 \, d\lambda, \qquad (5.1.37)$$

where

$$\int_{-1/2}^{1/2} H(\lambda)^2 \, d\lambda < \infty, \qquad (5.1.38)$$

do satisfy the Ergodic Principle! As a simple example consider

$$\int_{-1/2}^{1/2} |\zeta_N(\lambda)|^2 \, d\lambda = \frac{1}{N+1} \sum_0^N \sum_0^N \int_{-1/2}^{1/2} x_k x_m e^{2\pi i (k-m)\lambda} \, d\lambda$$

$$= \frac{1}{N+1} \sum_0^N x_k^2,$$

which we have seen does converge to

$$R(0) = \int_{-1/2}^{1/2} p(\lambda) \, d\lambda,$$

provided that (2.31) or the weaker condition is given by

$$\sum_{-\infty}^{\infty} R(k)^2 < \infty. \qquad (2.31a)$$

To prove this generally, we note first that we may assume

$$H(\lambda) = H(-\lambda)$$

without changing the values of (5.1.37). Let

$$\gamma_N = \int_{-1/2}^{1/2} H(\lambda)|\zeta_N(\lambda)|^2 \, d\lambda.$$

Then we can calculate that

$$\gamma_N = \frac{1}{2N+1} \sum_0^N \sum_0^N x_m x_n H_{m-n},$$

where $\{H_p\}$ denote the Fourier coefficients

$$H_p = \int_{-1/2}^{1/2} e^{2\pi i p \lambda} H(\lambda) \, d\lambda.$$

Note that

$$H_p \to 0 \qquad \text{as } p \to \infty,$$

since

$$\sum_{-\infty}^{\infty} H_p^2 = \int_{-1/2}^{1/2} H(\lambda)^2 \, d\lambda,$$

and this is what distinguishes the sum

$$E[\gamma_N^2] = \frac{1}{(N+1)^2} \sum_0^N \sum_0^N \sum_0^N \sum_0^N E[x_m x_n x_p x_q] H_{m-n} H_{p-q}$$

from (5.1.33). Again using the Gaussian four-product formula we obtain

$$E[\gamma_N^2] = \frac{1}{(N+1)^2} \sum_0^N \sum_0^N \sum_0^N \sum_0^N R(m-n)R(p-q)H(m-n)H(p-q)$$

$$+ \frac{1}{(N+1)^2} \sum_0^N \sum_0^N \sum_0^N \sum_0^N R(m-p)R(n-q)H(m-n)H(p-q)$$

$$+ \frac{1}{(N+1)^2} \sum_0^N \sum_0^N \sum_0^N \sum_0^N R(m-q)R(n-p)H(m-n)H(p-q).$$

$$(5.1.39)$$

The first sum

$$= \frac{1}{(N+1)^2} \left(\sum_0^N \sum_0^N R(m-n)H(m-n) \right)^2$$

and

$$\frac{1}{N+1}\sum_{0}^{N}\sum_{0}^{N} R(m-n)H(m-n)$$

as in Chapter 2 and can be expressed as

$$\sum_{-N}^{N}\left(1 - \frac{|n|}{N+1}\right) R(n)H(n)$$

and converges as $N \to \infty$ to

$$\sum_{-\infty}^{\infty} R(n)H(n) = \int_{-1/2}^{1/2} H(\lambda)p(\lambda)\, d\lambda \qquad (5.1.40)$$

under our assumption of (2.31) or (2.31a) and (5.1.38). We shall next show that both the second and third sums in (5.1.39) go to zero. It is readily seen that the two sums are equal by interchanging p and q in the third sum and using

$$H(q-p) = H(p-q).$$

Substituting

$$R(k) = \int_{-1/2}^{1/2} e^{2\pi i k\lambda} p(\lambda)\, d\lambda$$

$$H_k = \int_{-1/2}^{1/2} e^{2\pi i k\lambda} H(\lambda)\, d\lambda,$$

either sum can be expressed as

$$\int_{-1/2}^{1/2} \cdots \int_{-1/2}^{1/2} F_N(\lambda_1, \lambda_2, \sigma_1, \sigma_2) H(\sigma_1) H(\sigma_2) p(\lambda_1) p(\lambda_2)\, d\sigma_1\, d\sigma_2\, d\lambda_1\, d\lambda_2, \quad (5.1.41)$$

where

$$F_N(\lambda_1, \lambda_2, \sigma_1, \sigma_2) = \frac{1}{(N+1)^2} F_N(\lambda_1 + \sigma_1)\, \overline{F_N(\lambda_2 + \sigma_2)}\, F_N(\lambda_2 - \sigma_1)\, \overline{F_N(\lambda_1 - \sigma_2)},$$

where

$$F_N(\lambda) = \frac{1 - e^{-2\pi i \lambda(N+1)}}{1 - e^{-2\pi i \lambda}}.$$

Note that $F_N(\lambda) \to 0$ except at $\lambda = 0$, as $N \to \infty$, and hence $F_N(\lambda_1, \lambda_2, \sigma_1, \sigma_2) \to 0$ except for points where two of the four equalities hold:

$$\lambda_1 + \sigma_1 = 0, \qquad \lambda_2 + \sigma_2 = 0, \qquad \lambda_2 = \sigma_1, \qquad \lambda_1 = \sigma_2,$$

where $F_N(\lambda_1, \lambda_2, \sigma_1, \sigma_2)$ is actually nonnegative. On the other hand,

$$\int_{-1/2}^{1/2} \int_{-1/2}^{1/2} F_N(\lambda_1, \lambda_2, \sigma_1, \sigma_2) \, d\lambda_1 \, d\lambda_2 \, d\sigma_1 \, d\sigma_2 = \frac{1}{N+1},$$

using the fact that

$$\int_{-1/2}^{1/2} e^{2\pi i \lambda_1 m} e^{2\pi i \lambda_2 n} e^{2\pi i \sigma_1 p} e^{2\pi i \sigma_2 q} \, d\lambda_1 \, d\lambda_2 \, d\sigma_1 \, d\sigma_2 = 1, \qquad \text{if } m = n = p = q$$

$$= 0, \qquad \text{otherwise.}$$

These facts are enough to establish that (5.1.41) goes to zero as $N \to \infty$ but we omit the details.

$$\lim_N E[\gamma_N^2] = \left(\int_{-1/2}^{1/2} H(\lambda) p(\lambda) \, d\lambda \right)^2 = (E[\gamma_N])^2$$

or,

$$\gamma_N \to E[\gamma_N] \qquad \text{as } N \to \infty.$$

proving the Ergodic Principle.

5.1.6   Example: Delta Function Spectral Density

Next let us consider what happens when the spectral density consists of δ-functions. Thus let us specialize to the case

$$x_k = A \cos(2\pi\lambda_0 k + \phi), \qquad -1/2 \leq \lambda_0 \leq 1/2$$

corresponding to one term in (5.1.28) with

$$p(\lambda) = \frac{E[A^2]}{4} [\delta(\lambda - \lambda_0) + \delta(\lambda + \lambda_0)].$$

For any $\lambda \neq \pm\lambda_0$, we have

$$\zeta_N(\lambda) = \frac{1}{\sqrt{N+1}} \sum_0^N x_k e^{2\pi i \lambda k}$$

$$= \frac{A}{2\sqrt{N+1}} \left(\sum_0^N e^{2\pi i (\lambda + \lambda_0)k + i\phi} + \sum_0^N e^{2\pi i (\lambda - \lambda_0)k - i\phi} \right)$$

$$= \frac{A}{2\sqrt{N+1}} \left(\frac{e^{i\phi}(1 - e^{2\pi i (\lambda + \lambda_0)(N+1)})}{1 - e^{2\pi i (\lambda + \lambda_0) + i\phi}} + \frac{e^{-i\phi}(1 - e^{2\pi i (\lambda - \lambda_0)(N+1)})}{1 - e^{2\pi i (\lambda - \lambda_0) + i\phi}} \right)$$

$$\to 0 \qquad \text{as } N \to \infty. \tag{5.1.42}$$

Hence, as we have shown more generally,

$$|\zeta_N(\lambda)|^2 = \frac{1}{N+1}\left|\sum_0^N x_k e^{2\pi i\lambda k}\right|^2 \to 0 \qquad \text{as } N \to \infty \text{ for } \lambda \neq \pm\lambda_0.$$

On the other hand, for $\lambda = \pm\lambda_0$, one of the sums in (5.1.42) becomes $(N+1)$ and hence

$$\frac{1}{N+1}\left|\sum_0^N x_k e^{2\pi i\lambda k}\right|^2 \to \infty \qquad \text{as } N \to \infty, \lambda = \pm\lambda_0.$$

In fact we can see that

$$\zeta_N(\lambda) = \frac{A}{2}\left(e^{i\phi} f_N(\lambda - \lambda_0) + e^{-i\phi} f_N(\lambda + \lambda_0)\right),$$

where

$$f_N(\lambda) = \frac{1}{\sqrt{N+1}}\left(\sum_0^N e^{2\pi ik\lambda}\right),$$

and we saw in the proof of Theorem 2.8 in Chapter 2 that

$$|f_N(\lambda)|^2$$

is a δ-function sequence in N. Now

$$|\zeta_N(\lambda)|^2 = \frac{A^2}{4}\left(|f_N(\lambda - \lambda_0)|^2 + |f_N(\lambda + \lambda_0)|^2\right.$$

$$+ e^{2i\phi} f_N(\lambda - \lambda_0)\,\overline{f_N(\lambda + \lambda_0)}$$

$$\left. + e^{-2i\phi} f_N(\lambda + \lambda_0)\,\overline{f_N(\lambda - \lambda_0)}\right). \qquad (5.1.43)$$

For $\lambda_0 \neq 0$, the third and fourth terms in (5.1.43) go to zero and hence

$$|\zeta_N(\lambda)|^2$$

is a δ-function sequence, converging to

$$\frac{A^2}{4}\left(\delta(\lambda - \lambda_0) + \delta(\lambda + \lambda_0)\right).$$

For $\lambda_0 = 0$,

$$|\zeta_N(\lambda)|^2 = \frac{A^2}{4}|f_N(\lambda)|^2[2 + 2\cos\phi]$$

and converges to

$$\frac{A^2}{4}(1 + \cos \phi)\delta(\lambda).$$

Thus the Ergodic Principle fails at $\lambda = \pm\lambda_0$ for the spectral density estimate. It also fails for (5.1.37) in this case!

Finally we should note the difference between the discrete Fourier transform estimate (5.1.31) and the transform of a ("sample") covariance function obtained by taking time-average estimates

$$\sum_{-\infty}^{\infty} r_k \cos 2\pi\lambda k, \tag{5.1.44}$$

where r_k is the estimate for the covariance $R(k)$, and must perforce be taken to be zero for $|k| > N$, for some N depending on the data length. To illustrate this, $|\zeta_N(\lambda)|$ for $N = 1000$ and $N = 2000$ in Figures 5.1 and 5.2, respectively, for a Gaussian process $\{x_k\}$ with zero mean and covariance function

$$R(m) = (0.9)^{|m|}$$

We show the sample covariance $\{r_k\}$ for $k = 0, \ldots, 100$ obtained using 2000 samples and the true covariance function for comparison in Figure 5.3. Note the close agreement for k. Finally, Figure 5.4 shows the spectral density using the sample covariance setting

$$r_k = 0 \qquad \text{for } |k| > 100.$$

Note the close agreement with the true spectral density.

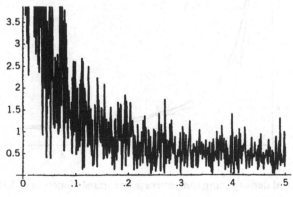

Figure 5.1. $|\zeta_N(\lambda)|$ based on $N = 1000$ samples. $\rho = 0.9$.

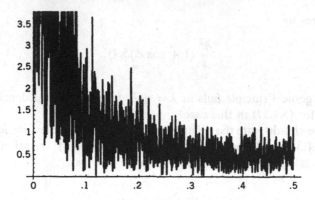

Figure 5.2. $|\zeta_N(\lambda)|$ based on $N = 2000$ samples. $\rho = 0$.

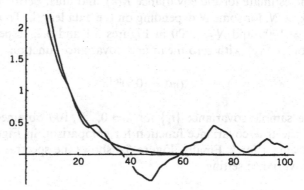

Figure 5.3. Time average covariance function based on 2000 samples. $\rho = 0.9$. Compare with the true covariance function (smooth curve).

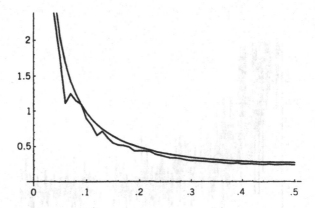

Figure 5.4. Spectral density using time-average covariance function of 5.10. True spectral density (smooth curve).

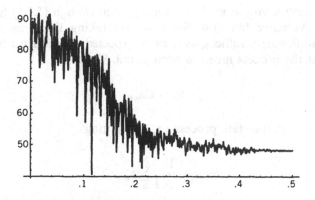

Figure 5.5. Curve for $20 \log_{10} |\zeta_N(\lambda)|$. Corresponding to Figure 1.8; $N = 1000$.

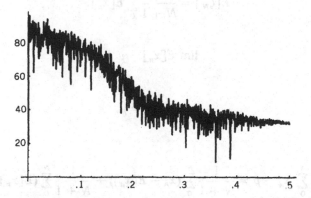

Figure 5.6. Curve for $20 \log_{10} |\zeta_N(\lambda)|$. Corresponding to Figure 1.8; $N = 2000$.

In many applications, extensive use is made of spectral density estimates of physically observed data ("real" data) known to be modellable (from physical theory) as sample paths of a stationary random process. Without going into the details of such procedures, we shall illustrate again the dependence on sample size using the laser beam scintillation data shown in Figure 1.8. Figures 5.5 and 5.6 plot $20 \log_{10} |\zeta_N(\lambda)|$ for $N = 1000$ and $N = 2000$, respectively (actually for $\lambda = n/N$, $0 \le n \le N/2$). The appearance is typical of "noiselike" data. We may make a similar estimate for the rainfall time series in Figure 1.9, but any conclusions drawn (any functionals calculated) must be tempered by the fact that the data length may not be adequate even if the stationarity label is reasonable.

5.1.7 Signal Generation Models

Let us now specialize to processes generated by state-space models—the Kalman signal generation model (3.2.23). As we recall, for a stable system such a process is only asymptotically stationary, even though (5.1.5) holds for the stationary covariance function. Since we are taking limits as $N \to \infty$, the asymptotic stationarity suffices, as may be expected. For illustrative purposes, let us look at the process mean in some detail. Since

$$v_n = Cx_n,$$

we may work with the state process $\{x_n\}$. Denoting

$$\frac{1}{N+1} \sum_0^N x_n$$

by ζ_N as before, we have

$$E[\zeta_N] = \frac{1}{N+1} \sum_0^N E[x_n].$$

Now letting

$$\lim_n E[x_n] = \mu$$

we see that

$$\lim_N E[\zeta_N] = \mu.$$

Moreover,

$$\frac{1}{N+1} \sum_0^N x_n - \mu = \frac{1}{N+1} \sum_0^N (x_n - E[x_n]) + \frac{1}{N+1} \sum_0^N (E[x_n] - \mu)$$

and the second sum on the right goes to zero, while the covariance of the first term

$$= \frac{1}{(N+1)^2} \sum_0^N \sum_0^N R(n, m). \tag{5.1.45}$$

Now in the notation of Chapter 3,

$$R(n, m) = R(n, n)A^{n-m} = R_\infty A^{n-m} + (R_\infty - R(n, n))A^{n-m}. \tag{5.1.46}$$

Here the average (5.1.45) using the first term in (5.1.46) corresponding to the stationary covariance goes to zero, while the norm of the second term for all n, m is given by

$$\|(R_\infty - R(n, n))A^{n-m}\|_0 \leq \|(R_\infty - R(n, n))\|_0 \|A\|_0^{|n-m|}$$
$$\to 0$$

for each n as $m \to \infty$, and for each m as $n \to \infty$. Hence the average also goes to zero. However, we see that the second term slows the convergence and depends on the covariance of the initial condition, and thus the "correlation time" based on the steady-state covariance is inadequate. Nevertheless, it is clear that the smaller the spectral radius of A, the faster the convergence.

5.2 CONTINUOUS-TIME MODELS

The results (as well as techniques) for continuous-time models parallel those for discrete time. For this reason, and for the reason that continuous-time integrals have to be approximated by discrete sums anyway, we shall be brief.

Thus let $x(t)$, $-\infty < t < \infty$, be a second-order stationary process with mean μ and covariance function $R(\cdot)$. Then the "time average" or "sample mean" is defined by

$$\zeta_T = \frac{1}{T} \int_0^T x(s)\, ds, \tag{5.2.1}$$

where, as in Chapter 4, we have the option of pathwise integrals or interpreting them in the mean square sense. Thus we will need to assume that the stationary covariance function $R(t)$ is continuous in t. In that case, however defined,

$$E[\zeta_T] = \frac{1}{T} \int_0^T E[x(s)]\, ds = \mu$$

and

$$E[(\zeta_T - \mu)(\zeta_T - \mu)^*] = \frac{1}{T^2} \int_0^T \int_0^T R(t - s)\, ds\, dt,$$

which, as in Chapter 3, we can rewrite as a single integral

$$\Lambda(T) = \frac{1}{T} \int_{-T}^T R(s)\left(1 - \frac{|s|}{T}\right) ds. \tag{5.2.2}$$

Hence we have mean square convergence of the time average to the phase average if (5.2.2) goes to zero as $T \to \infty$. Now $\Lambda(T)$ being nonnegative definite, we know (cf. Problem 2.16)) that $\Lambda(T)$ goes to zero if and only if

$$\text{Tr } \Lambda(T) \to 0$$

or, equivalently,

$$\frac{1}{T} \int_{-T}^T \text{Tr } R(s)\left(1 - \frac{|s|}{T}\right) ds \to 0.$$

Hence a sufficient condition for (5.2.2) to go to zero is that

$$\text{Tr } R(t) \to 0 \qquad \text{as } t \to \infty, \tag{5.2.3}$$

analogous to (5.1.5). Of course (5.2.3) is automatically satisfied if the stronger condition (3.2.23) holds. Because the analog of the considerations in the discrete case is transparent, we shall conclude here by stating merely the analog of the key Theorem 5.1.1.

Theorem 5.2.1. Let $x(t)$, $-\infty < t < \infty$, be a stationary Gaussian process with continuous covariance function $R(\cdot)$ for which (5.2.3) holds. Let $g(\cdot)$ be any function mapping $\mathbf{R}^n$ into $\mathbf{R}^1$ such that

$$E[g(x(t))^2] < \infty. \tag{5.2.4}$$

Then the Ergodic Principle holds in the mean square sense for $g(\cdot)$:

$$\frac{1}{T}\int_0^T g(x(t))\, dt \tag{5.2.5}$$

converges in the mean square sense to

$$E[g(x(t))]. \tag{5.2.6}$$

Proof. We shall omit the proof, since it is a paraphrase of the proof of Theorem 5.1.1.

As in the discrete-time case, pathwise convergence holds for (5.2.2) for Gaussian processes—see reference 1 for these mathematical refinements.

5.2.1 Signal Generation Models

If the process is defined by a state-space signal generation model (4.4.1) for a stable system, we have stationarity only asymptotically, even though the stationary process does satisfy (5.2.3). As in the discrete case, nevertheless, the time average (5.2.5) converges to the steady-state average

$$\lim_{t \to \infty} E[x(t)].$$

The details are the same as in the discrete-time case and are omitted.

Example 5.2.1. Let $\{x_n\}$ be a 1×1 Gaussian process with zero mean and stationary covariance function

$$R(n) = \rho^{|n|}, \qquad 0 \le \rho < 1.$$

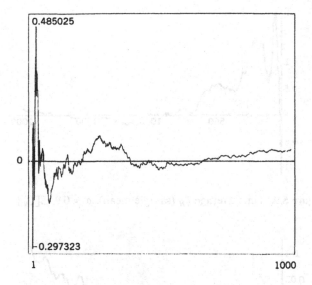

Figure 5.7. Time average ζ_N. $\rho = 0$, $E[\zeta_N] = 0$.

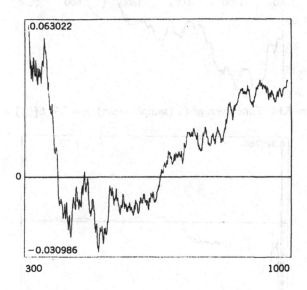

Figure 5.8. Time average ζ_N. $\rho = 0$, $E[\zeta_N] = 0$.

First consider the time average

$$\zeta_N = \frac{1}{N+1} \sum_0^N x_k.$$

We note that

$$\Pr\left[|\zeta_N| > 1/10\right] \leq 10^{-3}$$

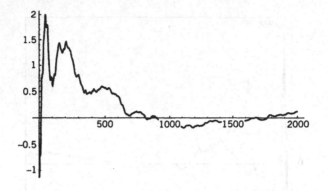

Figure 5.9. Time average ζ_N (sample mean). $\rho = 0.9$, $E[\zeta_N] = 0$.

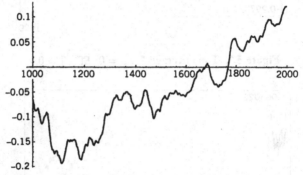

Figure 5.10. Time average ζ_N (sample mean). $\rho = 0.9$, $E[\zeta_N] = 0$.

Figure 5.11. Time average ζ_N. $\rho = 0$, $E[\zeta_N] = \rho$.

Figure 5.12. Time average ζ_N. $\rho = 0$, $E[\zeta_N] = \rho$.

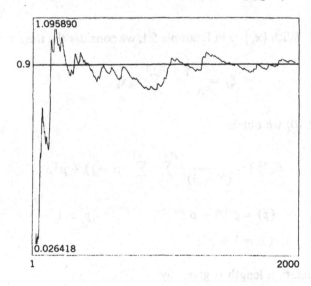

Figure 5.13. Time average ζ_N. $\rho = 0.9$, $E[\zeta_N] = \rho$.

for

$$N > \frac{900}{1 - \rho^2}.$$

The behaviour of ζ_N for various values of ρ is shown in Figures 5.7 and 5.8 for $\rho = 0$ and Figures 5.9 and 5.10 for $\rho = 0.9$. Figures 5.8 and 5.10 show the approach to the limit, and Figures 5.7 and 5.9 show the early transients.

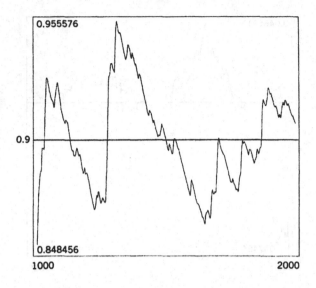

Figure 5.14. Time average ζ_N. $\rho = 0.9$, $E[\zeta_N] = \rho$.

Example 5.2.2. With $\{x_n\}$ as in Example 5.1, we consider the time average for ρ:

$$\zeta_N = \frac{1}{N+1} \sum_1^{N+1} x_k x_{k-1}.$$

Here from (5.1.10) we obtain

$$E[\zeta_N^2] = \frac{1}{(N+1)^2} \sum_1^{N+1} \sum_1^{N+1} r(i-j) + \rho^2,$$

where

$$r(p) = \rho^{2|p|} + \rho^{|p+1|+|p-1|}, \qquad |p| > 1.$$
$$r(0) = 1 + \rho^2,$$

Hence the correlation length is given by

$$\bar{N} = \frac{1}{r(0)} \sum_{-\infty}^{\infty} r(p)$$

$$= \frac{1}{(1+\rho^2)} \left(\sum_{-\infty}^{\infty} \rho^{2|p|} + \sum_1^{\infty} \rho^{2p} + \sum_{-\infty}^{-1} \rho^{2|p|} + \rho^2 \right)$$

$$= \frac{1 + 4\rho^2 - \rho^4}{1 - \rho^4}.$$

(For $\rho = 0$

$$E[\zeta_N^2] = \frac{1}{N+1}, \qquad E[\zeta_N] = 0.)$$

Figures 5.11 and 5.12 show the behavior of ζ_N for $\rho = 0$, and Figures 5.13 and 5.14 show it for $\rho = 0.9$. Since ζ_N is not Gaussian, we have to fall back on the Chebyshev inequality. Note that the value of N for a given accuracy is much larger in this case.

PROBLEMS

5.1. Show that we may rewrite the time average (5.1.1) in the "two-sided average" form

$$\zeta_N = \frac{1}{2N+1} \sum_{-N}^{N} x_k$$

and that the asymptotic properties would be identical.

5.2. Show that (5.1.4) goes to zero if (5.1.5) holds.
Hint:

$$\left\| \frac{1}{N+1} \sum_{-N}^{N} R(k) \left(1 - \frac{|k|}{N+1} \right) \right\|_0 \leq \frac{1}{N+1} \sum_{-N}^{N} \|R(k)\|_0.$$

Choose N_0 so that

$$\|R(k)\|_0 < \varepsilon \qquad \text{for } |N| > N_0.$$

Then

$$\frac{1}{1 + N_0 + p} \sum_{-N_0 - p}^{N_0 + p} \|R(k)\|_0 \leq \frac{1}{N_0 + p} \left(\sum_{-N_0}^{N_0} \|R(k)\|_0 + 2p\varepsilon \right).$$

5.3. Calculate the spectral density of the process defined by (5.1.9) in terms of that of the process $\{x_k\}$, and calculate the "correlation time."

5.4. Let $\{x_k\}$ be a stationary Gaussian process whose covariance function is nonsingular and satisfies (5.1.5). Let $g_1(\cdot), g_2(\cdot)$ be two step functions defined as in (5.1.12). Show that

$$E[g_1(x_k)g_2(x_{k+p})] \to E[g_1(x_k)]E[g_2(x_k)] \qquad \text{as } p \to \infty.$$

5.5. Given a (one-by-one) process $\{x_k\}$ such that

$$x_k = \mu + N_k, \qquad |\mu| > 0$$

and $\{N_k\}$ is white Gaussian with unit variance, determine how large N should be in order that the sample mean

$$\frac{1}{N}\sum_1^N x_k$$

has less than 5% error, with probability higher than $(1 - 0.001)$. Explain the dependence on the value of μ.

5.6. Calculate the correlation time for the process

$$y_n = e^{x_n},$$

where $\{x_n\}$ is stationary Gaussian with zero mean and stationary covariance function

$$R(m) = \rho^{|m|}.$$

Answer:

$$\bar{N} = \left(\frac{1}{e - 1}\right)\left(\sum_1^\infty \left(\frac{1 + \rho^n}{1 - \rho^n}\right)\frac{1}{n!}\right).$$

5.7. Calculate the correlation time for the process

$$y_n = x_n x_{n+k}$$

with $\{x_n\}$ as in Example 5.1, for arbitrary integer $k \geq 0$.
  Answer:

$$\frac{1 + \rho^2 + (2k + 1)\rho^{2k} - (2k - 1)\rho^{2k+2}}{(1 - \rho^2)(1 + \rho^{2k})}.$$

Let $\bar{N}(k)$ denote the correlation time as a function of k. Show that

$$\lim_{k \to \infty} \bar{N}(k) = \bar{N}(0).$$

A plot of $\bar{N}(k)$ for $\rho = 0.9$ is shown in Figure 5.15. Note the peak at $k = 6$.

5.8. Let $\{x_n\}$ be a 1×1 zero mean stationary Gaussian process with spectral density $p(\cdot)$. Show that the correlation time for the process

$$y_n = x_n x_{n-k}, \qquad k \text{ positive integer}$$

can be expressed as

$$\bar{N} = \frac{\displaystyle\int_{-1/2}^{1/2} (1 + \cos 4\pi k\lambda)p(\lambda)^2 \, d\lambda}{\displaystyle\left(\int_{-1/2}^{1/2} p(\lambda) \, d\lambda\right)^2 + \left(\int_{-1/2}^{1/2} \cos 2\pi k\lambda \, p(\lambda) \, d\lambda\right)^2}.$$

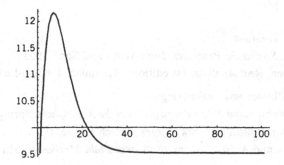

Figure 5.15. Correlation time $\bar{N}(k)$. $\rho = 0.9$, $\bar{N}(0) = 9.5$.

5.9. Show that the correlation time of the output of a time-invariant linear system with finite memory M excited by white noise is less than or equal to $(2M - 1)$ or $(2M + 1)$ depending on how memory is defined.

Hint: Use Schwarz inequality.

5.10. For the process $\{x_k\}$ as in Example 5.1, show how you would use the Ergodic Principle to calculate the probability

$$\Pr [x_k \leq 3]$$

from the process samples. How large a sample is needed to make the odds less than one in a thousand to have an error of more than 10 percent?

5.11. Construct an example of a process with covariance function $R(\cdot)$ such that

$$R(k) = 0, \qquad |k| \geq M$$

for which the correlation time is zero. What does it imply for the time-average estimate of the mean of the process?

NOTES AND COMMENTS

See references 3 and 4 for more on the use of ergodicity. For more on the correlation time concept see references 4 and 5. As befitting a course on second-order theory, we have limited our discussion of the Ergodic Principle to mean square convergence. Of course pathwise or probability one convergence is of great importance, but mathematical tools are beyond our scope and a mathematical treatise like reference 1 should be consulted for more on this aspect.

The approximation by simple functions is a standard topic in real analysis [2].

REFERENCES

Mathematical Treatises
1. J. L. Doob. *Stochastic Processes.* John Wiley and Sons, 1953.
2. R. L. Royden. *Real Analysis,* 1st edition. Macmillan, 1963. 3rd edition, 1988.

Applications: Physics and Engineering
3. J. W. Strohbehn. *Laser Beam Propagation in the Atmosphere.* Springer-Verlag, 1978.
4. Y. Weissman. *Optical Network Theory.* Artech House, 1992.
5. A. S. Monin and A. M. Yaglom. *Statistical Fluid Mechanics.* M.I.T. Press, 1971.

6

SAMPLING PRINCIPLE
AND INTERPOLATION

For any digital computer processing, all analog data have to be first sampled in time, usually at a fixed rate, however large. In other words, given a signal $x(t)$ definable in principle in continuous time—for every value of t in any interval—the A-D converter first takes discrete-time samples ("sample hold") every fixed interval or "sampling period"—call it Δ. Thus the computer only retains the discrete-time samples

$$x_n = x(n\Delta), \qquad n = 0, 1, 2, \ldots$$

(which are then quantized, but given modern processors; this further discretization is fine enough so that we may (and shall!) ignore it for our purposes). The first question then is: How faithful is this operation? What is the "degradation" due to the sampling process?

It was C. E. Shannon in 1948 [1] who gave the first answer to this question, introducing the notion of "band-limited" signals and incidentally laying the foundation for digital communications, and the "sampling principle" is named after him even though the mathematical theory ("cardinal series") predates his work (see reference 2).

Let us recall first the Sampling Principle for deterministic signals (see any standard text such as references 12 and 14 of the Review Chapter). Let $r(t)$, $-\infty < t < \infty$, be an $n \times 1$ function such that its Fourier transform

$$\psi(f) = \int_{-\infty}^{\infty} e^{-2\pi i f t} r(t) \, dt$$

vanishes outside $[-W, W]$:

$$\psi(f) = 0, \qquad |f| > W.$$

We say such a function is "band-limited" and the "bandwidth" is $2W$. Then the Shannon "Sampling Principle" is that the function is "determined" by the "samples" at $t = n/2W$, $n \in I$. More specifically, we have the "representation"

$$r(t) = \sum_{-\infty}^{\infty} a_n(t) r\left(\frac{n}{2W}\right), \qquad -\infty < t < \infty,$$

where

$$a_n(t) = \frac{\sin \pi(2Wt - n)}{\pi(2Wt - n)}.$$

In other words, the function values between samples may be "interpolated" from the values at the sample points.

Here we have to deal with signals which are modeled as random processes, and hence the Sampling Principle has to be recast accordingly. Recall that a stationary random process is band-limited to $[-W, W]$ if the spectral density vanishes outside this frequency interval. The "stochastic" Sampling Principle is embodied in the following theorem:

Theorem 6.1. Let $x(t)$, $-\infty < t < \infty$, denote an $n \times 1$ second-order stationary stochastic process which is band-limited to $[-W, W]$. Then

$$x(t) = \sum_{-\infty}^{\infty} a_n(t) x(n\Delta), \tag{6.1}$$

where

$$a_n(t) = \frac{\sin \pi(2Wt - n)}{\pi(2Wt - n)} \tag{6.2}$$

and

$$\Delta = \frac{1}{2W}. \tag{6.3}$$

The convergence of the series is in the mean square sense. That is,

$$E\left[\left(x(t) - \sum_{-N}^{N} a_n(t) x(n\Delta)\right)\left(x(t) - \sum_{-N}^{N} a_n(t) x(n\Delta)\right)^*\right] \tag{6.4}$$

goes to zero for each t as $N \to \infty$.

Proof. Let μ denote the mean and $R(\cdot)$ the stationary covariance function of the process. Let

$$\zeta_N(t) = x(t) - \sum_{-N}^{N} a_n(t) x(n\Delta). \tag{6.5}$$

Then

$$E[\zeta_N(t)] = \mu - \left(\sum_{-N}^{N} a_n(t)\right)\mu. \tag{6.6}$$

Now for each t expand $e^{2\pi i f t}$ as a function of f in $[-W, W]$ in a Fourier series. Thus

$$\frac{1}{2W}\int_{-W}^{W} e^{2\pi i f t - (i\pi n f /W)}\, df = \frac{\sin \pi(2Wt - n)}{\pi(2Wt - n)} = a_n(t). \tag{6.7}$$

Hence

$$e^{2\pi i f t} = \sum_{-\infty}^{\infty} a_n(t)e^{(i\pi n f /W)}. \tag{6.8}$$

Thus $a_n(t)$ are the Fourier coefficients of the function

$$e^{2\pi i f t}$$

over the interval $[-W, W]$. From (6.8) we see that, taking $f = 0$,

$$\sum_{-\infty}^{\infty} a_n(t) = 1$$

and hence

$$E[\zeta_N(t)] = \mu\left(1 - \sum_{-N}^{N} a_n(t)\right) \qquad \text{as } N \to \infty.$$

Hence we only need consider the covariance to be

$$\Lambda_N(t) = E[(\zeta_N(t) - E[\zeta_N(t)])(\zeta_N(t) - E[\zeta_N(t)])^*]$$

$$= E\left[\left(\tilde{x}(n\Delta) - \sum_{-N}^{N} a_n(t)\tilde{x}(n\Delta)\right)\left(\tilde{x}(n\Delta) - \sum_{-N}^{N} a_n(t)\tilde{x}(n\Delta)\right)^*\right], \tag{6.9}$$

where

$$\tilde{x}(n\Delta) = x(t) - E[x(t)].$$

Carrying out the multiplication in (6.9), we obtain

$$\Lambda_N(t) = R(0) + \sum_{-N}^{N}\sum_{-N}^{N} a_n(t)R((n - m)\Delta)a_m(t)$$

$$- \sum_{-N}^{N} a_n(t)(R(t - n\Delta) + R(t - n\Delta)^*). \tag{6.10}$$

We shall now assume that the spectral density of the process does not contain any δ-functions. (For a more general proof without assuming this, see reference 3.) Then

$$R(t) = \int_{-W}^{W} e^{2\pi i f t} P(f) \, df,$$

and after substituting this into (6.10) we have

$$\Lambda_N(t) = \int_{-W}^{W} \left(I + \sum_{-N}^{N} \sum_{-N}^{N} e^{2\pi i f(n-m)\Delta} a_n(t) a_m(t) \right.$$

$$\left. - \sum_{-N}^{N} a_n(t) (e^{2\pi i f(t-n\Delta)} + e^{-2\pi i f(t-n\Delta)}) \right) P(f) \, df \quad (6.11)$$

and the quantity in parentheses

$$= \left| e^{2\pi i f t} - \sum_{-N}^{N} e^{2\pi i f n \Delta} a_n(t) \right|^2$$

so that

$$\Lambda_N(t) = \int_{-W}^{W} \left| e^{2\pi i f t} - \sum_{-N}^{N} e^{2\pi i f n \Delta} a_n(t) \right|^2 P(f) \, df. \quad (6.12)$$

But from (6.8) we see that the integrand goes to zero for each f ("boundedly," a technicality) so that the limit of the integral is the integral of the limit and hence zero. This proves the theorem.

As in the deterministic case, we may look upon (6.1) as yielding an interpolation formula for "filling in data" between sampling points. For any t which modulo Δ can be expressed for some k as

$$t = k\Delta + \theta, \qquad 0 < \theta < \Delta,$$

we have that

$$x(k\Delta + \theta) = \sum_{n=-\infty}^{\infty} a_n(k\Delta + \theta) x(n\Delta)$$

and

$$a_n(k\Delta + \theta) = \frac{\sin \pi(2W\theta - (n-k))}{\pi(2W\theta - (n-k))}.$$

Hence

$$x(k\Delta + \theta) = \sum_{-\infty}^{\infty} a_n(\theta) x((n+k)\Delta) \quad (6.13)$$

and since

$$|a_n(\theta)| = \left| \frac{\sin \pi(2W\theta - n)}{\pi(2W\theta - n)} \right| \leq \left| \frac{1}{\pi(n - 2W\theta)} \right|$$

and decreases rapidly as $|n|$ increases, we see that (6.1) is a "two-sided" interpolation formula:

$$x(k\Delta + \theta) = \frac{\sin \pi 2W\theta}{\pi 2W\theta} x(k\Delta) + a_1(\theta)x((k+1)\Delta) + a_{-1}(\theta)x((k-1)\Delta) + \cdots$$

Also, if the process is Gaussian we can replace mean square convergence by sample path (probability one) convergence.

There are other interpolation formulas using the samples $\{x(n\Delta)\}$ such as the Newton–Gauss formula using successive differences—see reference 2.

The main point is that these interpolation formulas do not require any detailed knowledge of the process, such as the covariance or spectral density; only the bandwidth is required, assuming that the spectral density is band-limited.

Even if the process is *not* band-limited, it may be approximately so. In fact, from

$$R(0) = \int_{-\infty}^{\infty} P(f) \, df$$

we may pick a W such that $P(f)$ is essentially zero outside $[-W, W]$. In such a case it is useful to know what the error is in using the Shannon interpolation formula.

Theorem 6.2. Let $x(t)$, $-\infty < t < \infty$, be any (zero mean, for simplicity) stationary process with covariance function $R(\cdot)$ and spectral density $P(\cdot)$. Then the series

$$\hat{x}(t) = \sum_{-\infty}^{\infty} a_n(t) x\left(\frac{n}{2W}\right) \tag{6.14}$$

with $a_n(\cdot)$ given by (6.2) converges in the mean square for any choice of W. The process $\hat{x}(t)$, so defined, is stationary and band-limited to $[-W, W]$ with spectral density ("folded" or "aliasing"), denoted $\hat{P}(f)$, given by

$$= \sum_{k=-\infty}^{\infty} P(f + 2kW), \quad |f| < W$$

$$= 0, \quad |f| > W. \tag{6.15}$$

Moreover, the interpolation error-covariance ("aliasing" error) is given by

$$E[(x(t) - \hat{x}(t))(x(t) - \hat{x}(t))^*] \le 4 \int_W^\infty (P(f) + P(-f)) \, df$$

and the "normalized" mean square error is expressed as

$$\frac{E[\|x(t) - \hat{x}(t)\|^2]}{E[\|x(t)\|^2]} \le 4 \int_W^\infty \mathrm{Tr} \, P(f) \, df \bigg/ \int_0^\infty \mathrm{Tr} \, P(f) \, df. \tag{6.16}$$

Proof. To prove first the mean square convergence of (6.14) let us consider the "tails" for any $N > 0$:

$$E\left(\left\| \sum_N^\infty a_n(t) x\left(\frac{n}{2W}\right) \right\|^2 + \left\| \sum_{-\infty}^{-N} a_n(t) x\left(\frac{n}{2W}\right) \right\|^2 \right)$$

$$= \sum_N^\infty \sum_N^\infty + \sum_{-\infty}^{-N} \sum_{-\infty}^{-N} a_n(t) R\left(\frac{n}{2W} + \frac{m}{2W}\right) a_m(t)$$

$$= \int_{-\infty}^\infty P(f) \left\| \sum_N^\infty + \sum_{-\infty}^{-N} a_n(t) e^{(\pi i f n/W)} \right\|^2 df$$

$$= \sum_{-\infty}^\infty \int_{-W}^W P(f + 2kW) \left| \sum_N^\infty + \sum_{-\infty}^{-N} a_n(t) e^{(\pi i f n/W)} \right|^2 df$$

$$= \sum_{-\infty}^\infty \int_{-W}^W P(f + 2kW) \left| e^{2\pi i f t} - \sum_{-N+1}^{N-1} a_n(t) e^{(\pi i f n/W)} \right|^2 df.$$

From the convergence of the Fourier series on the right of (6.8), it follows as in (6.12) that the integrand goes to zero as $N \to \infty$. This is enough to establish the mean convergence of (6.14). Thus defined let us calculate the covariance function

$$E[\hat{x}(s + t)\hat{x}(s)^*]$$

$$= \sum_{-\infty}^\infty \sum_{-\infty}^\infty a_n(s + t) E[x(n\Delta) x(m\Delta)^*] a_m(s)$$

$$= \sum_{-\infty}^\infty \sum_{-\infty}^\infty a_n(s + t) R((n - m)\Delta) a_m(s)$$

$$= \int_{-\infty}^\infty \left(\sum_{-\infty}^\infty \sum_{-\infty}^\infty a_n(s + t) e^{2\pi i f(n-m)\Delta} a_m(s) \right) P(f) \, df$$

$$= \int_{-\infty}^\infty \left(\sum_{-\infty}^\infty a_n(s + t) e^{2\pi i n f/(2W)} \right) \overline{\left(\sum_{-\infty}^\infty a_m(s) e^{2\pi i f m/(2W)} \right)} P(f) \, df.$$

Now the factors in parentheses in the integrand are periodic in f, with period $2W$, and

$$= \sum_{-\infty}^{\infty} a_n(s)e^{2\pi i n f/(2W)} = e^{2\pi i f s}, \qquad -W < f < W.$$

Hence

$$E[\hat{x}(s+t)\hat{x}(s)^*] = \sum_{k=-\infty}^{\infty} \int_{-W}^{W} e^{2\pi i f(s+t)} e^{-2\pi i f s} P(f + 2kW)\, df$$

$$= \int_{-W}^{W} e^{2\pi i f t} \sum_{-\infty}^{\infty} P(f + 2kW)\, df.$$

Hence the process $\hat{x}(\cdot)$ is second-order stationary with zero-mean, band-limited, spectral density $\hat{P}(\cdot)$ given by (6.15), as required. Next let

$$z(t) = x(t) - \hat{x}(t). \tag{6.17}$$

Then

$$z(t) = 0 \qquad \text{for } t = \frac{n}{2W}.$$

More generally,

$$E[z(t)z(t)^*] = E[x(t)x(t)^*] + E[\hat{x}(t)\hat{x}(t)^*] - E[x(t)\hat{x}(t)^*] - E[x(t)^*\hat{x}(t)]. \tag{6.18}$$

Now

$$E[\hat{x}(t)\hat{x}(t)^*] = \int_{-W}^{W} \sum_{-\infty}^{\infty} P(f + 2kW)\, df$$

$$= \int_{-\infty}^{\infty} P(f)\, df = E[x(t)x(t)^*] \tag{6.19}$$

and

$$E[x(t)\hat{x}(t)^*] = \sum_{-\infty}^{\infty} a_n(t) R\left(t - \frac{n}{2W}\right)$$

$$= \int_{-\infty}^{\infty} e^{2\pi i f t} \left(\sum_{-\infty}^{\infty} a_n(t) e^{-2\pi i f n/W}\right) P(f)\, df$$

and

$$E[x(t)^*\hat{x}(t)] = \int_{-\infty}^{\infty} e^{-2\pi i f t} \left(\sum_{-\infty}^{\infty} a_n(t) e^{i\pi f n/(2W)}\right) P(f)\, df.$$

Hence

$$E[z(t)z(t)^*] = \int_{-\infty}^{\infty} \left(2 - e^{2\pi i f t}\psi(t, f) - e^{-2\pi i f t}\overline{\psi(t, f)}\right)P(f)\, df$$

$$= \int_{-\infty}^{\infty} |1 - e^{2\pi i f t}\psi(t, f)|^2 P(f)\, df,$$

where

$$\psi(t, f) = \sum_{-\infty}^{\infty} a_n(t)e^{-i\pi f n/W},$$

which is periodic in f for each t, with period $2W$, and

$$\psi(t, f) = e^{-2\pi i f t} = \psi(t, f + 2kW), \qquad -W < f < W.$$

Hence after breaking up the integral as in (2.38), we have

$$E[z(t)z(t)^*] = \int_{-W}^{W} \sum_{-\infty}^{\infty} |1 - e^{4\pi i k W t}|^2 P(f + 2kW)\, df$$

$$= 4\int_{-W}^{W} \sum_{-\infty}^{\infty} (\sin^2 2\pi k W t) P(f + 2kW)\, df. \qquad (6.20)$$

Hence

$$E[z(t)z(t)^*] \le 4\int_{W}^{\infty} (P(f) + P(-f))\, df$$

and

$$E[\|z(t)\|^2] \le 8\int_{W}^{\infty} \text{Tr}\, P(f)\, df,$$

as required, with (6.16) an immediate consequence.

We see from Theorem 6.2 that the Shannon interpolation formula breaks down if the process is not band-limited. This naturally raises the question as to whether there are other interpolation formulas which hold more generally. In order words, we may inquire whether we can find scalar, real-valued functions $a_n(\cdot)$ such that for t not of the form $m/2W$, the error

$$E\left[\left\|x(t) - \sum_{-\infty}^{\infty} a_n(t)x\left(\frac{n}{2W}\right)\right\|^2\right] \qquad (6.21)$$

is actually zero. As we shall show, the answer is in the negative.

Indeed, we can calculate that (6.21)

$$= \text{Tr}\left[R(0) + \sum_{-\infty}^{\infty}\sum_{-\infty}^{\infty} a_n(t)a_m(t)R\left(\frac{n}{2} - \frac{m}{2W}\right)\right.$$

$$\left. - \sum_{-\infty}^{\infty} a_n(t)\left(R\left(\frac{n}{2W} - t\right) + R\left(t - \frac{n}{2W}\right)\right)\right], \qquad (6.22)$$

which, in terms of the spectral density, becomes

$$= \int_{-\infty}^{\infty} (1 - |\psi(t, f)|^2 - \psi(t, f)e^{-2\pi i f t} - \overline{\psi(t, f)}e^{2\pi i f t}) \operatorname{Tr} P(f) \, df,$$

where, as before,

$$\psi(t, f) = \sum_{-\infty}^{\infty} a_n(t)e^{i\pi n f/W}$$

and the factor in parentheses in the integrand

$$= |1 - \psi(t, f)e^{-2\pi i f t}|^2$$

so that the mean square error

$$= \int_{-\infty}^{\infty} |1 - \psi(t, f)e^{-2\pi i f t}|^2 \operatorname{Tr} P(f) \, df.$$

Now this integral is zero if and only if

$$|1 - \psi(t, f)e^{-2\pi i f t}| = 0 \qquad (6.23)$$

when

$$\operatorname{Tr} P(f) \neq 0,$$

but

$$e^{2\pi i f t} = \psi(t, f) = \sum_{-\infty}^{\infty} a_n(t)e^{i\pi n f/W} \qquad ((6.24)$$

in any interval $-W < f < W$ already determines $a_n(t)$ and in that case (6.23) does not hold for $|f| > W$.

On the other hand it *is* possible to choose coefficients other than $a_n(\cdot)$ such that the error is smaller than (6.20), but the interpolated process will no longer be stationary! See Chapter 9 and Problem 9.13.

Now we can ask the reverse question: Given the samples $\{x_n\}$, a second-order stationary process with mean μ, and spectral density $p(\lambda)$, $-1/2 < \lambda < 1/2$, suppose we use the interpolation formula (6.1) and define

$$x(t) = \sum_{-\infty}^{\infty} a_n(t)x_n, \qquad (6.25)$$

where

$$a_n(t) = \frac{\sin \pi(2Wt - n)}{\pi(2Wt - n)}$$

and the limit is taken in the mean square sense for some W, $0 < W$. What can we say about the process so interpolated?

Theorem 6.3. The process $x(t)$, $-\infty < t < \infty$, defined by (6.25) is a second-order stationary process band-limited to $[-W, W]$ with spectral density given by

$$P(f) = \frac{1}{2W} p\left(\frac{f}{2W}\right) \qquad -W < f < W$$

$$= 0 \qquad\qquad |f| > W. \tag{6.26}$$

Proof. First of all

$$R(m) = E[x_{n+m} x_n^*] = \int_{-1/2}^{1/2} e^{2\pi i \lambda m} p(\lambda)\, d\lambda,$$

which by a change of variable

$$\lambda = \frac{f}{2W}$$

becomes

$$= \frac{1}{2W} \int_{-W}^{W} e^{2\pi i f m/(2W)} p\left(\frac{f}{2W}\right) df.$$

The mean square convergence of (6.24) follows by similar arguments as in the case of (6.14).

Moreover, thus defined,

$$E[x(t+s)x(s)^*] = \sum_{-\infty}^{\infty} \sum_{-\infty}^{\infty} a_n(t+s) E[x_n x_m^*] a_m(s)$$

$$= \sum_{-\infty}^{\infty} \sum_{-\infty}^{\infty} a_n(t+s) R(n-m) a_m(s)$$

$$= \frac{1}{2W} \int_{-W}^{W} \left(\sum_{-\infty}^{\infty} \sum_{-\infty}^{\infty} a_n(t+s) e^{\pi i f(n-m)} a_m(s) \right) p\left(\frac{f}{2W}\right) df$$

$$= \frac{1}{2W} \int_{-W}^{W} (e^{2\pi i f(t+s)} e^{-2\pi i f s}) p\left(\frac{f}{2W}\right) df$$

$$= \frac{1}{2W} \int_{-W}^{W} e^{2\pi i f t} p\left(\frac{f}{2W}\right) df,$$

proving all the statements of the theorem. In other words, every stationary discrete-parameter process can be considered as obtained by sampling a continuous-time process, with spectral density given by (6.26).

In conclusion we note that the primary virtue of the Shannon Sampling Principle is its universality—that the interpolation does not require any knowledge of the process statistics except the spectral band-limitedness.

It should also be noted that if the process is Gaussian, the series in (6.1) converges pathwise.

Example 6.1. Let us consider a 1×1 continuous-time process $x(t)$, $-\infty < t < \infty$, with spectral density

$$P(f) = \frac{2k}{k^2 + 4\pi^2 f^2}, \qquad -\infty < f < \infty.$$

This example has the virtue that we can calculate explicitly the folded spectral density $\hat{P}(\cdot)$ as well as the interpolation error bound (6.16). Since

$$P(f) = \frac{P(0)}{2},$$

at the 3-db "break frequency"

$$f_b = \frac{k}{2\pi},$$

we may study sampling rates

$$2W = 2f_b \cdot p$$

for any positive number p. The interpolation error bound given by (6.16)

$$= 8 \int_{pf_b}^{\infty} \frac{2k}{k^2 + 4\pi^2 f^2} \, df \qquad (6.27)$$

$$= 8 \left[\frac{1}{2} - \frac{1}{\pi} \tan^{-1} p \right]. \qquad (6.28)$$

Figure 6.1 is a plot of (6.28) in decibels as a function of p, $1 \le p$, and is less than zero for $p \ge 2.5$ approximately.

The corresponding folded spectral density is given by

$$\hat{P}(f) = \sum_{n=-\infty}^{\infty} P(f + 2nW) = \sum_{n=-\infty}^{\infty} \frac{2k}{k^2 + 4\pi^2 (f + 2npf_b)^2},$$

$$|f| < W = pf_b. \quad (6.29)$$

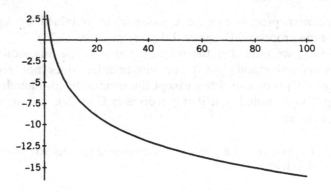

Figure 6.1. Interpolation error bound versus $\left(\dfrac{\text{sampling rate}}{2 \cdot \text{break frequency}}\right)$.

This can be evaluated explicitly using (2.26). Thus let $\hat{x}(\cdot)$ denote the interpolated process. We know that

$$\hat{x}\left(\frac{n}{2W}\right) = x\left(\frac{n}{2W}\right)$$

and hence the covariance function is expressed as

$$E\left[\hat{x}\left(\frac{n}{2W}\right)\hat{x}\left(\frac{n+m}{2W}\right)\right] = e^{-(|m|k)/2W} = \rho^{|m|}, \qquad \rho = e^{-k/2W}. \tag{6.30}$$

The corresponding spectral density is (cf. 2.12)

$$\frac{1-\rho^2}{1+\rho^2 - 2\rho\cos 2\pi\lambda}, \qquad |\lambda| < \frac{1}{2}.$$

Hence by (6.26) we have

$$\hat{P}(f) = \sum_{-\infty}^{\infty} P(f + 2nW) = \frac{1}{2W}\frac{1-\rho^2}{1+\rho^2 - 2\rho\cos\left(\dfrac{2\pi f}{2W}\right)}. \tag{6.31}$$

Hence the folded spectral density is given by

$$\hat{P}(f) = \frac{1}{2W}\frac{(1-e^{-k/W})}{1+e^{-k/W} - 2e^{-k/2W}\cos(\pi f/W)}, \qquad |f| < W \tag{6.32}$$

and can be expressed in terms of f_b and p as

$$\hat{P}(f) = \frac{1}{2pf_b}\frac{(1-e^{-2\pi/p})}{1+e^{-2\pi/p} - 2e^{-\pi/p}\cos(\pi f/pf_b)}, \qquad |f| < pf_b. \tag{6.33}$$

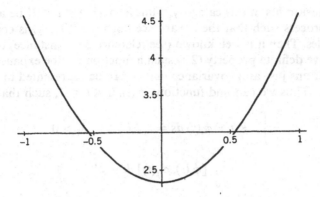

Figure 6.2. Curve for $10 \log_{10}(\hat{P}(f)|P(f))$. $p = 1 = f_b$.

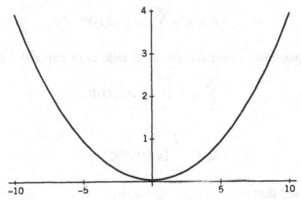

Figure 6.3. Curve for $10 \log_{10}(\hat{P}(f)|P(f))$. $p = 10$; $f_b = 1$.

A good measure of the relative shape of $\hat{P}(\cdot)$ with respect to $P(\cdot)$ is given by

$$10 \log_{10}(\hat{P}(f)|P(f)), \qquad |f| < W = pf_b.$$

This is plotted in Figure 6.2 for $f_b = 1$, $p = 1$ and in Figure 6.3 for $f_b = 1$, $p = 10$. Notice the increase as the frequency increases. As p increases, the effect is less, as we expect.

6.1 KARHUNEN-LOEVE EXPANSION

By "sampling" we may mean any technique which converts continuous-time data into discrete-time data (by a linear operation) which may then be reconstituted to retrieve the continuous-time data without error. The Karhunen-

Loeve Expansion fits in this category. Thus let $x(t)$, $0 \leq t \leq T$, be a zero-mean stochastic process such that the covariance function $R(t_1, t_2)$ is continuous in both variables. Then it is well known (see reference 3 for instance) that because of the positive definite property (2.3) such a function can be expanded in terms of eigenfunctions just as a covariance matrix can be represented in terms of its eigenvectors. Thus we can find functions $\phi_i(t)$, $0 \leq t \leq T$, such that

$$\int_0^T R(t, s)\phi_i(s)\, ds = \lambda_i \phi_i(t), \qquad \lambda_i > 0$$

$$\int_0^T [\phi_i(t), \phi_j(t)]\, dt = \delta_j^i$$

$$\int_0^T [R(t, s)\phi_i(s), \phi_j(s)]\, ds = \delta_j^i \lambda_i$$

and $\{\lambda_i\}$ can be arranged in a decreasing sequence; $\lambda_i \to 0$ and

$$R(t, s) = \sum_1^\infty \lambda_i \phi_i(t)\phi_i(s)^*$$

("Mercer Expansion"—see reference 5). It follows in particular that

$$\sum_1^\infty \lambda_i = \int_0^T \operatorname{Tr} R(t, t)\, dt.$$

Let

$$\zeta_i = \int_0^T [\phi_i(s), x(s)]\, ds.$$

Then we can see that

$$E[\zeta_i \zeta_j] = \int_0^T \int_0^T E[(\phi_i(t), x(t))(\phi_i(s), x(s))]\, ds\, dt$$

$$= \int_0^T \phi_i(t)^* \int_0^T R(t, s)\phi_j(s)\, ds\, dt$$

$$= \delta_j^i \lambda_i.$$

Let

$$r_N(t) = x(t) - \sum_1^N \zeta_i \phi_i(t).$$

Then

$$\int_0^T E[\|r_N(t)\|^2] = \int_0^T E\left[\left(x(t) - \sum_1^N \zeta_i \phi_i(t), x(t) - \sum_1^N \zeta_j \phi_j(t)\right)\right] dt$$

$$= \int_0^T \operatorname{Tr} R(t, t)\, dt - \sum_1^N \lambda_{ii}.$$

Hence

$$\int_0^T E[\|r_N(t)\|^2]\, dt \to 0 \qquad \text{as } N \to \infty.$$

In particular it follows that

$$E[\|r_N(t)\|^2] \to 0 \qquad 0 \le t \le T$$

or

$$x(t) = \sum_1^\infty \zeta_i \phi_i(t), \qquad 0 \le t \le T \tag{6.34}$$

in the mean square sense. Note that if $x(\cdot)$ is a Gaussian process, then $\{\zeta_i/\sqrt{\lambda_i}\}$ is a white-noise sequence, and in fact we can also show that the convergence is also pathwise with probability one.

The fact that the representation (6.27) (interpolation back from the samples to the process) is valid only in the finite interval $0 \le t \le T$ is not the main drawback of this sampling process, nor is finding the orthonormal sequence of functions $\phi_i(\cdot)$, but rather that it requires complete knowledge of the covariance function, unrealistic in practice.

6.2 DISCRETE-TIME MODELS

A further drawback is that the integrals defining $\{\zeta_i\}$ have to be finite sums in practice. In view of this, one may as well assume we are given a finite sequence of $n \times 1$ random variables $\{x_i\}$, $i = 1, \ldots, N$. In that case the K-L procedure is no more than finding the eigenvalues of the compound matrix

$$\Lambda = \{\Lambda_{ij}\}$$

$$\Lambda_{ij} = E[x_i x_j^*], \qquad i, j = 1, \ldots, N.$$

Let $\{\phi_i\}$ denote the $(nN) \times 1$ orthonormalized eigenvectors of Λ. Let x denote the $(nN) \times 1$ vector composed of $\{x_i\}$ as

$$\begin{vmatrix} x_1 \\ \vdots \\ x_N \end{vmatrix}.$$

Then

$$\zeta_i = [x, \phi_i]$$

and

$$x = \sum_1^{nN} \zeta_i \phi_i$$

$$E[\zeta_i^2] = \lambda_i$$

and $\{\lambda_i\}$ is monotone decreasing

$$\sum_1^{nN} \lambda_i = \text{Tr}\, \Lambda = \sum_1^N \text{Tr}\, \Lambda_{ii}.$$

One virtue of this representation is that it may be possible to retain only the first few ζ_i, say $i = 1, \ldots, p$ where

$$p \ll nN.$$

Let us explore this idea further. Let ζ denote an $n \times 1$ zero mean random variable with covariance matrix Λ. Given any $n \times 1$ vector ϕ, we first consider the problem of the best representation for ζ in terms of ϕ. For any $x \in \mathbf{R}^n$, such representation would be

$$a\phi$$

and the error is

$$x - a\phi$$

and we calculate that for any given x,

$$\|x - a\phi\|^2 = \|x\|^2 + a^2 \|\phi\|^2 - 2a[x, \phi]$$

which is a minimum when

$$a = \frac{[x, \phi]}{[\phi, \phi]}.$$

Hence our representation for ζ would be

$$\frac{[\zeta, \phi]}{[\phi, \phi]} \phi$$

where now we see that

$$[\zeta, \phi]$$

is a random variable, and further that the mean square error

$$E\left[\left\| x - \frac{[x, \phi]}{[\phi, \phi]} \phi \right\|^2 \right] = \text{Tr}\, \Lambda - \frac{[\Lambda \phi, \phi]}{[\phi, \phi]}.$$

Now the question is what is the best choice for ϕ, to minimize this mean square error, or equivalently to maximise

$$\frac{[\Lambda\phi, \phi]}{[\phi, \phi]}.$$

Representing ϕ in terms of the orthonormalized eigenvectors of Λ, we have

$$\phi = \sum_1^n [\phi, e_i]e_i; \qquad Ae_i = \lambda_i e_i, \qquad \lambda_1 \geq \lambda_2 \geq \cdots \geq \lambda_i \geq 0.$$

We see that

$$\frac{[\Lambda\phi, \phi]}{[\phi, \phi]} = \frac{\sum_1^n \lambda_i[\phi, e_i]^2}{\sum_1^n [\phi, e_i]^2}$$

which is

$$\leq \lambda_1,$$

and is

$$= \lambda_1$$

if we choose

$$\phi = e_1.$$

Hence if we are to choose just one vector, the best choice would be the eigenvector e_1 corresponding to the largest eigenvalue, with the representation

$$[\zeta, e_1]e_1.$$

The corresponding mean square error is

$$\text{Tr } \Lambda - \lambda_1.$$

It is not too difficult to see, continuing this procedure, that the best representation in terms of, say, m vectors would be

$$\sum_1^m [\zeta, e_i]e_i$$

where

$$\lambda_1 \geq \lambda_2 \geq \cdots \geq \lambda_m$$

—the first m eigenfunctions in order of decreasing eigenvalues—with corresponding error

$$\text{Tr } \Lambda - \sum_1^m \lambda_k.$$

For a continuous-time model, we have, similarly,

$$\sum_{1}^{m} \left(\int_{0}^{T} [x(t), \phi_i(t)] \, dt \right) \phi_i(t), \qquad 0 \le t \le T$$

with mean-square error

$$= \int_{0}^{T} \text{Tr} \, R(t, t) \, dt - \sum_{1}^{m} \lambda_k.$$

Example 6.2. Karhunen-Loeve Expansion. Let us consider an example of a 1×1 continuous time process $x(t)$, $0 < t < T$ with covariance function for which we can calculate the eigenfunctions explicitly. Thus let

$$R(t, s) = \min(t, s).$$

(The corresponding process is a Wiener process—see problem 1.2.5, Chapter 1.) In that case, for any $\phi(\cdot)$ we calculate

$$\int_{0}^{T} R(t, s)\phi(s) \, ds = \int_{0}^{t} s\phi(s) \, ds + t \int_{t}^{T} \phi(s) \, ds$$

and the eigenvalue problem becomes

$$\int_{0}^{t} s\phi(s) \, ds + t \int_{t}^{T} \phi(s) \, ds = \lambda\phi(t), \qquad 0 < t < T; \lambda \ge 0.$$

In particular

$$\lambda\phi(0) \equiv 0.$$

Differentiating with respect to t, we have

$$\int_{t}^{T} \phi(s) \, ds = \lambda\phi'(t)$$

and in particular

$$\lambda\phi'(T) = 0.$$

Differentiating once more with respect to t yields

$$\phi''(t) = \frac{-1}{\lambda} \phi(t)$$

and hence

$$\phi(t) = A \sin \frac{t}{\sqrt{\lambda}}, \qquad 0 < t < T.$$

Since

$$\phi'(T) = \frac{1}{\sqrt{\lambda}} \cos \frac{T}{\sqrt{\lambda}}$$

we see that

$$\frac{(2n+1)\pi}{2} = \frac{T}{\sqrt{\lambda}}$$

or

$$\lambda_n = \left(\frac{2T}{(2n+1)\pi}\right)^2, \qquad n = 0, 1, 2, \ldots$$

are the eigenvalues. The eigenfunctions are

$$\phi_n(t) = A_n \sin \frac{t}{\sqrt{\lambda_n}}, \qquad 0 \le t \le T$$

where A_n is to be determined from

$$A_n^2 \int_0^T \sin^2 \frac{t}{\sqrt{\lambda_n}} \, dt = 1.$$

Note that the eigenvalues decrease as $(1/n)^2$ and

$$\int_0^T t \, dt = \frac{T^2}{2} = \sum_0^\infty \left(\frac{2T}{(2n+1)\pi}\right)^2$$

is a by-product. Also

$$\min(t, s) = \sum_{n=0}^\infty \lambda_n A_n^2 \sin \frac{t}{\sqrt{\lambda_n}} \sin \frac{2}{\sqrt{\lambda_n}}, \qquad 0 < s, t < T$$

where

$$A_n^2 = \frac{1}{\displaystyle\int_0^T \sin^2(t/\sqrt{\lambda_n}) \, dt}.$$

Finally the Karhunen-Loeve expansion is

$$x(t) = \sum_1^\infty \zeta_k \phi_k(t),$$

where

$$\zeta_k = \int_0^T x(t)\phi_k(t)\,dt.$$

Note that while the process $x(\cdot)$ does not have a mean square derivative (compare problem 4.20), any finite sum approximation

$$x_N(t) = \sum_1^N \zeta_k \phi_k(t), \qquad 0 < t < T$$

does have a mean square derivative.

6.3 DISCRETE-TIME APPROXIMATION

We may also consider the discretized version where we take

$$t = n\Delta, \qquad n = 0, 1, \ldots, N-1; \; N\Delta = T.$$

The covariance matrix corresponding to the $N \times 1$ vector

$$x(0)$$
$$x(\Delta)$$
$$\vdots$$
$$x(\overline{N-1}\,\Delta)$$

is given by

$$\Lambda = \Delta\{\min(i,j)\}, \qquad 0 \le i,j \le N-1$$

$$= \begin{vmatrix} 0 & 0 & 0 & \cdots & 0 \\ 0 & 1 & 1 & \cdots & 1 \\ 0 & 1 & 2 & \cdots & 2 \\ \cdot & \cdot & \cdot & \cdot & \cdot & \cdot & \cdot \\ 0 & 1 & 1 & \cdots & N-1 \end{vmatrix}.$$

To allow for replacing the integral by a finite sum we need to multiply this matrix by Δ. Zero is an eigenvalue for all N. Numerical approximations are required to find all other eigenvalues and eigenvectors. The largest eigenvalue of $\Delta\Lambda$ should be close to

$$\frac{4N^2\Delta^2}{\pi^2}$$

for large N. We desist from further analysis. (See, however, Problem 6.12.) For $N = 6$, computer calculation yields

$$(12.345)\Delta^2$$

for the largest eigenvalue where the sum of the eigenvalues

$$= 15\Delta^2.$$

Hence the approximation using the eigenvector corresponding to the largest eigenvalue would be expected to be generally good, but would of course depend on the particular application!

PROBLEMS

6.1. Show that under the conditions of Theorem 6.1 we obtain

$$x(t + \tau) = \sum_{-\infty}^{\infty} a_n(t) x\left(\tau + \frac{n}{2W}\right)$$

$$R(t + \tau) = \sum_{-\infty}^{\infty} a_n(t) R\left(\tau + \frac{n}{2W}\right).$$

6.2. Show that a band-limited signal has mean square derivatives of all orders and that we can take derivatives on both sides of (6.1) to yield

$$x'(t) = \sum_{-\infty}^{\infty} a_n'(t) x\left(\frac{n}{2W}\right),$$

where the prime denotes the derivative (in the mean square sense). Show that

$$\sum_{-\infty}^{\infty} a_n'(t)^2 = \frac{1}{2W}\int_{-W}^{W} 4\pi^2 f^2\, df.$$

More generally,

$$\frac{d^p x(t)}{dt'} = \sum_{-\infty}^{\infty} \frac{d^p a_n(t)}{dt^p} x\left(\frac{n}{2W}\right)$$

for any p, and

$$\sum_{-\infty}^{\infty} \left|\frac{d^p a_n(t)}{dt^p}\right|^2 = \frac{1}{2W}\int_{-W}^{W} (4\pi^2 f^2)^p\, df.$$

6.3. Let $\{s_n\}$ be a stationary Gaussian signal with spectral density $p(\lambda)$. Define (for fixed W) the continuous-time stochastic process by

$$s(t) = \sum_{-\infty}^{\infty} s_n \frac{\sin \pi(2Wt - n)}{\pi(2Wt - n)}, \qquad -\infty < t < \infty,$$

where the convergence of the infinite series is taken in the mean square sense. Find the covariance of the process

$$s(k\Delta), \qquad -\infty < t < \infty$$

for fixed Δ, $2W\Delta \leq 1$. Specialize to the case where the spectral density of $\{s_n\}$ is such that

$$p(\lambda) = 1, \qquad -1/2 < \lambda < 1/2.$$

6.4. Show that the process defined by (6.17) is not stationary, unless $x(\cdot)$ is band-limited to $[-W, W]$. Calculate the covariance function.

6.5. Let $x(t)$, $-\infty < t < \infty$, be band-limited white noise with spectral density

$$P(f) = D, \qquad |f| \leq W$$
$$= 0, \qquad |f| > W.$$

Let this process be sampled at the rate $2Wp$ samples/second, and let the samples be interpolated by the Shannon formula corresponding to the bandwidth $2W(p)$, with

$$x_n = x\left(\frac{n}{2Wp}\right).$$

Find the folded spectrum for $p = 1/4$ and $p = 4$, and more generally for any p.

6.6. Let $\{x_n\}$, $n \in I$, denote a stationary Gaussian process with spectral density $P(\cdot)$. For interpolation between sample points, let

$$x(t) = \sum_{-\infty}^{\infty} x_n \frac{\sin \pi(2Wt - n)}{\pi(2Wt - n)}, \qquad -\infty < t < \infty,$$

and define

$$y_n = x(n\Delta) \qquad \text{where } \Delta = \frac{\gamma}{2W}, \quad n \in I.$$

Find the spectral density of the process $\{y_n\}$, for γ of the form $1/p$, p an integer.

6.7. In sampling (periodic) a signal $x(\cdot)$, one has the choice of band-limiting the signal first (by passing it through a linear time-invariant system whose transfer function ideally is equal to unity in the pass band and zero outside) and then sampling, or sampling the pristine signal. Show that the mean square interpolation error corresponding to the first operation is a constant but not so for the second. Which is better?

Hint: In (6.20), for any integer k we have

$$\sin^2 2\pi k W t = 1$$

if we choose t such that $4kWt = (2m + 1)$ for some integer m.

6.8. Response of linear systems to band-limited input processes: Let $\psi(\cdot)$ denote the transfer function of a linear time-invariant system, with weighting function $W(\cdot)$. Let

$$H(t) = \int_{-W}^{W} e^{2\pi i f t} \psi(f)\, df.$$

Let $x(\cdot)$ be the stationary input process band-limited to $[-W, W]$. Show that the output process has the representation

$$y(t) = \sum_{-\infty}^{\infty} a_n(t) y\left(\frac{n}{2W}\right),$$

where

$$y\left(\frac{n}{2W}\right) = \sum_{-\infty}^{\infty} H\left(\frac{m}{2W}\right) x\left(\frac{n-m}{2W}\right).$$

Note that $H(\cdot)$ is not "physically realizable" in that $H(t) \neq 0$ for $t < 0$.

Hint: Spectral density of $y(\cdot)$

$$
\begin{aligned}
&= P_x(f)|\psi(f)|^2, && |f| \le W \\
&= 0, && |f| > W
\end{aligned}
$$

$$y\left(\frac{n}{2W}\right) = \int_0^\infty W(\sigma) x\left(\frac{n}{2W} - \sigma\right) = \int_0^\infty W(\sigma) \sum_{-\infty}^{\infty} a_m(\sigma) x\left(\frac{n-m}{2W}\right) d\sigma$$

$$\int_0^\infty W(\sigma) a_m(\sigma)\, d\sigma = \frac{1}{2W} \int_{-W}^{W} \psi(f) e^{2\pi i f m/2W}\, df = H\left(\frac{m}{2W}\right).$$

6.9. Let $x(t)$, $-\infty < t < \infty$, be a 1×1 stationary process with spectral density

$$P(f) = e^{-k|f|}, \qquad k > 0.$$

Let the process be sampled at rate $2W$ samples/second and interpolated by the corresponding Shannon formula

$$\hat{x}(t) = \sum_{-\infty}^{\infty} a_n(t) x\left(\frac{n}{2W}\right),$$

where $a_n(t)$ is given by (6.2.). Find the "folded" spectral density of the continuous-time process $\hat{x}(\cdot)$.

Answer:

$$\frac{e^{-k|f|} + e^{k|f|-2Wk}}{1 - e^{-2kW}}, \qquad |f| < W.$$

Estimate the aliasing error.

Answer:

$$\leq \frac{8}{k} e^{-kW}.$$

6.10. Let ζ denote a $2N \times 1$ zero-mean Gaussian with covariance matrix

$$\Lambda = \left\{ c_1 + c_2 \cos \frac{(j-k)\pi}{N} \right\}, \qquad 1 \leq j, k \leq 2N$$

where c_1 and c_2 are nonnegative constants. Find the Karhunen-Loeve expansion for ζ.

Hint: The nonzero eigenvalues are $2Nc_1, Nc_2, Nc_2$. Use

$$\sum_{1}^{2N} \exp \frac{ik\pi}{N} = 0.$$

6.11. Given the $N \times N$ matrix

$$\Lambda = \{\min (j, k)\}_{1 \leq j, k \leq N}.$$

Show that the eigenvectors are of the form

$$\alpha_n = |1 \quad 0| A^{n-1} \begin{vmatrix} 1 \\ 0 \end{vmatrix}, \qquad \alpha_1 = 1$$

where

$$A = \begin{vmatrix} 2 - (1/\lambda) & -1 \\ 1 & 0 \end{vmatrix}$$

where the eigenvalue λ is determined by the root of

$$\sum_{0}^{N-1} |1 \quad 0| A^k \begin{vmatrix} 1 \\ 0 \end{vmatrix} - \lambda = 0.$$

NOTES

For recent extensions of the Sampling Principle see references 4 and 5 and the references therein.

REFERENCES

Classic Papers and Treatises

1. C. E. Shannon. "A Mathematical Theory of Communication," *Bell System Technical Journal* (August 1948).
2. J. M. Whittaker. *Interpolating Function Theory*. Cambridge University Press, 1935.
3. R. Courant and D. Hilbert. *Methods of Mathematical Physics*, Volume 1. Wiley-Interscience, 1955.

Recent Publications

4. J. P. Gillis. Multidimensional Point Processes and Random Sampling. Ph.D. thesis, UCLA, 1988.
5. I. Bilinskis and A. Mikelsons. *Randomized Signal Processing*. Prentice-Hall, 1992.

7

SIMULATION OF RANDOM PROCESSES

Computer simulation of system performance is an integral part of the design process in communications and control, as in any application where stochastic models are employed. In this chapter we study techniques for digital computer simulation of random processes.

7.1 IID SEQUENCES: WHITE NOISE

Any random number generator, on a personal computer for example, purports to generate an IID sequence with uniform distribution: In the decimal system, the integers 0 to 9 are generated with equal probability of (1/10). A basic method of generation is the "linear congruence method." It will take us too far afield to go into any further discussion of generating uniform random numbers. See references 1 and 2 for detailed treatments, and see references 3 and 4 for algorithms. We may note that a desirable property of any technique, apart from passing tests for "randomness" for the desired number of samples, is "repeatability" not only on a given computer but on a class of computers of given size.

Nearly all tools for generating (discrete-time!) Gaussian random processes (like the signal generation model (3.3.1)) requires that we begin with white noise—an IID Gaussian sequence. In principle they could be generated from an IID uniform sequence based on this simple idea. If ζ is any 1×1 random variable, with a continuous distribution function $F(\cdot)$, then the random variable

$$\eta = F(\zeta)$$

is uniformly distributed between 0 and 1. If, for example, $F(\cdot)$ is the Gaussian distribution,

$$F(x) = \frac{1}{\sqrt{2\pi}} \int_{-\infty}^{x} \exp -\frac{y^2}{2} \, dy,$$

then we can define the inverse mapping

$$\zeta = F^{-1}(\eta); \tag{7.1.1}$$

and if we begin with a random variable η uniformly distributed in $[0, 1]$, ζ given by (7.1.1) produces a Gaussian. See pages 17–20 in the Review Chapter. Thus

$$N_n = F^{-1}(Z_n),$$

where $\{Z_n\}$ is an IID uniform sequence, suitably scaled and $\{N_n\}$ will yield an IID Gaussian sequence. Here each uniform sample yields a Gaussian sample, but the function (7.1.1) is complicated. Given a pair of independent uniformly distributed samples, we can create a Gaussian sample using the "polar" method (see reference 2), which has the advantage that the functions are simpler.

Thus let η_1, η_2 denote a pair of independent uniform random variables $[0, 1]$. Define

$$A = F^{-1}(\eta_1),$$

where

$$F(x) = 1 - \exp\left(\frac{-x^2}{2}\right), \qquad x > 0,$$

$$F'(x) = x \exp\left(\frac{-x^2}{2}\right), \qquad x > 0.$$

In this case

$$A = \sqrt{-2 \log (1 - \eta_1)},$$

where log is to the base e. Then A is Rayleigh. Define

$$\phi = (2\pi)\eta_2$$

so that ϕ is uniformly distributed in $[0, 2\pi]$. Then

$$\zeta = (\sqrt{-2 \log (1 - \eta_1)}) \sin 2\pi\eta_2$$

is Gaussian with mean zero and unit variance—see pages 17–20 in the Review Chapter.

If we can spare several uniform samples, one may opt for the simpler operation of averaging a suitable number of them each, relying on the law of large numbers or the central limit theorem (see References to the Review Chapter). There are of course still other methods—but it is beyond our scope to provide a detailed review.

We shall simply assume therefore that we have at our disposal a generator, for generating as many samples as needed of IID (0, 1) Gaussian variables. In particular therefore we may assume as available an $n \times 1$ white-noise sequence $\{N_k\}$ with zero mean and identity covariance, for given n.

We can then generate independent Gaussian variables $\{x_k\}$ with any given $(n \times n)$ covariance Λ by defining

$$x_k = LN_k,$$

where L is a lower triangular matrix such that

$$\Lambda = LL^*.$$

Such a factorization of Λ can be accomplished in many ways (Gram-Schmidt orthogonalization or other means—see any standard text on linear algebra (compare the references for the Review Chapter).

7.2 SIGNAL-GENERATION MODELS

Given the capability to generate a white-noise sequence, we can simulate stationary Gaussian processes in many ways, depending largely on how the process is specified. We study the Rice model and the Kalman model.

7.2.1 Rice Model

The earliest such technique is due to S. O. Rice (see reference 2 of Chapter 1), and we shall call it the Rice model. Here the spectral density is specified, and the mean is zero. Let $p(\lambda)$, $-1/2 < \lambda < 1/2$, denote the spectral density where we assume that it is continuous in λ. Then the covariance function

$$R(m) = \int_{-1/2}^{1/2} e^{2\pi i m \lambda} p(\lambda) \, d\lambda \qquad (7.2.1)$$

can be approximated by the partial sums

$$R_M(m) = \sum_{k=-M}^{M} e^{2\pi i m \lambda_k} p(\lambda_k) \Delta\lambda, \qquad k \neq 0$$

$$\frac{k-1}{2M} < \lambda_k < \frac{k}{2M}, \qquad k = 1, \ldots, M; \lambda_{-k} = -\lambda_k$$

$$\Delta\lambda = \frac{1}{2M}$$

and for each n

$$\| R_M(n) - R(n) \| \to 0 \qquad \text{as } M \to \infty. \qquad (7.2.2)$$

Fixing M, we generate a sequence of vectors x_k of zero-mean $n \times 1$ Gaussians such that

$$E[x_k \ x_k^*] = p(\lambda_k)\left(\frac{1}{2M}\right) \qquad (7.2.3)$$

for each k as above. Let $y_k, k = -M, \ldots, -1, 1, \ldots, M$, be a similar zero-mean $n \times 1$ Gaussian vector sequence such that

$$E[y_k \ y_k^*] = E[x_k \ x_k^*]$$

$$E[x_k \ y_j^*] = 0 \qquad \text{for all } j, k.$$

Then in the Rice model we generate the process

$$x_M(n) = \sum_{k=-M}^{M} (x_k \sin 2\pi\lambda_k n + y_k \cos 2\pi n\lambda_k), \qquad k \neq 0. \qquad (7.2.4)$$

This, in other words, is our simulation of a zero-mean Gaussian with spectral density which approximates the given spectral density depending on the choice of M—how large it is. For a given M, we may, in principle, generate as many samples of the process as needed, indexed by the integer n. In practice there is of course an upper bound. This is because for fixed $\{x_k, y_k\}$, $1 \leq |k| \leq M$, (7.2.4) is actually periodic in n, since the $\{\lambda_k\}$ perforce have to be rational numbers. The period then is the maximum number of samples we can use. The process is of course "trivial" in the sense that it is a function of the $4M$ variables $\{x_k, y_k\}$. If we use the polar methods to generate them, we would need $4M$ independent random variables, each $n \times 1$. If we choose $\lambda_k = (2k - 1)/4M$, the period is $4M$, so that we generate at most $4M$ samples using $8M$ "seed" variables, each $n \times 1$.

Since we are only interested in the second-order properties, we shall examine only the mean and spectral density of the process specified by (7.2.4). A distinction has to be made here between "time average" and "ensemble average." Thus under the assumption that the $\{x_k, y_k\}$ are independent Gaussians, we can calculate that

$$E[x_M(n)] = 0$$

$$R_M(n) = E[x_M(n)x_M(n + m)^*] = \sum_{-M}^{M} \frac{1}{2M} p(\lambda_k) \cos 2\pi\lambda_k m, \qquad k \neq 0.$$

In particular, we see that for fixed M, property (5.1.5) does not hold, eventhough it holds for the truth model (7.2.1). On the other hand, given the simulated data defined by (7.2.4)—one sample path—we can only calculate time average, and hence we must rely on the fact that we can make M in (7.2.4) large, as necessary, exploiting the ergodic properties of the sequences $\{x_k\}, \{y_k\}$.

Let us first consider the time average for the mean: Let

$$\zeta_N = \frac{1}{N+1} \sum_0^N x_M(n) = \sum_{-M}^M x_k a_k(N) + \sum_{-M}^N y_k b_k(N), \qquad (7.2.5)$$

where

$$a_k(N) = \frac{1}{N+1} \sum_0^N \sin 2\pi\lambda_k n$$

$$b_k(N) = \frac{1}{N+1} \sum_0^N \cos 2\pi\lambda_k n.$$

Note that (7.2.5) is a sum of independent zero-mean Gaussian variables, and we can calculate (compare Chapter 5) the covariance

$$E[\zeta_N \zeta_N^*] = \frac{1}{N+1} \sum_{-N}^N R_M(n)\left(1 - \frac{|n|}{N+1}\right),$$

which can be expressed as

$$= \frac{1}{N+1} \sum_{-N}^N R(n)\left(1 - \frac{|n|}{N+1}\right) + \frac{1}{N+1} \sum_{-N}^N (R_M(n) - R(n))\left(1 - \frac{|n|}{N+1}\right).$$

$$(7.2.6)$$

As in Chapter 5, the first term can be approximated by

$$\frac{\bar{N}R(0)}{N+1}$$

and can be made as small as desired by choosing N large enough.

We shall show next that M can be chosen large enough to make the second term in (7.2.6) small. Thus

$$R_M(n) - R(n) = 2\sum_1^M \frac{p(\lambda_k)}{2M} \cos 2\pi n\lambda_k - 2\int_0^{1/2} \cos 2\pi n\lambda p(\lambda)\, d\lambda$$

$$(7.2.7)$$

$$= 2\sum_1^M \int_{(k-1)/2M}^{k/2M} (p(\lambda_k) \cos 2\pi n\lambda_k - p(\lambda) \cos 2\pi n\lambda)\, d\lambda.$$

Now

$$\| p(\lambda_k) \cos 2\pi n\lambda_k - p(\lambda) \cos 2\pi n\lambda \|$$

$$\leq \| p(\lambda_k) - p(\lambda) \| + \| p(\lambda) \| |\cos 2\pi n\lambda_k - \cos 2\pi n\lambda|,$$

where

$$|\cos 2\pi n \lambda_k - \cos 2\pi n \lambda| \le 2|\sin \pi n(\lambda - \lambda_k)|$$

$$\le \frac{2\pi n}{2M}, \qquad \text{for } |\lambda - \lambda_k| < \frac{1}{2M}$$

$$\le \frac{2\pi N}{2M},$$

$$\le \pi 2^{-\nu}, \qquad \text{for } M \ge 2^{\nu} N$$

where ν is a positive integer. Hence the second term in (7.2.6) can be made smaller than

$$2 \sum_1^M \int_{(k-1)/2M}^{k/2M} \| p(\lambda_k) - p(\lambda) \| \, d\lambda + 2 \left(\int_0^{1/2} \| p(\lambda) \| \, d\lambda \right) \pi 2^{-\nu},$$

where the first term goes to zero as $M \to \infty$. Hence, invoking the Chebyshev inequality as usual, the time average (7.2.5) can be made as as small as desired with as large a probability as desired by choosing M large enough.

Next let us consider the time-average estimate for the covariance $R(m)$:

$$\zeta_N(m) = \frac{1}{N+1} \sum_0^N x_M(n) x_M(n+m)^*, \tag{7.2.8}$$

where m is a fixed positive integer. We shall show that $\zeta_N(m)$ converges to $R(m)$ as $N, M \to \infty$. As in Chapter 5, let

$$y_M(n) = x_M(n) x_M(n+m)^* \tag{7.2.9}$$

so that (7.2.8) is recognized as the time-average estimate of the process defined by (7.2.9). Now

$$E[y_M(n)] = R_M(m). \tag{7.2.10}$$

As follows from (7.2.7),

$$\| R_M(m) - R(m) \|$$

$$\le 2 \sum_1^M \int_{(k-1)/2M}^{k/2M} \| p(\lambda_k) - p(\lambda) \| \, d\lambda + 2 \int_0^{1/2} \| p(\lambda) \| \, d\lambda \cdot \left(\frac{2\pi m}{2M} \right) \tag{7.2.11}$$

and can be made as small as desired by making M large enough, m being fixed. Hence it is enough to consider the convergence of (7.2.8) to its mean given by (7.2.10). We are thus led to examine the covariance function of the process

$y_M(\cdot)$, and as in the case of the mean, we have to allow M to be large as well. Since $y_M(n)$ is now a square matrix for each n, much of the notational complexity can be avoided by considering each term of the matrix, and this in turn means that it is enough to consider the one-dimensional version where $\{x_k\}, \{y_k\}$ are both one-dimensional. Thus

$$y_M(n) = x_M(n)x_M(n + m)$$

and the covariance function of this process is given by

$$r_M(p) = E[x_M(n)x_M(n + m)x_M(n + p)x_M(n + p + m)] - R_M(m)^2$$
$$= R_M(p)^2 + R_M(m + p)R_M(m - p). \tag{7.2.12}$$

This does not go to zero as $p \to \infty$ if M is fixed, so we allow M to go to infinity as well. Thus, as in Chapter 5,

$$E[(\zeta_N(m) - R_M(m))^2] = \frac{1}{N + 1}\left(\sum_{-N}^{N} r_M(p)\left[1 - \frac{|p|}{N + 1}\right]\right)$$
$$= \frac{1}{N + 1}\sum_{-N}^{N} r(p)\left[1 - \frac{|p|}{N + 1}\right]$$
$$+ \frac{1}{N + 1}\sum_{-N}^{N}(r_M(p) - r(p))\left[1 - \frac{|p|}{N + 1}\right]\right), \tag{7.2.13}$$

where

$$r(p) = R(p)^2 + R(m + p)R(m - p).$$

Let $\bar{N}_y$ denote the correlation time of the process $\{y_M(n)\}$. Then the first term in (7.2.13)

$$\approx \frac{r(0)\bar{N}_y}{N + 1}$$

for large N, and it can be made as small as desired by choosing N large enough. Next let us consider the second term in (7.2.13):

$$r_M(p) - r(p) = (R_M(p)^2 - r(p)^2) + R_M(m + p)R_M(m - p)$$
$$- r(m + p)r(m - p). \tag{7.2.14}$$

As we have seen (specializing (7.2.7) to the one-dimensional case) for $p \le (N + m)$,

$$|R_M(p) - r(p)| \le 2\sum_{1}^{M}\int_{(k-1)/2M}^{k/2M}|p(\lambda_k) - p(\lambda)|\,d\lambda + R(0)\pi 2^{-\nu}$$

for

$$M \geq 2^v(N + m),$$

and hence (7.2.14) and in turn the second term in (7.2.13) can be made as small as desired by choosing v, and hence M, large enough. Thus the time average (7.2.8) converges to $R(m)$ in the sense that the error can be made as small as desired with probability as high as desired, as in the case of the mean.

7.2.2 Kalman (State-Space) Models

The Kalman signal generation model (3.3.1) is "ready-made" for recursive digital computer programs. Of course the state-space model has to be specified. Often, as in control engineering, such a discrete-time model as in the case of the flight data in Chapter 4 may be deduced from a continuous-time dynamic system model. As we have seen, this constrains the spectral density of the process to be rational and is thus less general than the Rice model. Given a rational spectral density (satisfying the Paley-Wiener condition; see Chapter 4) it is possible to derive a corresponding state-space model, using techniques from state-space theory. It is beyond our scope to go into the details here.

7.2.3 Weighting Pattern Models

Simulating processes with a prescribed spectral density using weighting pattern models is computer memory intensive, but because advances in computer technology continue and memory size is less of a problem, the advantages over other methods may make this option attractive.

We shall begin with the general case and illustrate with a simple example. Thus let $P(\lambda)$, $-1/2 \leq \lambda \leq 1/2$, the spectral density matrix, be given. The first step is to factorize it as

$$P(\lambda) = \psi(\lambda)\psi(\lambda)^*,$$

which, because $P(\lambda)$ is nonnegative definite, can be done in many ways—we may invoke the covariance factorization technique (see Chapter 9), for example. In the one-dimensional case we may simply take the square root. As we have seen in Chapter 3, we place no requirement on the spectral density: that it be physically realizable or rational. We define

$$
\begin{aligned}
W_k &= \int_{-1/2}^{1/2} e^{2\pi i k \lambda} \psi(\lambda)\, d\lambda \\
&= 2 \int_{0}^{1/2} \cos 2\pi k \lambda \, \psi(\lambda)\, d\lambda, \qquad k \geq 0,
\end{aligned}
\tag{7.2.15}
$$

which is evaluated numerically, and then use

$$S_n = \sum_0^{2M} \tilde{W}_k N_{n-k},$$ (7.2.16)

$$\tilde{W}_k = W_{k-M},$$

where, of course, by (7.2.15)

$$W_{-k} = W_k.$$

As in the Rice model, the process defined by (7.2.16) is stationary for $n > 2M$. To produce N process samples we need $(N + 2M)$ Gaussian white noise samples.

Example 7.1. We consider a simple one-dimensional illustrative example. Let the discrete-time process be one-dimensional with corresponding spectral density

$$p(\lambda) = \frac{1 - \rho^2}{1 + \rho^2 - 2\rho \cos 2\pi\lambda}, \qquad -1/2 < \lambda < 1/2.$$ (7.2.17)

We will now simulate it in two ways, taking $\rho = 0.9881$.
First using the Rice model (7.5) with

$$\lambda_k = \frac{2k - 1}{4M}, \qquad k = 1, \ldots, M,$$

using the polar method for generating the Gaussian variables. To see how large M should be, we see first of all that for the approximation for the integral of $p(\cdot)$,

$$2\left(\frac{1}{M}\right)\sum_1^M p(\lambda_k),$$

we have

$$= 0.838 \qquad \text{for } M = 100$$
$$= 0.990 \qquad \text{for } M = 200$$
$$= 1.007 \qquad \text{for } M = 300.$$

The true value is, of course, 1. Sample paths for $M = 50, 100$ and 200 are shown in Figures 7.1, 7.2 and 7.3, respectively. Note the periodicity with period $4M$, illustrated in Figure 7.1. Computational time and memory size increase as M increases, as is to be expected. For large M, it is more efficient to use Fast Fourier transform techniques, recognizing that (7.2.4) can be interpreted as a discrete Fourier transform. Figure 7.4 shows a simple path so generated for $M = 512$.

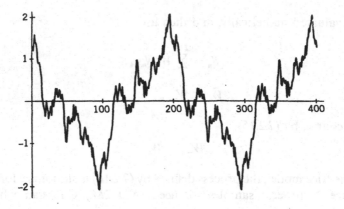

Figure 7.1. Rice model. $M = 50$.

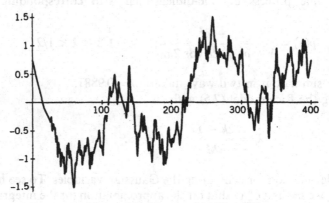

Figure 7.2. Rice model. $M = 100$.

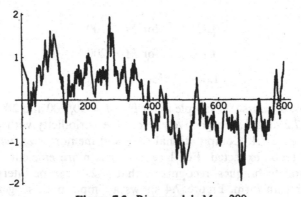

Figure 7.3. Rice model. $M = 200$.

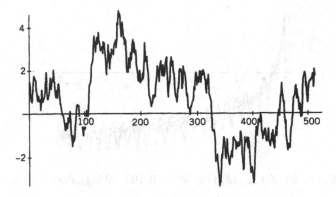

Figure 7.4. $M = 512.$ $\rho = 0.9881.$

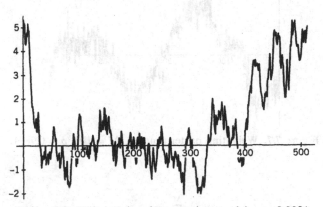

Figure 7.5. Kalman signal generation model. $\rho = 0.9881.$

Next we simulate the same process using the Kalman signal generation model:

$$s_{n+1} = \rho s_n + \sqrt{1 - \rho^2}\, N_n, \qquad \rho = 0.9881,$$

where $\{N_n\}$ is white Gaussian (0.1). Here 800 samples were generated with $s_0 = 0$ and the sample path obtained using samples $n = 300$ to $n = 700$ is shown in Figure 7.5. The steady-state variance of s_n being equal to 1, the peaks close to ± 2 are reasonable.

To simulate process with the weighting pattern model, let

$$W_k = 2 \int_0^{1/2} \sqrt{P(\lambda)}\, \cos 2\pi k\lambda = W_{-k}$$

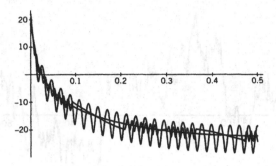

Figure 7.6. Curve for $20 \log_{10} |\psi_M(\lambda)|$. $M = 50, 100$. $20 \log_{10} P(\lambda)$: The smooth curve.

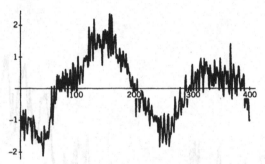

Figure 7.7. Weighting pattern model. $M = 100$, $\rho = 0.9881$.

and let

$$\psi_M(\lambda) = \sum_{-M}^{M} W_k \cos 2\pi k \lambda.$$

To illustrate how large M should be we plot

$$10 \log_{10} |\psi_M(\lambda)|^2$$

for $M = 50, 100$ in Figure 7.6, along with

$$10 \log_{10} P(\lambda).$$

It is seen that $M = 100$ is well within 3db. Figure 7.7 shows a sample path of 400 points of the simulated process using $M = 100$.

In terms of comparison of these methods, we note that with the Rice model, Figure 7.2 with $M = 200$ compares visually well with Figure 7.5. The number of white noise samples used ($4M$) in the Rice model yields $4M$ process samples; however, in the Kalman model we need to use n white noise samples for n

process samples and n must be larger than 300 for steady state. In the weighting pattern model we used 600 white noise samples. The value of $M = 100$ would appear to be inadequate, judging by the larger high-frequency components in relation to the Kalman model. We could compare the sample spectral density and covariance estimates following Chapter 5. A more detailed comparative evaluation would depend on many other factors not only purely computational but also the eventual use to be made of the simulated process.

7.3 CONTINUOUS-TIME PROCESSES

How do we generate a continuous-time process $x(t)$ on a digital computer? The direct answer is: No way. The best we can do is create a sequence such that

$$x_n = x(n\Delta).$$

The values are those of the continuous-time process at sampling times.

Let us illustrate this for the case of interest in control applications where the continuous-time process is defined by a Kalman signal generation model (4.4.1)

$$\dot{x}(t) = Ax(t) + BN(t)$$
$$v(t) = Cx(t), \tag{7.3.1}$$

where $N(\cdot)$ is white Gaussian noise with spectral density D. Let x_0 denote the initial condition at $t = 0$.

Theorem 7.1. Given the process $v(\cdot)$ defined by the state-space description (4.4.1) and a sampling period Δ, we can generate

$$v_n = v(n\Delta)$$
$$x_n = x(n\Delta)$$

for $n \geq 0$, by the discrete-time signal generation model:

$$v_n = Cx_n$$
$$x_{n+1} = e^{A\Delta}x_n + B_\Delta N_n, \tag{7.3.2}$$

where $\{N_n\}$ is a white Gaussian sequence with identity spectral density and

$$B_\Delta B_\Delta^* = \int_0^\Delta e^{A\sigma}BDB^*e^{A^*\sigma}\,d\sigma$$

and initial condition x_0.

Proof. We note that

$$x(\overline{n+1}\,\Delta) = e^{A\Delta}x(n\Delta) + \int_0^{\Delta} e^{A(\Delta-\sigma)}BN(n\Delta + \sigma)\,d\sigma. \qquad (7.3.3)$$

Let

$$\zeta_n = \int_0^{\Delta} e^{A(\Delta-\sigma)}BN(n\Delta + \sigma)\,d\sigma. \qquad (7.3.4)$$

Then

$$E[\zeta_n\,\zeta_m^*] = 0 \qquad\qquad\qquad n \neq m$$

$$= \int_0^{\Delta} e^{A(\Delta-\sigma)}BDB^*e^{A^*(\Delta-\sigma)}\,d\sigma. \qquad n = m.$$

Thus $\{\zeta_n\}$ is white Gaussian with covariance matrix Λ given by

$$\Lambda = \int_0^{\Delta} e^{A\sigma}BDB^*e^{A^*\sigma}\,d\sigma. \qquad (7.3.5)$$

Factorizing Λ as

$$\Lambda = LL^*$$

in one of many ways, we can write

$$\zeta_n = LN_n,$$

where $\{N_n\}$ is white Gaussian with identity covariance. Defining

$$x_n = x(n\Delta)$$

we have

$$x_{n+1} = e^{A\Delta}x_n + LN_n$$

$$v_n = v(n\Delta) = Cx_n,$$

which is of the required form (7.3.2). We note that Λ can be obtained without integration by using the linear equation

$$0 = A\Lambda + \Lambda A^* + BDB^* - e^{A\Delta}BDB^*e^{A^*\Delta} \qquad (7.3.6)$$

(see Problem 4.7). It should be emphasized that (7.3.2) generates the process $\{x(n\Delta)\}$ without any approximation whatever. Note that the system (7.3.2) is stable if the continuous-time model (7.3.1) is, since the eigenvalues of $e^{A\Delta}$ are

$$e^{\Delta(\text{eigenvalues of } A)}.$$

Moreover, the corresponding steady-state covariance is given by

$$E[x_n x_m^*] = e^{A\Delta(n-m)} R_\infty, \qquad n > m.$$

7.3.1 Approximation

Usually we want Δ to be small (but not too small, otherwise too many samples are needed!) so that the integral

$$\int_0^\Delta e^{A\sigma} BDB^* e^{A^*\sigma}\, d\sigma$$

can be approximated by

$$\Delta BDB^*. \qquad (7.3.7)$$

In that case we may define

$$\zeta_n = B_\Delta N_n.$$

where

$$B_\Delta = \sqrt{\Delta}\, B \qquad (7.3.8)$$

and $\{N_n\}$ is white Gaussian noise with covariance $D =$ spectral density of the white Gaussian in the continuous-time model. In practice (7.3.7) suffices to yield a good enough approximation.

Example 7.2. As an example let us consider simulating the response to white noise of a second-order system of Examples 4.5.1:

$$\frac{d^2 v}{dt^2} + 2b\frac{dv}{dt} + cv(t) = N(t). \qquad (7.3.9)$$

Rather than converting this directly into a difference equation, we first go into the state-space form (4.5.2) and thus to (7.3.2) where now

$$\Lambda = \int_0^\Delta e^{At} BDB^* e^{A^*t}\, dt,$$

where e^{At} is given explicitly by (4.5.7). Assuming that Δ is small enough, let us use the approximation

$$B_\Delta = \sqrt{\Delta}\, B$$

and use

$$x_{n+1} = e^{A\Delta} x_n + B_\Delta N_n$$

$$v_n = Cx_n, \qquad (7.3.10)$$

where $\{N_n\}$ is white Gaussian with identity (unit) covariance, being 1×1 in this example.

Let us compare the steady-state covariance between the discrete-time and continuous-time models. From (7.3.10) we obtain

$$R = e^{A\Delta} R e^{A^*\Delta} + BDB^*\Delta. \qquad (7.3.11)$$

If Δ is small enough so that

$$e^{A\Delta} \doteq I + A\Delta,$$

substitution in (7.3.11) yields

$$R = (I + A\Delta)R(I + A^*\Delta) + BDB^*D$$

or

$$0 = \Delta(AR + RA^* + BDB^*), \qquad (7.3.12)$$

which is the Liapunov equation for the steady-state covariance in the continuous-time model. Hence we need Δ small enough so that the approximation

$$e^{A\Delta} \doteq I + A\Delta$$

is reasonable. Or

$$e^{\lambda_i\Delta} \doteq 1 + \lambda_i\Delta$$

for every eigenvalue λ_i of A, or minimally we need that

$$\frac{r^2\Delta^2}{2} < 1,$$

where

$$r = \max_i |\lambda_i| = \text{spectral radius of } A.$$

Hence

$$\Delta < \frac{1}{r} \qquad (7.3.13)$$

is a minimal requirement. In this particular example, we obtain

$$\lambda_1 = -b + \sqrt{b^2 - c}$$
$$\lambda_2 = -b - \sqrt{b^2 - c}.$$

If

$$c < b^2,$$

we need

$$\Delta < \frac{1}{b + \sqrt{b^2 - c}}.$$

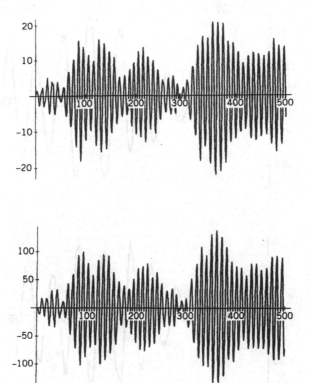

Figure 7.8. Simulation: Example 7.2. Case 1, $\Delta = 0.1$.

On the other hand, if

$$c > b^2$$

we have that

$$r = \sqrt{c}.$$

Hence

$$\Delta < \frac{1}{\sqrt{c}}.$$

Note that in this case

$$\sqrt{c} = 2\pi f_c$$

is the natural (angular) frequency of the system and

$$\Delta < \frac{1}{2\pi f_c}$$

is consistent with the Shannon Sampling Principle. Thus the spectral density

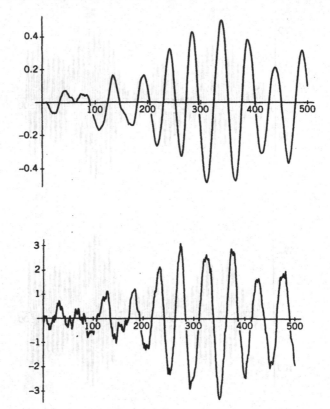

Figure 7.9. Simulation: Example 7.2. Case 1, $\Delta = 0.02$.

$(c > b^2)$ of the continuous-time process $v(\cdot)$ is (compare Chapter 6)

$$\frac{1}{|(-4\pi^2 f^2 + 4\pi i b f + c|^2}$$

$$= \frac{1}{|2\pi i f + b + i\sqrt{c - b^2}|^2 + |2\pi i f + b - i\sqrt{c - b^2}|^2}$$

$$= \frac{1}{(b^2 + (\sqrt{c - b^2} - 2\pi f)^2)(b^2 + (\sqrt{c - b^2} + 2\pi f)^2)}$$

peaks at (has a maximum for positive frequencies) at

$$2\pi f = \sqrt{c - b^2}$$

decreasing rapidly thereafter. The spectral density is 3 db down from the peak

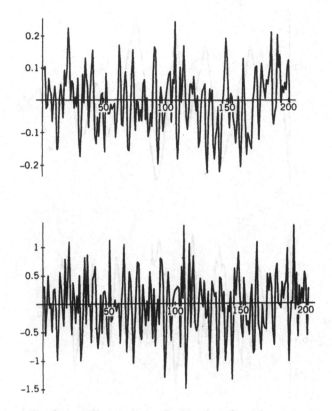

Figure 7.10. Simulation: Example 7.2. Case 2, $\Delta = 0.3$.

at approximately

$$f = \frac{b + \sqrt{c - b^2}}{2\pi}$$

and the Shannon Sampling Principle would suggest

$$\Delta < \frac{2\pi}{2(b + \sqrt{c - b^2})}. \qquad (7.3.14)$$

Again this is especially good when damping ($\approx b$) is small as is common in modeling oscillatory systems. In general, one may work with the diagonal terms in the spectral density matrix to get a reasonably optimal sampling interval since too small a sampling interval would entail too many samples to work with.

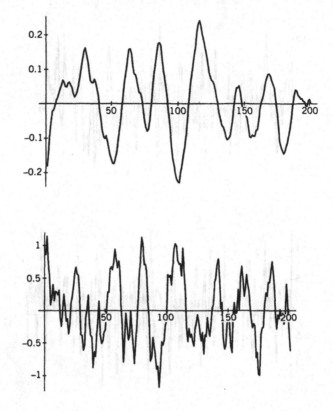

Figure 7.11. Simulation: Example 7.2. Case 2, $\Delta = 0.5$.

7.3.2   Simulation

Case 1: Low Damping

$$\sqrt{c} = 6.2832, \qquad b = 0.01$$

$$\frac{\sqrt{c - b^2}}{2\pi} \approx 1.$$

(7.3.14) yields

$$\Delta = \frac{1}{2},$$

whereas (7.3.13) yields

$$\Delta < \frac{1}{6.2832}.$$

Sample paths (for components 1 and 2) using $\Delta = 1/10$ and $\Delta = 1/50$ are shown in Figures 7.8 and 7.9 respectively. Note that increasing sampling rate improves fidelity, but for the same number N of samples generated the actual time interval is less. In Figure 7.8 we have $N\Delta = 50$ seconds, or 50 periods at the center frequency of 1 Hz, whereas in Figure 7.9, $N\Delta = 10$ seconds or 10 periods only.

Case 2: High Damping Case (taken from Problem 4.1)

$$c = 13.818, \qquad b = 1.59, \qquad \sqrt{c - b^2} = 3.36$$

(7.3.14) yields

$$\Delta < 0.6,$$

while (7.3.13) yields

$$\Delta < 0.3.$$

The period corresponding to the angular frequency of 3.36 being 1.9 seconds, Δ of 0.3 yields 6 samples per period. Figure 7.10 shows sample paths obtained for $\Delta = 0.3$ for $N = 200$. The sampling rate is clearly inadequate. Figure 7.11 shows the improvements with $\Delta = 0.5$, but the sampling rate is still not quite adequate for $\dot{v}(t)$. Thus (7.3.14) and (7.3.13) are at best only lower bounds for Δ.

PROBLEMS

7.1. Simulate a band-limited white noise process sampled at half the Shannon Rate. Simulate the same process using the Rice approximation. Compare sample paths.

7.2. Simulate the continuous-time processes with spectral densities given in Problem 2.20—cases (i) and (iii) using an appropriate sampling period. Compare the Kalman signal generation model with the Rice model. Compare the discrete Fourier transforms (absolute values).

7.3. Simulate the process in Problem 4.1 with $\Delta = 0.02$. Using time averages to check fidelity.

NOTES AND COMMENTS

The Rice model was in vogue in the days of analog computers. Books on simulation (e.g., reference 2) deal with simulation of IID sequences with given distribution rather than general processes.

REFERENCES

1. D. E. Knuth. *The Art of Computer Programming*. Chapter 3, Volume 2. Addison-Wesley, 1969.

2. P. Bradley, B. L. Fox, and L. E. Schrage. *A Guide to Simulation.* Springer-Verlag, 1983.

3. G. Marsaglia and A. Zaman. Toward a Universal Random Number Generator. Florida State University Report, FSU-SCRI-87-50, 1987.

4. A. I. McLeod. "Remark AS 458. A Remark on Algorithm AS 182. An Efficient and Portable Pseudo-random Number Generator," *Applied Statistics*, vol. 34 (1985), pp. 189–200.

8

RANDOM FIELDS

Random field models play an essential role in processing spatially distributed data. Two major areas of application we can mention are: (i) geophysical data of various kinds (gravitational anomaly is but one example) and (ii) turbulence (in particular the refractive index field as it affects electromagnetic/acoustic wave propagation).

In this chapter we present an elementary account of the second-order theory: correlation and spectrum of homogeneous fields, drawing on and extrapolating from the random process theory developed in the previous chapters. No essentially new tools are needed.

We shall consider random processes $x(t)$, $t \in \mathbf{R}^2$ (Euclidean two-space), and $t \in \mathbf{R}^3$ (Euclidean three-space).

We begin with $T = \mathbf{R}^3$ and distinguish the processes by adopting the name:

8.1 RANDOM FIELDS IN THREE-SPACE

We shall use r in place of t just to remind us that we are not dealing with time any more as the index. Also we shall use the notation $f(r)$ in place of $x(t)$; for each r the random variable $f(r)$ moreover will be restricted to be 1×1 (one-dimensional), since our concern is primarily with what the spatial dependence entails. Also we shall use

$$r = ix + jy + kz \tag{8.1.1}$$

in the usual coordinate component notation, or, when appropriate,

$$r = ix_1 + jx_2 + kx_3 \tag{8.1.2}$$

$$\lambda = i\lambda_1 + j\lambda_2 + k\lambda_3 \qquad (8.1.3)$$

$$r = |r| = \sqrt{x_1^2 + x_2^2 + x_3^2}; \qquad \lambda = |\lambda| = \sqrt{\lambda_1^2 + \lambda_2^2 + \lambda_3^2}.$$

We shall use

$$|dr|, \qquad |d\lambda|$$

to denote the volume element in three-space. We shall continue to use the inner product notation

$$[\lambda, r] = x_1\lambda_1 + x_2\lambda_2 + x_3\lambda_3.$$

Give any unit vector e we may use spherical coordinates with respect to e defined by

$$x = r \sin \phi \cos \theta$$

$$y = r \sin \phi \sin \theta \qquad (8.1.4)$$

$$z = r \cos \phi,$$

where

$$[r, e] = r \cos \phi,$$

and by the corresponding volume element

$$r^2 \sin \phi \, d\phi \, d\theta \, dr, \qquad 0 \le \phi \le \pi; 0 \le \theta \le 2\pi; 0 \le r < \infty. \qquad (8.1.5)$$

8.1.1 Homogeneous Fields

The main statistical quantities of interest to us are the mean

$$m(r) = E[f(r)] \qquad (8.1.6)$$

and the covariance

$$R(r_1, r_2) = E[f(r_1)f(r_2)] - m(r_1)m(r_2). \qquad (8.1.7)$$

As before we are concerned primarily with stationary (or second-order stationary) processes. Thus we call a random field stationary (or "homogeneous," a term reserved for random fields exclusively) if

$$m(r) = \text{constant}$$

and

$$R(r_1, r_2) = R(r_1 - r_2). \qquad (8.1.8)$$

Examples of Homogeneous Fields   Here are some elementary examples, beginning perhaps with the least interesting: Given three independent stationary processes

$$x_i(t), \qquad -\infty < t < \infty, i = 1, 2, 3$$

we define

$$f(r) = x_1(x)x_2(y)x_3(z)$$

with r given by (8.1.1). The field is clearly homogeneous and

$$R(r) = R_1(x)R_2(y)R_3(z), \tag{8.1.9}$$

where $R_i(\cdot)$ is the covariance function of the process $x_i(\cdot)$. Also the Fourier transform of $R(\cdot)$ is given by

$$\phi(\lambda) = \int_{-\infty}^{\infty} e^{2\pi i[\lambda, r]} R(r) |dr| = p_1(\lambda_1)p_2(\lambda_2)p_3(\lambda_3), \qquad \lambda \in \mathbf{R}^3 \tag{8.1.10}$$

with λ as in (8.1.3), where $p_i(\cdot)$ are the spectral densities corresponding to $R_i(\cdot)$.

Our next example is modeled on Example 1.1.3—the Rice model. Let A_k, ϕ_k be as in Example 1.1.3, and let

$$f(r) = \sum_{1}^{N} A_k \sin(2\pi[\lambda_k, r] + \phi_k), \tag{8.1.11}$$

where λ_k are fixed elements in R^3. The field is homogeneous, with covariance function

$$R(r) = \sum_{1}^{N} \sigma_k^2 \cos(2\pi[\lambda_k, r]). \tag{8.1.12}$$

As in the random process case, we call a field Gaussian if the joint distributions of $R(r_i)$. $i = 1, \ldots, N$, are Gaussian. The field defined by (8.1.11) is clearly Gaussian.

From the definition it is immediate that

$$R(r) = R(-r). \tag{8.1.13}$$

Also, just as in the case of stationary processes, the mean feature of the stationary covariance function is the "positive definite" property (cf. (2.3)):

$$\sum_{1}^{N} \sum_{1}^{N} a_i R(r_i - r_j) \bar{a}_j \geq 0 \tag{8.1.14}$$

as is immediate from the fact that the left side

$$= E \left| \sum_{1}^{N} a_i f(r_i) \right|^2.$$

We therefore expect the Bochner theorem (cf. Chapter 2) to hold. It does and, as in Chapter 2, we shall state a version paralleling Theorem 2.5 that is adequate for our purposes.

Theorem 8.1.1. Let $R(r)$, $r \in \mathbf{R}^3$, be a continuous positive definite function satisfying (8.1.13) and (8.1.14), such that

$$\int_{\mathbf{R}^3} |R(r)| \, |dr| < \infty. \tag{8.1.15}$$

Then we can find a continuous nonnegative function $\phi(\lambda)$, $\lambda \in \mathbf{R}^3$, such that

$$R(r) = \int_{\mathbf{R}^3} e^{2\pi i[\lambda, r]} \phi(\lambda) |d(\lambda)|. \tag{8.1.16}$$

Proof. We shall give a brief outline of the proof only, since it parallels completely the proof in Theorem 2.5. Thus we define

$$\phi_T(\lambda) = \frac{1}{T^3} \int_{B_T} \int_{B_T} e^{2\pi i[\lambda, r-s]} R(r-s) \, |dr| \, |ds|, \tag{8.1.17}$$

where B_T is the box in R^3 defined by

$$0 \leq x_i \leq T, \quad i = 1, 2, 3.$$

By the positive definite property we have

$$\phi_T(\lambda) \geq 0,$$

and by a change of variable, as in Chapter 2, we obtain

$$\phi_T(\lambda) = \int_{E_T} e^{2\pi i[\lambda, \tau]} R(\tau) \left(1 - \frac{|\tau_1|}{T}\right)\left(1 - \frac{|\tau_2|}{T}\right)\left(1 - \frac{|\tau_3|}{T}\right) d\tau_1 \, d\tau_2 \, d\tau_3, \tag{8.1.18}$$

where

$$E_T = [-T \leq \tau_i \leq T, i = 1, 2, 3] \subset \mathbf{R}^3$$

$$\tau = i\tau_1 + j\tau_2 + k\tau_3.$$

As $T \to \infty$, $\phi_T(\lambda)$ converges to a function $\phi(\lambda)$ for each λ in $\mathbf{R}^3$, satisfying (8.1.16), by arguments paralleling those in the proof of Theorem 2.5.

As before, the function $\phi(\cdot)$ in (8.1.16) is called the "spectral density" of the process and is the Fourier transform of the covariance function:

$$\phi(\lambda) = \int_{\mathbf{R}^3} e^{-2\pi i[\lambda, r]} R(r)|dr|, \tag{8.1.19}$$

from which we see that $\phi(\cdot)$ is continuous because of (8.1.15).

Remark. As in the case of continuous-time process, we extend spectral densities to include delta functions; thus the Fourier transform of the covariance function (8.1.12) is the sum of (now three-dimensional) delta functions.

Isotropic Fields It turns out that fields of physical interest, at the primordial level as in turbulence [1], have more structure than implied by homogeneity alone. The new feature, peculiar to fields, is that of isotropy. Thus we say that a random field is isotropic if it is homogeneous and, further, the stationary covariance $R(\cdot)$ is such that

$$R(r) = B(r), \qquad r = |r|. \tag{8.1.20}$$

The covariance does not depend on direction either—only on magnitude. (The field defined by (8.1.11) is *not* isotropic!)

We can give a simple characterization of the covariance function of isotropic fields. We assume (8.1.15) which in turn implies that

$$\int_{\mathbf{R}^3} \phi(\lambda)|d\lambda| < \infty, \tag{8.1.21}$$

being equal in fact to $R(0)$ by Theorem 8.1.1.

Theorem 8.1.2. Suppose (8.1.20) holds. Then the spectral density satisfies

$$\phi(\lambda) = Q(\lambda), \qquad \lambda = |\lambda|. \tag{8.1.22}$$

Conversely, if the spectral density satisfies (8.1.22), then the field is isotropic; that is, (8.1.20) holds.

Proof. We shall actually prove a little bit more than the statement of the theorem. Thus suppose (8.1.20) holds. Then

$$\phi(\lambda) = \int_{\mathbf{R}^3} e^{-2\pi i[\lambda, r]} B(r)|dr|. \tag{8.1.23}$$

For fixed λ, Let

$$e = \frac{\lambda}{\lambda}$$

so that

$$[\lambda, r] = \lambda r \cos \phi.$$

Choose spherical coordinates as in (8.1.4), with z-axis along e. Then

$$\phi(\lambda) = \int_0^\infty \int_0^{2\pi} \int_0^\pi B(r) e^{-2\pi i \lambda r \cos \phi} r^2 \sin \phi \, d\phi \, d\theta \, dr$$

$$= 2\pi \int_0^\infty B(r) r^2 \left(\int_0^\pi e^{-2\pi i \lambda r \cos \phi} \sin \phi \, d\phi \right) dr.$$

Now for $\lambda > 0$,

$$\int_0^\pi e^{-2\pi i \lambda r \cos \phi} \sin \phi \, d\phi = \frac{\sin 2\pi\lambda r}{\pi\lambda r}.$$

Hence

$$\phi(\lambda) = 2 \int_0^\infty \frac{\sin 2\pi\lambda r}{\lambda} B(r) r \, dr = Q(\lambda) \tag{8.1.24}$$

or $\phi(\lambda)$ is a function of λ only, proving the first part of the theorem. Conversely, if (8.1.22) holds, we have

$$R(r) = \int_{\mathbf{R}^3} e^{2\pi i [\lambda, r]} Q(\lambda) |d\lambda|.$$

And by the same arguments in dealing in (8.1.23), or directly from (8.1.24), the right side

$$= 2 \int_0^\infty \left(\frac{\sin 2\pi\lambda r}{r} \right) \lambda Q(\lambda) \, d\lambda = d\lambda = B(r). \tag{8.1.25}$$

Or the field is homogeneous. Note that

$$B(0) = 4\pi \int_0^\infty \lambda^2 Q(\lambda) \, d\lambda < \infty.$$

We have proved a little more in that we have now the representations (8.1.24) and (8.1.25), which we shall exploit next.

Let

$$P(\lambda) = 2 \int_0^\infty \cos 2\pi t \lambda B(t) \, dt \tag{8.1.26}$$

$$= \int_{-\infty}^\infty e^{-2\pi i t \lambda} B(t) \, dt \tag{8.1.27}$$

defining

$$B(-|t|) = B(|t|).$$

Now we can readily see that, thus defined, $B(t)$ is a covariance function in $-\infty < t < \infty$. Indeed

$$\sum \sum a_i B(t_k - t_j)\bar{a}_j = \sum \sum a_i R(t_k e - t_j e)\bar{a}_j \geq 0, \qquad (8.1.28)$$

where e is a unit vector in $\mathbf{R}^3$. Hence

$$P(\lambda) \geq 0.$$

Or $P(\lambda)$ is a spectral density in $-\infty < \lambda < \infty$. Also, from (8.1.15) we obtain

$$\int_0^\infty r^2 |B(r)| \, dr < \infty. \qquad (8.1.29)$$

Hence

$$\int_1^\infty r |B(r)| \, dr < \infty$$

and if $B(\cdot)$ is continuous in $0 \leq t < \infty$, which follows from the continuity of $R(\cdot)$, we have that

$$\int_0^\infty r |B(r)| \, dr < \infty.$$

Hence in (8.1.26), we can differentiate with respect to λ inside the integral to obtain

$$P'(\lambda) = -2 \int_0^\infty 2\pi(\sin 2\pi\lambda r) r B(r) \, dr = -2\pi\lambda\phi(\lambda)$$

or

$$\phi(\lambda) = \frac{-P'(\lambda)}{2\pi\lambda}, \qquad 0 < \lambda, \qquad (8.1.30)$$

yielding a representation for the spectral density of any homogeneous and isotropic field in three-space. We can use it to generate such spectral densities starting from a spectral density in one variable. From (8.1.26) we also have that

$$B(r) = 2 \int_0^\infty \cos \pi\lambda r \, P(r) \, d\lambda. \qquad (8.1.31)$$

In an entirely similar—or "symmetric"—fashion we have a representation for the covariance function of an isotropic field:

$$R(r) = \frac{-C'(r)}{2\pi r}, \qquad (8.1.32)$$

where $C(t)$, $-\infty < t < \infty$, is a continuous stationary covariance function given by

$$C(r) = 2 \int_0^\infty (\cos 2\pi\lambda r) Q(\lambda) \, d\lambda, \qquad r > 0 \qquad (8.1.33)$$

and which is twice continuously differentiable.

We can use (8.1.30) to generate examples of spectral densities of isotropic fields in three-space. Here are some which appear in applications.

1. For

$$P(\lambda) = \frac{1}{(k^2 + 4\pi^2\lambda^2)^\nu}, \qquad \nu \geq 1$$

(ν not necessarily an integer!) we have

$$\phi(\lambda) = \frac{4\pi\nu}{(k^2 + 4\pi^2\lambda^2)^{\nu+1}}, \qquad (8.1.34)$$

$$B(r) = \frac{1}{\Gamma(\nu + (1/2))} \left(\frac{r}{2k}\right)^{\nu-(1/2)} K_{\nu-(1/2)}(kr) \qquad (8.1.35)$$

where $K_n(\cdot)$ is the modified Hankel function of order n (see reference 9, p. 172; see also reference 15 of the Review Chapter).

2. For

$$P(\lambda) = e \exp\left(\frac{-\sigma^2 4\pi^2\lambda^2}{2}\right)$$

$$\phi(\lambda) = 2\pi\sigma^2 \exp\left(-2\sigma^2\pi^2\lambda^2\right)$$

$$B(r) = \frac{1}{\sqrt{2\pi}\,\sigma} \exp\left(\frac{-r^2}{2\sigma^2}\right).$$

Here is an example in which $P(\lambda)$ in (8.1.26) is not defined. Let $k > 0$ and define

$$B(r) = \frac{1}{4\pi r} \exp\left(-kr\right), \qquad 0 < r.$$

Note that $B(r)$ is not continuous at $r = 0$. But we can use (8.1.24) to obtain

$$\phi(\lambda) = Q(\lambda) = \frac{1}{4\pi\lambda} \int_0^\infty 2 \sin 2\pi\lambda r \exp\left(-kr\right) dr = \frac{1}{k^2 + 4\pi^2\lambda^2}.$$

Note that

$$\int_0^\infty \lambda^2 Q(\lambda) \, d\lambda = \infty$$

and of course $B(0+)$ is not defined.

Structure Functions A useful construct for isotropic fields is the Kolmogorov structure function (see reference 1). Because of its importance in applications (adaptive optics, see reference 5) we shall discuss it, however briefly.

Given an isotropic field $f(r)$, $r \in \mathbf{R}^3$, the structure function $D(r)$ is defined to be

$$D(r) = E[(f(r) - f(0))^2], \quad r \in \mathbf{R}^3$$

or, equivalently, for arbitrary r_0 in $\mathbf{R}^3$:

$$= E[(f(r_0 + r) - f(r_0))^2].$$

Now it is readily seen that

$$D(r) = 2[R(0) - R(r)]$$
$$D(r) = 2[B(0) - B(r)]. \tag{8.1.36}$$

We can also evaluate $D(\cdot)$ from the spectral density function, using (8.1.25):

$$D(r) = 8\pi \int_0^\infty \left(1 - \frac{\sin 2\pi \lambda r}{2\pi \lambda r}\right) \lambda^2 Q(\lambda)\, d\lambda. \tag{8.1.37}$$

More importantly, however, we can express the spectral density in terms of the structure function. In fact, we have the following formula due to Tatarski [1]:

$$\phi(\lambda) = \frac{1}{2\pi\lambda^2} \int_0^\infty \frac{\sin 2\pi \lambda r}{2\pi \lambda r} \frac{d}{dr} [r^2 D'(r)]\, dr. \tag{8.1.38}$$

Deriving this formula is straightforward. Assuming

$$\int_0^\infty \lambda^3 Q(\lambda)\, d\lambda < \infty$$

we can do the differentiations inside the integral sign in (8.1.37). Thus

$$r^2 D'(r) = 8\pi \int_0^\infty \left(-r \cos 2\pi \lambda r + \frac{\sin 2\pi \lambda r}{2\pi \lambda}\right) \lambda^2 Q(\lambda)\, d\lambda$$

$$\frac{d}{dr}(r^2 D'(r)) = 8\pi \int_0^\infty (2\pi \lambda r)(\sin 2\pi \lambda r) \lambda^2 Q(\lambda)\, d\lambda.$$

Hence

$$2 \int_0^\infty (\sin 2\pi \lambda r) \lambda^3 Q(\lambda)\, d\lambda = \frac{1}{8\pi^2} \frac{1}{r} \frac{d}{dr}(r^2 D'(r)).$$

Hence, inverting the sine-transform (considering it as the Fourier transform of an odd function, see Problem 8.1) we have

$$\lambda^3 Q(\lambda) = 2 \int_0^\infty \sin 2\pi\lambda r \, \frac{1}{8\pi^2 r} \frac{d}{dr} (r^2 D'(r)) \, dr$$

and hence $Q(\cdot)$ is given by (8.1.38).

The main interest in the structure function is for small r. We can readily see that for the spectral density given by (8.1.34), we obtain

$$D(r) = 2(B(0) - B(r)) = 4 \int_0^\infty \frac{1 - \cos 2\pi\lambda r}{(k^2 + 4\pi^2\lambda^2)^\nu} \, d\lambda.$$

By a change of variable, the integral

$$= 2 \frac{r^{2\nu-1}}{\pi} \int_0^\infty \frac{1 - \cos t}{(t^2 + r^2 k^2)^\nu} \, dt.$$

Hence for $2\nu > 1$,

$$D(r) \le \left(\frac{2}{\pi} \int_0^\infty \frac{1 - \cos t}{t^{2\nu}} \, dt \right) r^{2\nu-1}$$

and as $r \to 0+$

$$r^{1-2\nu} D(r) \to \left(\frac{2}{\pi} \int_0^\infty \frac{1 - \cos t}{t^{2\nu}} \, dt \right).$$

For $\nu = 5/6$ (corresponding to the Kolmogorov turbulence spectrum [1]) we thus have

$$D(r) \approx (\text{constant}) r^{2/3} \qquad \text{for small } r.$$

Note, however, that as $r \to \infty$, $D(r) \to B(0)$.

We can also express $P(\lambda)$ in terms of the structure function. Integrating by parts in (8.1.26) and noting that

$$-B'(r) = D'(r)$$

we have

$$\lambda P(\lambda) = \frac{1}{\pi} \int_0^\infty D'(t) \sin 2\pi\lambda t \, dt,$$

which by a change of variable in the integral, can be expressed

$$\lambda^2 P(\lambda) = \frac{1}{\pi} \int_0^\infty D'\left(\frac{x}{\lambda}\right) \sin 2\pi x \, dx.$$

We see that the spectral density as $\lambda \to \infty$ is determined by $D(r)$ as $r \to 0$. For

$$D(r) = r^\nu \qquad \text{as } r \to 0$$

and assuming for example that $D'(r)$ is positive, nondecreasing, and goes to zero as $r \to \infty$ (a mathematical technicality to ensure the limit of the integral is the integral of the limiting integrand), we have

$$\lambda^{1+\nu} P(\lambda) \to \frac{\nu}{\pi} \int_0^\infty x^{\nu-1} \sin 2\pi x \, dx,$$

with the integral on the right interpreted as

$$\lim_{L \to \infty} \int_0^L x^{\nu-1} \sin 2\pi x \, dx.$$

Or

$$P(\lambda) = \frac{\text{constant}}{\lambda^{1+\nu}} \qquad \text{as } \lambda \to \infty$$

and by (8.1.31) we have

$$Q(\lambda) = \frac{\text{constant}}{\lambda^{3+\nu}} \qquad \text{as } \lambda \to \infty.$$

It should be noted that

$$Q(\lambda) = \frac{1}{\lambda^{3+\nu}} \qquad \lambda > 1$$

$$= 0 \qquad \lambda < 1$$

will yield the same structure function for small r as the Kolmogorov spectrum (8.1.34).

It should be noted that the Kolmogorov ("power") spectral density (8.1.34) for $\nu = 11/6$ is such that the corresponding "line" spectral density

$$P(\lambda) = \frac{1}{(k^2 + 4\pi^2\lambda^2)^{5/6}}$$

violates the condition

$$\int_0^\infty \lambda^2 P(\lambda) \, d\lambda < \infty.$$

which is necessary for the existence of a mean square derivative (gradient of the field). If on the other hand this condition holds, the structure function is given by

$$D(r) = 2 \int_0^\infty (1 - \cos 2\pi\lambda r) P(\lambda) \, d\lambda = 4 \int_0^\infty \sin^2 \pi\lambda r P(\lambda) \, d\lambda$$

$$\leq r^2 \int_0^\infty 4\pi^2 \lambda^2 P(\lambda) \, d\lambda$$

and hence is of the order of r^2 as $r \to 0$! See Problem 8.16 for more on this.

Response of Linear Homogeneous Systems   By a linear homogeneous (translation invariant) system we may mean the "input-output" relations given by a convolution transform

$$g(r) = \int_{\mathbf{R}^3} W(r - r')f(r')|dr'|, \qquad r \in \mathbf{R}^3, \qquad (8.1.39)$$

where $W(\cdot)$ is a "weighting-pattern," $f(\cdot)$ is the "input" field, and $g(\cdot)$ is the "output" field. We can see that $g(\cdot)$ is a homogeneous field if the input field is. Moreover, the corresponding spectral density is given by (analogous to (4.2.7))

$$\phi_g(\lambda) = |\psi(\lambda)|^2 \phi(\lambda),$$

where

$$\psi(\lambda) = \int_{\mathbf{R}^3} e^{-2\pi i[\lambda, r]} W(r)|dr|. \qquad (8.1.40)$$

In particular we can define a "white noise" field in three-space, imitating our procedure in Chapter 4. Thus if $f(\cdot)$ is white noise in three-space, for any continuous function $h(\cdot)$ in the box B: $-\infty < a_i \le \lambda_i \le b_i < \infty$, we must have

$$E\left(\int_B h(r)f(r)|dr| \right)^2 = \int_B |h(r)|^2 |dr|.$$

The corresponding covariance function is a delta function in three-space. We may create a white-noise process imitating the procedure in Chapter 4: Take any homogeneous process $f(\cdot)$ with spectral density $\phi(\cdot)$, such that

$$\phi(0) \ne 0.$$

Define

$$f_n(r) = n^{3/2} f(nr).$$

Then

$$E\left[\left(\int_B h(r)f_n(r)|dr| \right)^2 \right] = \int_{\mathbf{R}^3} |\psi(\lambda)|^2 \phi\left(\frac{\lambda}{n} \right) |d\lambda|$$

$$\to \phi(0) \int_B |h(r)|^2 |dr|.$$

Or we have white noise with spectral density $\phi(0)$ in the limit.

It is interesting to note that white noise is isotropic:

$$\phi(\lambda) = Q(\lambda) = \text{constant}.$$

On the other hand, if we take a line scan, the corresponding one-dimensional spectral density is $+\infty$, or is not defined! See Problem 8.6. It should be noted that there is no analog of the "physical realizability" condition on weighting functions.

If $W(r)$ is also "isotropic" in the sense that

$$W(r) = B(r),$$

the Fourier transform (8.1.40) takes the form

$$\psi(\lambda) = 2 \int_0^\infty \frac{\sin 2\pi\lambda r}{\lambda} B(r) r \, dr. \qquad (8.1.41)$$

In particular in (8.1.39), if $f(\cdot)$ is white noise we have that $g(\cdot)$ is isotropic if the weight function $W(\cdot)$ and the spectral density of $g(\cdot)$ is given by

$$Q(\lambda) = |\psi(\lambda)|^2. \qquad (8.1.42)$$

In fact for given isotropic spectral density $Q(\cdot)$, we may define

$$\psi(\lambda) = \sqrt{Q(\lambda)}$$

and correspondingly

$$W(r) = \int_{R^2} e^{2\pi i[\lambda, r]} \sqrt{Q(\lambda)} \, |d\lambda| = 2 \int_0^\infty \frac{\sin 2\pi\lambda r}{r} \lambda \sqrt{Q(\lambda)} \, d\lambda. \qquad (8.1.43)$$

For example, if

$$Q(\lambda) = 2\pi\sigma^2 \exp\left(-2\sigma^2\pi^2\lambda^2\right)$$

we can see that

$$W(r) = H(r) = \frac{1}{\pi} \exp\left(\frac{-r^2}{2\pi^2\sigma^2}\right).$$

For

$$Q(\lambda) = \frac{1}{(k^2 + 4\pi^2\lambda^2)^{2\gamma}}, \qquad \gamma \geq 1,$$

$W(r)$ is the covariance function corresponding to the spectral density

$$\frac{1}{(k^2 + 4\pi^2\lambda^2)^\gamma}$$

and can be calculated from (8.1.35). In particular, we see that every homogeneous random field can be considered as the response of white noise to a linear translation-invariant system. Moreover, the covariance function can be expressed as

$$R(r) = \int_{\mathbf{R}^3} W(r + r') W(r') |dr'|, \tag{8.1.44}$$

where

$$W(r) = \int_{\mathbf{R}^3} e^{2\pi i[\lambda, r]} \sqrt{\phi(\lambda)} \, |d\lambda|. \tag{8.1.45}$$

Signal Generation Models   Random field theory is distinguished by lack of useful signal generation models, such as the Kalman model in continuous-time processes. This is because of the fundamental difference between space and time—that unlike time there is no (natural) arrow in space to yield "past" and "future." We may consider partial differential equations with white-noise forcing terms, but no general representation theory is available at the present time. Thus "evolutionary" simulation of random fields in three-space via Kalman-type models is largely an open problem.

Line Scans   Of course if we consider a "line scan"—that is, the field along a line

$$r = r_0 + te,$$

where e is a fixed unit vector and r_0 is a fixed vector then

$$f(r) = g(te)$$

yields a random process in $-\infty < t < \infty$ which is stationary if the field is homogeneous. The corresponding spectral density does not depend on e if moreover the field is isotropic. Thus the problem is reduced to one where the random process theory we have developed can be brought to bear. In particular the Ergodic Principle and the Sampling Principle apply. The spectral density of the process $f(te)$, $-\infty < t < \infty$, is given by $P(\lambda)$, defined by (8.1.26). Conversely, the line spectral density determines the field spectral density through (8.1.30). From (8.1.30) we see in particular that

$$P'(\lambda) \le 0$$

or that the spectral distribution must be convex for $\lambda \ge 0$. See also Problem 8.3.

Gradient   The gradient $\nabla f(r)$ of a scalar field $f(r)$ is defined by

$$[\nabla f(r), e] = \frac{d}{dt} f(r + te)\Big|_{t=0}, \qquad |e| = 1.$$

Thus defined $\nabla f(r)$ is a 3×1 vector field. If the field $f(r)$ is homogeneous, so is the gradient. In fact it is readily verified that the covariance matrix is given by

$$E[(\nabla f(r_1))(\nabla f(r_2))^*] = 4\pi^2 \int_{\mathbf{R}^3} e^{2\pi i[\lambda, r_2 - r_1]} \lambda \lambda^* \phi(\lambda) \, d|\lambda|, \quad (8.1.46)$$

where

$$\lambda = \begin{vmatrix} [\lambda, i] \\ [\lambda, j] \\ [\lambda, k] \end{vmatrix}$$

and $\phi(\cdot)$ is the spectral density of the field $f(\cdot)$. Hence the gradient field is also homogeneous, with spectral density matrix

$$4\pi^2 \lambda \phi(\lambda) \lambda^*, \quad \lambda \in \mathbf{R}^3.$$

In particular,

$$E[\nabla f(r_1), (\nabla f(r_2))] = \text{Tr } E[(\nabla f(r_1))(\nabla f(r_2))^*]$$

$$= \int e^{2\pi i[\lambda, r_2 - r_1]} 4\pi^2 |\lambda|^2 \phi(\lambda) \, d|\lambda| \quad (8.1.47)$$

$$= -\nabla^2 R(r_2 - r_1), \quad (8.1.48)$$

generalizing (4.6.2). If the field $f(\cdot)$ is isotropic in addition, we can readily verify that the spectral density matrix in (8.1.47) is diagonal and, in fact, a multiple of the identity matrix. Moreover, (8.1.48) can be expressed as

$$= 16\pi^3 \int_0^\infty \left(\frac{\sin 2\pi \lambda r}{2\pi \lambda r} \right) \lambda^4 Q(\lambda) \, d\lambda, \quad (8.1.49)$$

where $r = |r_2 - r_1|$. In particular, we have that

$$E[|\nabla f(r)|^2] = 16\pi^3 \int_0^\infty \lambda^4 Q(\lambda) \, d\lambda, \quad (8.1.50)$$

where the integral has to be finite in order that the gradient is definable in the mean square sense, analogous to the derivative in the one-dimensional case discussed in Chapter 4. Also we can express (8.1.49) in terms of the spectral density $P(\cdot)$, corresponding to any line scan, as

$$= 12\pi^2 \int_{-\infty}^\infty \lambda^2 P(\lambda) \, d\lambda \quad (8.1.51)$$

using (8.1.31). Indeed we can verify that for any unit vector e,

$$E[|[\nabla f(r), e]|^2] = 4\pi^2 \int_{-\infty}^{\infty} \lambda^2 P(\lambda)\, d\lambda \tag{8.1.52}$$

from the fact that the derivative

$$\frac{d}{dt} f(r - te)$$

has the spectral density (cf. Chapter 4)

$$4\pi^2 \lambda^2 P(\lambda).$$

8.2   RANDOM FIELDS IN TWO-SPACE

Let us next consider random fields in two-space, concentrating on aspects different from the theory developed for three-space. Thus let $f(r)$, $r \in \mathbf{R}^2$, denote the field and otherwise follow the notation of Section 8.1. The concept of stationarity and the Bochner theorem relating spectral density to the covariance function can be carried through as in Section 8.1.

8.2.1   Isotropic Fields

If in addition, the field is isotropic so that

$$R(r) = B(r), \qquad r = ix + jy, r = \sqrt{x^2 + y^2}, \tag{8.2.1}$$

we would have for the spectral density

$$\lambda = i\lambda_1 + j\lambda_2$$

$$\lambda = \sqrt{\lambda_1^2 + \lambda_2^2}$$

$$\phi(\lambda) = \int_{-\infty}^{\infty} \int_{-\infty}^{\infty} e^{-2\pi i(\lambda_1 x + \lambda_2 y)} B(r)\, dx\, dy$$

$$= \int_{0}^{\infty} \int_{0}^{2\pi} e^{-2\pi i \lambda r \cos\theta} B(r) r\, d\theta\, dr$$

$$= \int_{0}^{\infty} 2\pi J_0(2\pi \lambda r) B(r) r\, dr, \dagger \tag{8.2.2}$$

† Sometimes known as a Hankel transform of zero order of $B(\cdot)$—see reference 9 for examples.

where $J_0(\cdot)$ is the Bessel function of order zero. It follows that

$$\phi(\lambda) = Q(\lambda),$$

where now $Q(\lambda)$ has the representation (8.2.2).

Examples For

$$B(r) = e^{-kr}$$

we have

$$Q(\lambda) = 2\pi \int_0^\infty J_0(2\pi\lambda r)e^{-kr}r\,dr$$

$$= \frac{6k\pi}{(k^2 + 4\pi^2\lambda^2)^{3/2}} \tag{8.2.3}$$

For

$$B(r) = e^{-r^2\sigma^2}$$

we have

$$Q(\lambda) = \frac{\pi}{\sigma^2}\exp\left(\frac{-\pi^2\lambda^2}{\sigma^2}\right). \tag{8.2.4}$$

See reference 2 for other examples.

For an isotropic field we also have that

$$R(r) = B(r) = \int_0^\infty 2\pi J_0(2\pi\lambda r)Q(\lambda)\lambda\,d\lambda. \tag{8.2.5}$$

Spectral Density of Line Scan As in the case of the field in three-space, we can consider the spectral density of a line scan. Thus let

$$r = r_0 + te,$$

where e is a fixed unit vector and r_0 any fixed vector, and consider the line scan

$$f(r) = g(te)$$

as a function of t, $-\infty < t < \infty$. This is a stationary process if the field is homogeneous. If, in addition, the latter is isotropic, then we can, as before (cf. (8.1.31) and Problem 8.2) relate the corresponding spectral density given by

$$P(\lambda) = 2\int_0^\infty B(r)\cos 2\pi\lambda r\,dr$$

to the two-space spectral density $Q(\cdot)$. Using (8.2.5) and substituting for $B(\cdot)$

in terms of $Q(\cdot)$, we have

$$P(\lambda) = 4\pi \int_0^\infty \int_0^\infty J_0(2\pi\sigma r)Q(\sigma)\sigma \, d\sigma \cos 2\pi\lambda r \, dr.$$

Reversing the order of integration and noting that (see reference 9), we obtain

$$4\pi \int_0^\infty J_0(2\pi\sigma r) \cos 2\pi\lambda r \, dr = \frac{2}{\sqrt{\sigma^2 - \lambda^2}}, \qquad \lambda < \sigma$$

$$= 0, \qquad\qquad \lambda > \sigma.$$

We obtain the following for the line scan spectral density:

$$P(\lambda) = 2 \int_\lambda^\infty \frac{\sigma Q(\sigma)}{\sqrt{\sigma^2 - \lambda^2}} \, d\sigma, \qquad 0 < \lambda < \infty. \tag{8.2.6}$$

Conversely, given the line scan spectral density $P(\cdot)$, we can determine the (isotropic) field spectral density $Q(\cdot)$ using (8.2.2), substituting for $B(\cdot)$ therein the integral form

$$B(r) = 2 \int_0^\infty \cos 2\pi\lambda r P(\lambda) \, d\lambda,$$

yielding

$$Q(\lambda) = 4\pi \int_0^\infty \int_0^\infty r J_0(2\pi\lambda r) \cos 2\pi\lambda r P(f) \, df \, dr.$$

We can simplify this to yield a single integral, noting that $P(\cdot)$ is differentiable and

$$P'(\lambda) = -2 \int_0^\infty 2\pi r B(r) \sin 2\pi\lambda r \, dr,$$

$$\int_0^\infty r|B(r)| \, dr < \infty,$$

which is necessary for the integral in (8.2.2) to be defined. Or we can use

$$-\pi r B(r) = \int_0^\infty \sin 2\pi\lambda r P'(\lambda) \, d\lambda,$$

and substituting this into (8.2.2) we obtain

$$Q(\lambda) = -2 \int_0^\infty \int_0^\infty J_0(2\pi\lambda r) \sin 2\pi f r P'(f) \, dr \, df.$$

Now from reference 9 we have that

$$2\pi \int_0^\infty J_0(2\pi\lambda r) \sin 2\pi f r \, dr = 0, \qquad\qquad f < \lambda$$

$$= \frac{1}{\sqrt{f^2 - \lambda^2}}, \qquad f > \lambda.$$

Hence

$$Q(\lambda) = \frac{1}{\pi} \int_\lambda^\infty \frac{-P'(f)}{\sqrt{f^2 - \lambda^2}} \, df, \qquad\qquad (8.2.6a)$$

which then determines the field spectral density in terms of the line scan spectral density. This is more complicated than the corresponding formula for the three-space version (8.1.30). Note that

$$Q(0) = \frac{1}{\pi} \int_0^\infty \frac{-P'(f)}{f} \, df = 2\pi \int_0^\infty r B(r) \, dr.$$

Structure Function   The structure function for an isotropic field is defined just as in the three-space case. Thus

$$D(r) = 2(B(0) - B(r))$$

and can be expressed in terms of the spectral density $Q(\cdot)$ as

$$D(r) = 4\pi \int_0^\infty (1 - J_0(2\pi\lambda r)) Q(\lambda) \lambda \, d\lambda.$$

For more on the structure function, see reference 2, p. 697.

8.2.2   Sampling Principle

We shall say that a homogeneous random field is "band-limited" if the spectral density is given by

$$\phi(\lambda) = 0, \qquad |\lambda_1| > W_1, |\lambda_2| > W_2. \qquad\qquad (8.2.7)$$

We shall see how the sampling principle (Chapter 6) for continuous-time processes extends to band-limited homogeneous fields in two space.
    Let

$$r_{m,n} = i \frac{m}{2W_1} + j \frac{n}{2W_2}. \qquad\qquad (8.2.8)$$

We refer to $r_{m,n}$ as "lattice points," or "grid points." The key to the extension is the Fourier series expansion for functions in two variables. Actually we only need to recall that (cf. (6.8))

$$e^{2\pi i \lambda_1 x} = \sum_{-\infty}^{\infty} a_m(x) e^{\frac{im\pi x}{W_1}}, \qquad -W_1 < x < W_1$$

$$e^{2\pi i \lambda_2 y} = \sum_{-\infty}^{\infty} a_n(y) e^{\frac{in\pi y}{W_2}}, \qquad -W_2 < y < W_2,$$

where

$$a_m(x) = \frac{\sin \pi(2W_1 x - m)}{\pi(2W_1 x - m)}$$

$$a_n(y) = \frac{\sin \pi(2W_2 y - n)}{\pi(2W_2 y - n)}.$$

Hence

$$e^{2\pi i[\lambda, r]} = \sum_{-\infty}^{\infty} \sum_{-\infty}^{\infty} a_m(x) a_n(y) e^{i2\pi[\lambda, r_{m,n}]},$$

where

$$\lambda = i\lambda_1 + j\lambda_2$$

$$r = ix + jy.$$

Then, as in Chapter 6, defining

$$\zeta_{M,N}(r) = \sum_{-M}^{M} \sum_{-N}^{N} a_m(x) a_n(y) f(r_{m,n}) \tag{8.2.9}$$

we can prove that

$$E[\zeta_{M,N}(r)] \to E[f(r)] \qquad \text{as } M, N \to \infty$$

and

$$E[|\zeta_{M,N}(r) - f(r)|^2] \to \infty$$

or

$$f(r) = \sum_{-\infty}^{\infty} \sum_{-\infty}^{\infty} a_m(x) a_n(y) f(r_{m,n}), \tag{8.2.10}$$

the series converging in the "mean square" sense—thus being the Shannon principle for random fields in two-space. Once again, no specific knowledge of the spectral density is needed other than band-limitedness and W_1, W_2. The fact that (8.2.10) is a two-sided interpolation formula is less of a problem since most data reduction problems for fields are "off line"—all the data are available at once. Because of the possibility of other coordinate systems (still orthogonal), interpolation formulas of other kinds are possible. See reference 4 of Chapter 6.

8.2.3 Response of Linear Systems

Analogous to (8.1.39) we may consider an input-output relation:

$$g(r) = \int_{\mathbf{R}^2} W(r - r') f(r') \, dr', \qquad (8.2.11)$$

where $f(\cdot)$ is the "input" field and $g(\cdot)$ is the "output" field (arising perhaps as a solution to a partial differential equation with constant coefficients—see the Vening-Meinesz formula in reference 7 for optical image processing). If $f(\cdot)$ is homogeneous, so is $g(\cdot)$. Moreover, the spectral density of $g(\cdot)$ is given by

$$\phi_g(\lambda) = |\psi(\lambda)|^2 \phi_f(\lambda),$$

where $\phi_f(\cdot)$ is the spectral density of $f(\cdot)$ and $\psi(\lambda)$ is the Fourier transform of $W(\cdot)$:

$$\psi(\lambda) = \int_{\mathbf{R}^2} e^{2\pi i[\lambda, r]} W(r) \, |dr|. \qquad (8.2.12)$$

We can also go on to define a "white noise" field in two-space as in the case of three-space, again following the development in Chapter 4. Indeed every homogeneous random field can be considered as the response of a linear translation-invariant system to white noise, via (8.2.11). Here

$$\psi(\lambda) = \sqrt{\phi_g(\lambda)} \qquad (8.2.13)$$

and the corresponding weighting function is given by

$$W(r) = \int_{\mathbf{R}^2} e^{2\pi i[\lambda, r]} \sqrt{\phi_g(\lambda)} \, |d\lambda|.$$

If the field is isotropic, then

$$W(r) = H(r)$$

and is given by the zero-order Hankel transform (cf. 8.2.6)

$$H(r) = \int_0^\infty 2\pi J_0(2\pi\lambda r) \sqrt{Q(\lambda)} \lambda \, d\lambda. \qquad (8.2.14)$$

Also the covariance function has the representation

$$R(r) = \int_{\mathbf{R}^2} W(r + r') W(r') \, |dr'|. \qquad (8.2.15)$$

White noise is isotropic and the spectral density along a line scan is not defined. Also, "physical realizability" of the weighting pattern $W(\cdot)$ has no meaning since there is no natural ordering into past and future. We may define an ordering with desirable features, as in "lexicographic" ordering, for prediction theory (see reference 9)

8.3 FIELDS ON LATTICES (GRIDS)

8.3.1 Homogeneous Fields

Analogous to "discrete-time" processes or time series, we may consider fields defined on lattices (grids)—indexed by pairs of integers (m, n), $m \in \mathbf{I}$, $n \in \mathbf{I}$. We have already seen an example arising from sampling of band-limited fields in two-space. Here we shall consider the general case. Let

$$f_{m,n}, \qquad m \in \mathbf{I}, n \in \mathbf{I}$$

denote the field. Let

$$M_{m,n} = E[f_{m,n}] \tag{8.3.1}$$

denote the mean function and let

$$R(m_1, n_1; m_2, n_2) = E[(f_{m_1, n_1} - M_{m_1, n_1})(f_{m_2, n_2} - M_{m_2, n_2})]. \tag{8.3.2}$$

We shall call the field homogeneous if the mean is a constant and

$$R(m_1, n_1; m_2, n_2) = R(m_1 - m_2, n_1 - n_2). \tag{8.3.3}$$

As a simple example of such a field we may consider (analogous to (8.1.11))

$$\sum_1^N A_k \sin (2\pi m \lambda_{1,k} + 2\pi n \lambda_{2,k} + \phi_k), \tag{8.3.4}$$

where A_k and ϕ_k are as in Example 1.1.3, and

$$-1/2 \leq \lambda_{i,k} \leq 1/2, \qquad i = 1, 2; k = 1, \ldots, N$$

are fixed. This is a Gaussian field with covariance function:

$$\sum_1^N \sigma_k^2 \cos (2\pi m \lambda_{1,k} + 2\pi n \lambda_{2,k}). \tag{8.3.5}$$

As an illustrative example, let $N = 4$ and

$$\lambda_{11} = \lambda_{21} = \frac{1}{8}$$

$$\lambda_{12} = \frac{3}{8}, \quad \lambda_{22} = \frac{1}{8}$$

$$\lambda_{13} = \frac{3}{8} = \lambda_{23}$$

$$\lambda_{14} = \frac{1}{8}, \quad \lambda_{24} = \frac{3}{8}.$$

A contour plot of (8.3.4) with interpolation between points for $1 \le m \le 16$, $1 \le n \le 16$, is given in Figure 8.1, where of course the periodicity (period 8) is noticeable, but less noticeable in a 3-D plot of the same in Figure 8.2. The field is clearly not isotropic.

We have of course the analog of Bochner's theorem which we shall now prove for the case

$$\sum_{-\infty}^{\infty} \sum_{-\infty}^{\infty} |R(m, n)| < \infty \tag{8.3.6}$$

$$R(m, n) = R(-m, -n)$$

as in Theorem 2.8.

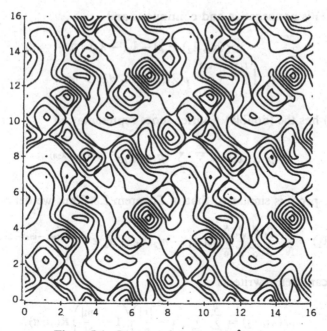

Figure 8.1. Contour plot. $N = 4$, $\sigma_k^2 = 1$.

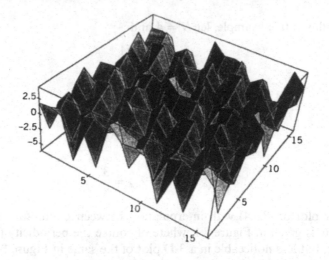

Figure 8.2. A 3-D plot corresponding to Figure 8.1.

Theorem 8.3.1. Suppose that the stationary covariance function satisfies (8.3.6). Then it has the representation:

$$R(m, n) = \int_{-1/2}^{1/2} \int_{-1/2}^{1/2} e^{2\pi i(\lambda_1 m + \lambda_2 n)} p(\lambda_1, \lambda_2)\, d\lambda_1\, d\lambda_2, \qquad (8.3.7)$$

where $p(\cdot, \cdot)$ is nonnegative and continuous and

$$p(\lambda_1, \lambda_2) = p(-\lambda_1, -\lambda_2)$$
$$\int_{-1/2}^{1/2} \int_{-1/2}^{1/2} p(\lambda_1, \lambda_2)\, d\lambda_1\, d\lambda_2 > \infty.$$

Also $p(\cdot, \cdot)$ has the (double) Fourier series representation

$$p(\lambda_1, \lambda_2) = \sum_{-\infty}^{\infty} \sum_{-\infty}^{\infty} R(m, n) e^{-2\pi i(\lambda_1 m + \lambda_2 n)}. \qquad (8.3.8)$$

Proof. The proof is similar to that of Theorem 2.8. Thus we define

$$p_N(\lambda_1, \lambda_2) = \frac{1}{N^2} \sum_0^N \sum_0^N \sum_0^N \sum_0^N R(m_1 - m_2, n_1 - n_2) e^{-2\pi i\lambda_1(m_1 - m_2) - 2\pi i\lambda_1(n_1 - n_2)},$$

which we can then rewrite as

$$p_N(\lambda_1, \lambda_2) = \sum_{-N}^{N} \sum_{-N}^{N} \left(1 - \frac{|n|}{N}\right)\left(1 - \frac{|m|}{N}\right) R(m, n). \qquad (8.3.9)$$

As in Theorem 2.8,

$$0 \le p_N(\lambda_1, \lambda_2) \to p(\lambda_1, \lambda_2) \qquad \text{as } N \to \infty,$$

which satisfies (8.3.7). We omit the details. The function $p(\lambda_1, \lambda_2)$, $-1/2 \le \lambda_1, \lambda_2 \le 1/2$, is, of course, called the "spectral density" of the process. To take care of our example (8.3.4), we have to extend the spectral density to include two-dimensional δ-functions, as in Chapter 2.

Since the lattice may arise from different coordinate systems, the notion of an isotropic field probably does not carry the same level of significance. Nevertheless, we may define it formally: A homogeneous field is isotropic if

$$R(m, n) = B(\sqrt{m^2 + n^2}). \qquad (8.3.10)$$

If the field on the lattice arises from sampling a band-limited field as in Section 8.2.2, we obtain

$$f_{m,n} = f(r_{m,n}), \qquad m \in \mathbf{I}, n \in \mathbf{I}. \qquad (8.3.11)$$

Then the spectral density of the field $\{f_{m,n}\}$ is readily expressed in terms of that of $f(\cdot)$ from

$$R(m, n) = \int_{-W_1}^{W_1} \int_{-W_2}^{W_2} \cos [\lambda, r_{m,n}] \phi(\lambda) |d\lambda|$$

$$= 4W_1 W_2 \int_{-1/2}^{1/2} \int_{-1/2}^{1/2} \phi(i2\lambda_1 W_1 + j2\lambda_2 W_2)$$

$$\cdot \cos (2\pi(\lambda_1 m + \lambda_2 n)) \, d\lambda_1 \, d\lambda_2.$$

Thus

$$p(\lambda_1, \lambda_2) = 4W_1 W_2 \phi(2i\lambda_1 W_1 + 2j\lambda_2 W_2). \qquad (8.3.12)$$

If the original field is band-limited and isotropic, so that we have

$$Q(\lambda) = \phi(\lambda) = 0 \qquad \text{for } \lambda > W$$

and thus taking $W_1 = W_2 = W$, we do have that

$$p(\lambda_1, \lambda_2) = 4W^2 Q(2W \sqrt{\lambda_1^2 + \lambda_2^2}), \qquad -1/2 \le \lambda_1, \lambda_2 \le 1/2$$

and curiously

$$= 0 \qquad \text{for } 1/4 < \lambda_1^2 + \lambda_2^2 < 1/2.$$

8.3.2   Simulation of Fields on Two-Dimensional Grids

Rice Model   We can extend the Rice model in Chapter 7 to simulate fields on two-dimensional grids using the Bochner theorem:

$$R(m, n) = \int_{-1/2}^{1/2} \int_{-1/2}^{1/2} e^{2\pi i[\lambda_1 m + \lambda_2 n]} p(\lambda_1, \lambda_2)\, d\lambda_1\, d\lambda_2$$

$$= \int_{-1/2}^{1/2} \int_{-1/2}^{1/2} \cos 2\pi(\lambda_1 m + \lambda_2 n) p(\lambda_1, \lambda_2)\, d\lambda_1\, d\lambda_2.$$

We approximate the integral on the right by Riemann sums. Thus for given M, we take elementary squares with sides $\Delta = 1/2M$ and form

$$f_M(m, n) = \sum_{-M}^{M} \sum_{-M}^{M} A_{jk} \sin(2\pi\lambda_j m + 2\pi\lambda_k n + \phi_{jk}), \qquad j \neq 0, k \neq 0 \quad (8.3.13)$$

$$\lambda_j = \frac{2j - 1}{4M}, \qquad j = 1, \ldots, M$$

$$\lambda_{-j} = -\lambda_j, \qquad j = 1, \ldots, M$$

$$\sigma_{jk}^2 = E[A_{jk}^2] = \frac{1}{4M^2} p(\lambda_j, \lambda_k)$$

$$p(-\lambda_j, -\lambda_k) = p(\lambda_j, \lambda_k)$$

A_{jk} Rayleigh (with second moment σ_{jk}^2);

ϕ_{jk} uniform in $[0, 2\pi]$;

all independent.

Note that the approximation process (8.3.13) is not isotropic even though the limit as $M \to \infty$ may be. How large M should be for the sample covariances to converge to the value may be estimated as in Chapter 7.

For illustrative purposes we consider the spectral density

$$p(\lambda_1, \lambda_2) = \frac{1}{(1 + 4\pi^2\lambda_1^2 + 4\pi^2\lambda_2^2)^{3/2}}, \qquad -1/2 \leq \lambda_1, \lambda_2 \leq 1/2. \quad (8.3.14)$$

For $M = 16$, we show two 3-D plots (sample paths) in Figures 8.3 and 8.4 and show corresponding contour plots in Figures 8.5 and 8.6. The latter clearly show the periodicity with period 8. The two realizations are of course different in appearance. For higher values of M it is more efficient to use fast Fourier transform techniques. See Figure 8.7 for a 3-D plot of the case $M = 64$.

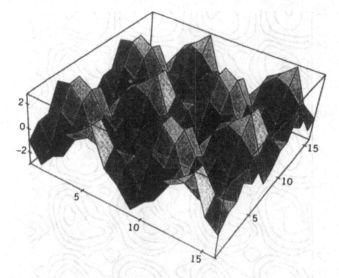

Figure 8.3. $\rho = 0.6$; $M = 16$.

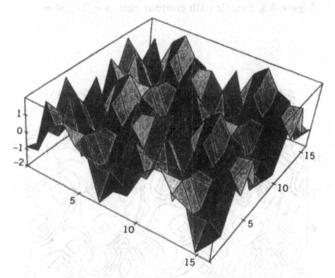

Figure 8.4. $\rho = 0.6$; $M = 16$.

Karhunen-Loeve Expansion   The Karhunen-Loeve expansion of random processes treated in Chapter 6 is readily extended to random fields. It is probably more useful for fields in two-space, especially as a simulation tool because often the field is required only in some rectangular area (wavefront in a planar aperture in optics, for example). Thus let $f(r)$ denote the field (with mean zero).

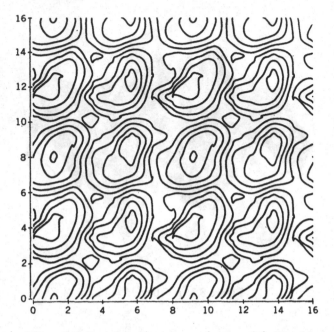

Figure 8.5. Sample path contour plot. $\rho = 0.6$; $M = 16$.

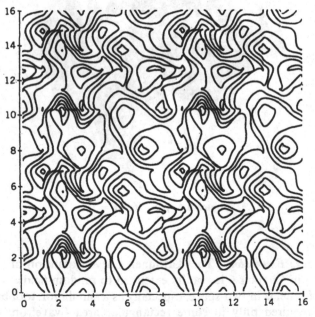

Figure 8.6. Sample path contour plot. $\rho = 0.6$; $M = 16$.

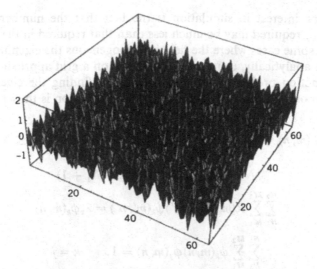

Figure 8.7. A 3-D plot. $\rho = 0.6$; $M = 64$.

For $r \in B$, where B is a rectangle,

$$a_1 \leq x \leq a_2, \qquad b_1 \leq y \leq b_2, \qquad r = ix + jy,$$

we have

$$f(r) = \sum_1^\infty \zeta_k \phi_k(r) \tag{8.3.15}$$

$$\zeta_k = \int_B f(r)\phi_k(r)\,|dr|,$$

where $\phi_k(\cdot)$ are the orthonormal eigenfunctions

$$\int_B R(r, r')\phi_k(r')\,|dr'| = \lambda_k \phi_k(r), \qquad r \in B$$

$$\int_B \phi_k(r')\,|dr| = 1,$$

where the $\{\lambda_k\}$ are positive and can be arranged in decreasing order, $\lambda_k \to 0$:

$$\sum_1^\infty \lambda_k = \int_B \int_B R(r, r')\,|dr|\,|dr'|.$$

The $\{\phi_k\}$ are independent zero-mean Gaussians, and

$$E[\zeta_k^2] = \lambda_k.$$

Of particular interest in simulation is the fact that the number of random variables $\{\zeta_k\}$ required may be much less than that required in the Rice model.

While in some cases where the field is homogeneous the eigenfunctions may be obtained analytically, usually one has to go on a grid approximation. Thus the eigenfunction problem is reduced to that of finding the eigenvalues and eigenvectors of a positive definite matrix. The expansion is thus

$$f(m, n) = \sum_1^N \zeta_k \phi_k(m, n), \qquad M_1 \le m \le M_2, N_1 \le n \le N_2, \qquad (8.3.16)$$

$$N = (M_2 - M_1 + 1)(N_2 - N_1 + 1)$$

$$\sum_{N_1}^{N_2} \sum_{M_1}^{M_2} R(m, n; m', n') \phi_k(m', n') = \lambda_k \phi_k(m, n) \qquad (8.3.17)$$

$$\sum_{N_1}^{N_2} \sum_{M_1}^{M_2} \phi_k(m, n) \phi_j(m, n) = 1, \qquad k = j$$

$$= 0, \qquad k \ne j$$

$$\zeta_k = \sum_{N_1}^{N_2} \sum_{M_1}^{M_2} f(m, n) \phi_k(m, n),$$

where $R(;)$ is the covariance function

$$R(m, n; m', n') = E[f(m, n) f(m', n')].$$

Again of special interest would be the homogeneous case. In contrast to the Rice model, we now need to specify the covariance function rather than the spectral density.

Example 8.3.1. As an illustrative example we shall consider the covariance function:

$$R(m, n; m', n') = \rho^{\sqrt{(m - m')^2 + (n - n')^2}}, \qquad 0 < \rho < 1, \qquad (8.3.18)$$

corresponding to sampling with period Δ an isotropic field in two-space with covariance function

$$B(r) = e^{-kr}$$

so that

$$\rho = e^{-k\Delta}.$$

Let the grid B be the square

$$B = [0 \le m, n \le M].$$

The covariance matrix whose eigenvalues and eigenvectors we need to determine

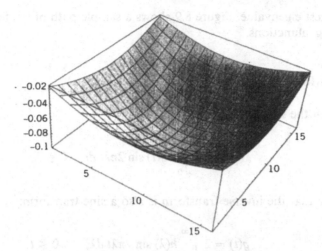

Figure 8.8. $\rho = 0.6$, $M = 16$. Eigenfunction corresponding to largest eigenvalue.

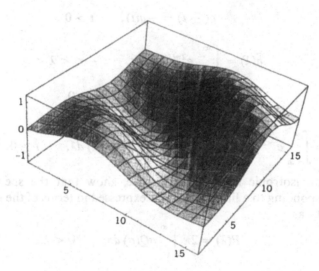

Figure 8.9. $\rho = 0.6$, $M = 16$. Karhunen-Loeve expansion using the first ten terms.

is then of size $(M + 1)^2 \times (M + 1)^2$. Since analytical solution is not possible, we have to be content with some general qualitative observations and a numerical example. As $\rho \to 0$, the matrix becomes the identity matrix, or the eigenvalues are nearly equal for small ρ. As $\rho \to 1$, the eigenvalues tend to be small except for the largest which approaches the matrix which is $\mathrm{Tr} = (M + 1)^2$. The number of random variables needed is $(M + 1)^2$. To minimize problems of purely numerical nature we consider $M = 15$, so that the matrix is 256×256. For $\rho = 0.6$, Figure 8.8 shows a 3-D plot of the eigenfunction corresponding to the largest eigenvalue. Figure 8.9 shows a sample path of the field using the first ten eigenfunctions.

PROBLEMS

8.1. Given the sine transform

$$h(\lambda) = w \int_0^\infty g(t) \sin 2\pi\lambda t \, dt, \quad 0 \le \lambda,$$

show that the inverse transform is also a sine transform:

$$g(t) = 2 \int_0^\infty h(\lambda) \sin 2\pi\lambda t \, d\lambda, \quad 0 \le t.$$

Hint: Define

$$g(-t) = -g(t), \quad t > 0$$

and

$$\tilde{h}(\lambda) = \int_{-\infty}^\infty e^{2\pi i\lambda t} g(t) \, dt, \quad -\infty < \lambda < \infty$$

$$= 2ih(\lambda) \quad \lambda \ge 0.$$

Then

$$g(t) = \int_{-\infty}^\infty e^{-2\pi i\lambda} \tilde{h}(\lambda) \, d\lambda = 2 \int_0^\infty \sin 2\pi\lambda t \, h(\lambda) \, d\lambda, \quad t \ge 0.$$

8.2. For an isotropic field in three-space, show that the spectral density corresponding to a line scan can be expressed in terms of the field spectral density as

$$P(\lambda) = 2\pi \int_\lambda^\infty \sigma Q(\sigma) \, d\sigma, \quad 0 < \lambda.$$

Calculate $P(\cdot)$ for $Q(\cdot)$ of the form

$$Q(\lambda) = \frac{1}{(k^2 + 4\pi^2\lambda^2)^\nu}.$$

8.3. Let $R(t)$, $-\infty < t < \infty$, denote a stationary covariance function with spectral density $p(\cdot)$:

$$R(t) = \int_{-\infty}^\infty e^{2\pi i\lambda t} P(\lambda) \, d\lambda,$$

where

$$P(\lambda) \ge 0.$$

Show that

$$B(r) = R(r), \qquad r > 0$$

is the covariance function of an isotropic field in three-space if and only if $P(\lambda)$ is differentiable and

$$P'(\lambda) \leq 0 \qquad \text{for } 0 < \lambda.$$

This condition is violated, for instance, by

$$P(\lambda) = \lambda^{-5/3} e^{-1/\lambda}, \qquad 0 < \lambda.$$

In particular,

$$B(r) = e^{-\sigma r} \cos 2\pi f_c r, \qquad r > 0, \sigma > 0$$

cannot be the covariance function of an isotropic field in three-space.

8.4. Let

$$R(r) = \left(1 - \frac{|x|}{T}\right)\left(1 - \frac{|y|}{T}\right)\left(1 - \frac{|z|}{T}\right), \qquad \frac{|x|}{T}, \frac{|y|}{T}, \frac{|z|}{T} < 1$$

$$= 0, \qquad\qquad\qquad\qquad\qquad\qquad\qquad \text{otherwise}$$

$$r = ix + jy + kz$$

denote the stationary covariance function of a random field im three-space. Find the corresponding spectral density. Denoting it $\phi_T(\lambda)$, calculate

$$\lim_{T \to \infty} \int_{\mathbf{R}^2} h(\lambda) \phi_T(\lambda) |d\lambda| = h(0),$$

where $h(\cdot)$ is a continuous function, defined on $\mathbf{R}^3$.

8.5. The spectral density of a line scan of an isotropic field in two-space has the form

$$P(\lambda) = \frac{1}{1 + 4\pi^2 \lambda^2}.$$

Find the spectral density of the field.

8.6. The covariance function of a line scan of an isotropic field in three-space has the form

$$B(r) = (1 + r)e^{-kr}, \qquad r > 0.$$

Find the corresponding spectral density. Repeat for two-space.

8.7. Let

$$f_{m,n}$$

be a random field on the lattice:

$$m \in \mathbf{I}, \quad n \in \mathbf{I}.$$

Extend the definition of "stationary" ("homogeneity") fields on lattices— a "discrete-parameter" process. Define the process for $r \in \mathbf{R}^2$ using (8.2.10) and show that the process is homogeneous if the field $\{ f_{m,n} \}$ is.

8.8. Let $f(r)$, $r \in \mathbf{R}^3$, denote an isotropic field. Let

$$g_n(t) = n^{3/2} f(nte), \qquad -\infty < t < \infty,$$

where e is a unit vector $\mathbf{R}^3$. Find the spectral density of $g_n(\cdot)$. What is the limit as $n \to \infty$?

Repeat the problem for $\mathbf{R}^2$ in place of $\mathbf{R}^3$. and

$$g_n(t) = nf(nte).$$

8.9. Let $f(r)$, $r \in \mathbf{R}^3$, denote an isotropic field, with

$$r = ix + jy + kz.$$

Show that the field obtained by taking $z =$ constant (say, zero), namely

$$f(jx + jy)$$

which is then defined on $\mathbf{R}^2$, is also isotropic, and find the corresponding spectral density. Calculate it for the three-space spectral density given by (8.1.34). Show that the structure functions have the same form.

Answer:

$$2 \int_0^\infty Q(\sqrt{\lambda^2 + x^2})\, dx$$

$$\frac{\text{constant}}{(k^2 + 4\pi^2 \lambda^2)^{2\nu + 1}}.$$

8.10. In the usual notation

$$\lambda = i\lambda_1 + j\lambda_2 + k\lambda_3$$
$$r = ix + jy + kz$$

let $R(r)$ and $\phi(\lambda)$ denote the covariance function and spectral density of a stationary field in three-space. Then

$$\phi(i\lambda_1 + j\lambda_2)$$

is a spectral density in two-space. Show that the corresponding covariance function is given by

$$R_2(ix + jy) = \int_{-\infty}^{\infty} R(ix + jy + kz)\, dz.$$

Specialize to the case where the three-space field is also isotropic, and show that

$$R_2(ix + jy) = \int_{-\infty}^{\infty} \int_{-\infty}^{\infty} e^{i2\pi(\lambda_1 x + \lambda_2 y)} \phi(i\lambda_1 + j\lambda_2)\, d\lambda_1\, d\lambda_2$$

$$= 2\pi \int_{0}^{\infty} J_0(2\pi\rho\lambda)\lambda Q(\lambda)\, d\lambda,$$

where

$$\rho = \sqrt{x^2 + y^2}$$

$$Q(\lambda) = \phi(\lambda).$$

In particular, show that

$$R_2(0) = 2\pi \int_{0}^{\infty} \lambda Q(\lambda)\, d\lambda.$$

Calculate $R_2(0)$ for the three-space spectral density given by (8.1.34).

8.11. Show that the "folded" spectrum of a homogeneous random field in two-space sampled at grid points

$$\left(\frac{n}{2W_1}, \frac{m}{2W_1} \right)$$

and interpolated by the Shannon formula is

$$= \sum_{k=-\infty}^{\infty} \sum_{j=-\infty}^{\infty} p(\lambda_1 + 2kW_1, \lambda_2 + 2jW_2), \qquad |\lambda_1| < W_1, |\lambda_2| < W_2,$$

$$= 0, \qquad\qquad\qquad\qquad\qquad\qquad \text{otherwise,}$$

analogous to (6.15).

8.12. For an isotropic random field in two-space with spectral density

$$Q(\lambda) = \frac{1}{(k^2 + 4\pi^2\lambda^2)^3}$$

determine the weighting function of the translation-invariant linear system such that the process can be realized as the response of this system to white noise input. Find also the corresponding formula for the covariance function following (8.2.15).

8.13. Show that for an isotropic field in two-space with spectral density given by

$$Q(\lambda) = \frac{1}{(k^2 + 4\pi^2\lambda^2)^\nu}$$

the structure function $D(r)$ defined by (8.2.6) has the form

$$(\text{constant})r^{2(\nu-1)}$$

for small r.

8.14. Higher-order structure-functions: The structure function

$$D(r) = E[(f(r_0 + r) + f(r_0))^2]$$

corresponds to a first-order difference and can be generalized to higher orders by defining the "operator" (see Problem 3.18):

$$\delta(f(r)) = f(r_0 + r) - f(r_0)$$

and

$$\delta^n(f(r)) = \delta^1(\delta^{n-1}(f(r)))$$

and correspondingly

$$D^n(r) = E[(\delta^n(f(r)))^2].$$

Show that for an isotropic field in three-space

$$D^n(r) = \int_{-\infty}^{\infty} 2^n(1 - \cos 2\pi\lambda r)^n P(\lambda)\, d\lambda, \qquad r = |r|$$

$$= \int_{-\infty}^{\infty} 2^n(1 - \cos 2\pi\lambda)^n P\left(\frac{\lambda}{r}\right) \cdot \frac{1}{r}\, d\lambda.$$

In particular, for the Kolmogorov spectrum (8.1.34) show that

$$D^n(r) = (\text{constant depending on } n) \cdot r^{2\nu-1}, \qquad \text{as } r \to 0$$

for all n.

8.15. The spectral density of a homogeneous field in three-space has the form

$$\phi(\lambda) = Q(\lambda_1, \sqrt{\lambda_2^2 + \lambda_3^2}).$$

Show that the covariance matrix of the gradient field is diagonal.

8.16. The Kolmogorov power spectrum

$$Q(\lambda) = \frac{1}{(k^2 + 4\pi^2\lambda^2)^{11/6}}$$

(taking $v = 5/6$ in (8.1.34)) violates the condition that

$$\int_0^\infty \lambda^2 Q(\lambda)\, d\lambda < \infty.$$

This is remedied (see reference 1) by modifying at high frequencies. Thus we define

$$Q(\lambda) = \frac{1}{(k^2 + 4\pi^2\lambda^2)^{11/6}} \exp(-\alpha\lambda^2), \qquad \alpha > 0.$$

Use (8.1.37) to show that while now the structure function $D(\cdot)$ is such that

$$D(r) \le r^2 \qquad \text{as } r \to 0,$$

there is a region where the approximation

$$D(r) \sim r^{2/3}$$

is valid. This starts at r_0, where

$$D''(r_0) = 0.$$

8.17. Let $f(r)$, $r \in \mathbf{R}^2$ denote an isotropic field. For fixed $r_1, r_2, \mathbf{v} \in \mathbf{R}^2$, define the process

$$D(t) = f(r_1 - \mathbf{v}t) - f(r_2 - \mathbf{v}t), \qquad -\infty < t < \infty.$$

Show that this is a stationary process and calculate $E[D(t)^2]$. Calculate the spectral density for $(r_2 - r_1)$ parallel to $\mathbf{v}$; $|\mathbf{v}| = 1$.

NOTES AND COMMENTS

Within the scope of an elementary exposition it has not been possible to include material on many areas of recent applications, such as image processing—see reference 8 and the references in Chapter 6. Of course no mention has been made either of much work on the pure mathematical side, such as that concerning Markovian fields, it being clearly out of range.

REFERENCES

Classic Treatises

1. V. I. Tatarski. *Wave Propagation in a Turbulent Medium*. McGraw-Hill, 1961.
2. A. S. Monin and A. M. Jaglom. *Statistical Fluid Mechanics: Mechanics of Turbulence*. M.I.T. Press, 1971.

Reviews

3. J. W. Strohbehn, ed. *Laser Beam Propagation in the Atmosphere*. Topics in Applied Physics Series. Springer-Verlag, 1977.

Recent Publications

4. M. I. Yadrenko. *Spectral Theory of Random Fields*. Optimization Software Publications, 1983.
5. R. K. Tyson. *Principles of Adaptive Optics*. Academic Press, 1991.
6. R. W. Preisendorfer. *Principal Component Analysis in Meteorology and Oceanography*. Elsevier, 1988.

Review Papers

7. R. S. Nash, Jr. and S. K. Jordan. "Statistical Geodesy—An Engineering Perspective," *Proceedings of the IEEE*, vol. 66, no. 5 (1978), pp. 532–550.

Image Processing

8. J. S. Lim. *Two-Dimensional Signal and Image Processing*. Prentice-Hall, 1990.

Lexicographic Ordering/Prediction Theory

9. H. Helson and D. Lowdenslager. "Multidimensional Prediction Theory," *Acta Mathematica*, vol. 99, no. 3 (1958), pp. 165–202.

General

10. G. N. Watson. *A Treatise on the Theory of Bessel Functions*. Cambridge University Press, 1944.

9

LINEAR FILTERING THEORY

The theory of Filtering of Signals in Noise is too extensive and too far-ranging in scope to be covered adequately in an introductory treatment of random processes. We can however present a representative across-the-board view (with no claim to be exhaustive!) of the basic backbone (as well as the bread-and-butter!) of this theory: namely *Linear Filtering*—since it draws only on the second order properties of the processes—and hence equivalently the theory of filtering Gaussian signals in Gaussian noise.

We begin with a canonical abstract formulation of the linear filtering problem.

9.1 BEST LINEAR ESTIMATE

Let x be an $m \times 1$ random variable which we need to estimate from y, an $n \times 1$ random variable. The class of estimators is restricted to be the class of linear estimates—that is to say, of the form

$$Ay + c, \qquad (9.1.1)$$

where A is an $m \times n$ matrix and c is an $m \times 1$ vector. The error in the estimate is

$$x - (Ay + c)$$

and the (linear) filtering problem is that of determining A and c so as to minimize the "mean square error":

$$E[\|x - (Ay + c)\|^2]. \qquad (9.1.2)$$

To proceed, we first "calculate" (9.1.2) for any A and c:

$$E[\|x - (Ay + c)\|^2] = \|\mu_x - (A\mu_y + c)\|^2$$
$$+ \operatorname{Tr} [R_{xx} + AR_{yy}A^* - 2R_{xy}A^*] \quad (9.1.3)$$

where

$$E[x] = \mu_x, \qquad E[y] = \mu_y$$
$$\tilde{x} = x - \mu_x, \qquad \tilde{y} = y - \mu_y$$
$$E[\tilde{x}\tilde{x}^*] = R_{xx}, \qquad E[\tilde{y}\tilde{y}^*] = R_{yy}$$
$$E[\tilde{x}\tilde{y}^*] = R_{xy}.$$

We see that the mean square error (9.1.2) is thus

$$\geq \operatorname{Tr} [R_{xx} + AR_{yy}A^* - 2R_{xy}A^*]. \quad (9.1.4)$$

Let us first minimize (9.1.4). Let

$$q(A) = \operatorname{Tr} [R_{xx} + AR_{yy}A^* - 2R_{xy}A^*], \quad (9.1.5)$$

defining a nonnegative function on the space of $m \times n$ matrices. Assume now that we can find a matrix A_0 such that it satisfies

$$R_{xy} = A_0 R_{yy}. \quad (9.1.6)$$

Then writing

$$A = A_0 + A - A_0$$

and substituting in (9.1.5), we have

$$q(A) = \operatorname{Tr} [R_{xx} + A_0 R_{yy} A_0^* - 2R_{xy}A_0^* + (A - A_0)R_{yy}A_0^* - R_{xy}(A - A_0)^*$$
$$+ A_0 R_{yy}(A - A_0)^* - R_{xy}(A - A_0)^* + (A - A_0)R_{yy}(A - A_0)^*]. \quad (9.1.7)$$

Replacing R_{xy} by $A_0 R_{yy}$ in (9.1.7) and noting that

$$\operatorname{Tr} R_{xy}(A - A_0)^* = \operatorname{Tr} A_0 R_{yy}(A - A_0)^* = \operatorname{Tr} (A - A_0)R_{yy}A_0^*$$

we see that

$$q(A) = q(A_0) + \operatorname{Tr} (A - A_0)R_{yy}(A - A_0)^*, \quad (9.1.8)$$

where the second term is nonnegative. Hence A_0 is an optimal solution for our minimization problem, and in particular the minimum

$$= \operatorname{Tr} [R_{xx} - R_{xy}A_0^*]. \quad (9.1.9)$$

Let us now get back to finding A_0. We see that A_0 is a solution of the equation

$$R_{xy} = AR_{yy}. \quad (9.1.10)$$

This equation (in its full generality) is referred to as a "Wiener-Hopf" equation. If R_{yy} is nonsingular, it has a unique solution, namely

$$A_0 = R_{xy}(R_{yy})^{-1}, \qquad (9.1.11)$$

and the second term in (9.1.8) is zero if and only if

$$A - A_0 = 0.$$

Let us consider next what happens if R_{yy} is singular. Let ϕ_i denote the orthonormalized eigenvectors of R_{yy} corresponding to nonzero eigenvalues λ_i, $i = 1, \ldots, p$ (see Eq. (1.6) on page 6). Then

$$\sum_1^p \lambda_i \phi_i \phi_i^*$$

is nonsingular. Hence let

$$A_0 = R_{xy}\left(\sum_1^p \lambda_i \phi_i \phi_i^*\right)^{-1}. \qquad (9.1.12)$$

Then for any ϕ_k, $k = 1, \ldots, p$:

$$R_{xy}\phi_k = A_0\left(\sum_1^p \lambda_i \phi_i \phi_i^*\right)\phi_k = A_0[\lambda_k \phi_k] = A_0 R_{yy}\phi_k.$$

For any eigenvector ψ corresponding to eigenvalue zero of R_{yy},

$$[R_{yy}\psi, \psi] = E[(y^*\psi)^2] = 0$$

or

$$y^*\psi = 0.$$

Hence

$$R_{xy}\psi = E[xy^*\psi] = 0 = A_0 R_{yy}\psi.$$

Hence it follows that A_0 is a solution of the Wiener-Hopf equation. But then so is

$$A_0 + \psi\psi^*, \qquad \text{if } R_{yy}\psi = 0,$$

since in that case

$$(A_0 + \psi\psi^*)R_{yy} = A_0 R_{yy}$$

since

$$\psi^*R_{yy} = (R_{yy}\psi)^* = 0.$$

Furthermore in (9.1.8) we obtain

$$q(A_0 + \psi\psi^*) = q(A_0) + \text{Tr}\,[\psi\psi^*R_{yy}\psi\psi^*] = q(A_0).$$

Thus the Wiener-Hopf equation has always a solution and any solution minimizes $q(\cdot)$.

Returning now to (9.1.3), we only need to set

$$c = \mu_x - A_0\mu_y.$$

The optimal estimate is thus given by

$$A_0 y + c = \mu_x + A_0(y - \mu_y), \tag{9.1.13}$$

where A_0 is any solution of (9.1.10) and $q(A_0)$ is the corresponding minimal mean square error. Note that

$$E[A_0 y + c] = E[x].$$

9.1.1 Minimizing the Error Covariance Matrix

For any choice of A and c, the covariance matrix of the corresponding error is given by

$$E[(x - (Ay + c))(x - (Ay + c))^*]. \tag{9.1.14}$$

Recall (Review Chapter) that given any two covariance matrices M and N we say that

$$M \leq N$$

if the matrix $(M - N)$ which is symmetric is nonpositive definite:

$$M - N \leq 0.$$

Hence we may consider the problem of minimizing the error covariance matrix (9.1.14), rather than the mean square error which is recognized as its trace. The optimal solution is exactly the same as before, and the arguments are the same, *mutatis mutandis*. Thus (9.1.14) can be calculated as

$$[(\mu_x - (A\mu_y + c))(\mu_x - (A\mu_y + c))^*] + Q(A),$$

where

$$Q(A) = R_{xx} + A R_{yy} A^* - R_{xy} A^* - A R_{yx}$$

and (9.1.14) is

$$\geq Q(A).$$

Now for A_0 satisfying the Wiener-Hopf equation, we can verify that

$$Q(A) = Q(A - A_0 + A_0) = Q(A_0) + (A - A_0) R_{yy} (A - A_0)^*.$$

Hence it follows that the minimum of (9.1.14) is given by

$$Q(A_0) = R_{xx} - A_0 R_{yx}, \tag{9.1.14a}$$

which is then the minimal error covariance matrix.

The fact that the error covariance matrix is minimized implies that if we wish to estimate any linear transformation of x, say Lx, where L is any $p \times m$ matrix:

$$\text{Best estimate of } Lx = L \text{ (Best estimate of } x\text{)}.$$

This is immediate from

$$E[L(x - (Ay + c))](L(x - (Ay + c)))^*$$
$$= LE[(x - (Ay + c))(x - (Ay + c))^*]L^*$$
$$\geq LQ(A_0)L^*.$$

Finally we note that the Wiener-Hopf equation (9.1.10) can be written

$$E[(\tilde{x} - A\tilde{y})y^*] = 0, \tag{9.1.15}$$

where we recognize

$$\tilde{x} - A\tilde{y}$$

as the estimate error (referred to as "residual" in statistical terminology) and hence (9.1.15) is equivalent to saying that "the residual is uncorrelated with the data" and may often by invoked in this form to determine the solution.

9.1.2 Extension to Continuous-Time Processes

Let us now extend the problem to the case where the $m \times 1$ variable x has to be estimated from a continuous-time $n \times 1$ process $y(t)$, $0 \leq t \leq T < \infty$. We assume the process has zero mean (otherwise we may consider the centered process, centered about the mean). Let

$$R(s, t) = E[y(s) y(t)^*]$$

and we assume that $R(s, t)$ is continuous is $0 \leq s, t \leq T$. The process is then mean-square continuous, so that for any $m \times p$ function $W(t)$, $0 \leq t \leq T$, such that

$$\int_0^T \| W(t) \|^2 \, dt < \infty \tag{9.1.16}$$

we have (cf. Chapter 4)

$$E\left[\int_0^T W(s)y(s)\,ds\right] = 0$$

and

$$E\left[\left(\int_0^T W(s)y(s)\,ds\right)\left(\int_0^T W(s)y(s)\,ds\right)^*\right] = \int_0^T\int_0^T W(s)R(s,t)W(t)^*\,ds\,dt.$$

The linear estimation problem is then to determine $W(\cdot)$ which minimizes the error covariance matrix

$$E\left[\left(\tilde{x} - \int_0^T W(s)y(s)\,ds\right)\left(\tilde{x} - \int_0^T W(s)y(s)\,ds\right)^*\right]. \qquad (9.1.17)$$

We may as well assume that x has zero mean. Let $W_0(\cdot)$ denote an optimal estimate. Then extending (9.1.15), we would expect that $W_0(\cdot)$ will satisfy

$$E\left[\left(x - \int_0^T W_0(s)y(s)\,ds\right)y(t)^*\right] = 0, \qquad 0 \le t \le T. \qquad (9.1.18)$$

This is equivalent to

$$R_{xy}(t) = \int_0^T W_0(s)R(s,t)\,ds, \qquad 0 \le t \le T, \qquad (9.1.19)$$

where

$$R_{xy}(t) = E[x\,y(t)^*], \qquad 0 \le t \le T.$$

We note that (9.1.19) is a Wiener-Hopf equation. Assume now that we have found $W_0(\cdot)$ satisfy (9.1.18). Then it is easy to see that for any $W(\cdot)$ (satisfying (9.1.16)) we obtain

$$x - \int_0^T W(s)y(s)\,ds = x - \int_0^T W_0(s)y(s)\,ds + \int_0^T (W_0(s) - W(s))y(s)\,ds$$

and

$$E\left[\left(x - \int_0^T W_0(s)y(s)\,ds\right)\left(\int_0^T (W_0(s) - W(s))y(s)\,ds\right)^*\right]$$

$$= \int_0^T \left(R_{xy}(t) - \int_0^T W_0(t)R(s,t)\,ds\right)(W_0(t) - W(t)^*)\,dt$$

$$= 0. \qquad (9.1.20)$$

Hence the error covariance matrix (9.1.17)

$$= E\left[\left(x - \int_0^T W_0(s)y(s)\,ds\right)\left(x - \int_0^T W_0(s)y(s)\,ds\right)^*\right]$$

$$+ E\left[\left(\int_0^T (W_0(s) - W(s))y(s)\,ds\right)\left(\int_0^T (W_0(s) - W(s))y(s)\,ds\right)^*\right]$$

$$\geq E\left[\left(x - \int_0^T W_0(s)y(s)\,ds\right)\left(x - \int_0^T W_0(s)y(s)\,ds\right)^*\right],$$

which is then the minimal error matrix, and denoting it by P we have

$$P = E[xx^*] - \int_0^T W_0(s)R_{xy}(s)^*\,ds. \qquad (9.1.21)$$

The question of whether we can find a function $W_0(\cdot)$ satisfying the Wiener-Hopf equation (9.1.19) is slightly more involved than in the "finite-dimensional" case (9.1.15), and we shall not go into it any further since in any computer solution the data would be converted discrete-time samples anyway, and we are back to (9.1.15).

Example 9.1.1. The Matched Filter. Before we leave the continuous-time process models, let us illustrate the theory by an example of interest in radar detection theory [5]—the concept of the matched filter. Thus let

$$y(t) = aS(t) + N(t), \qquad 0 \leq t \leq T, \qquad (9.1.22)$$

where $S(\cdot)$ is known and is 1×1 and the noise $N(\cdot)$ is 1×1 white noise with spectral density d. The problem is to estimate the parameter "a." Here we consider the model in which it is Gaussian with zero mean and variance σ^2 and is independent of the noise. We may assume (with no loss in generality) that

$$\int_0^T S(t)^2\,dt = 1.$$

The optimal $W_0(\cdot)$ is now 1×1. Because $N(\cdot)$ is white noise, we go back to the form (9.1.20) for the Wiener-Hopf equation, where we use $h(t)$ to denote

$$W_0(t) - W(t).$$

Then we note that $W_0(\cdot)$ must satisfy

$$E\left[\left(a - \int_0^T W_0(s)y(s)\,ds\right)\left(\int_0^T h(t)y(t)\,dt\right)\right] = 0 \qquad (9.1.23)$$

for arbitrary $h(\cdot)$. But we can calculate the integrals as

$$\int_0^T W_0(s)y(s)\,ds = a\int_0^T W_0(t)S(t)\,dt + \int_0^T W_0(s)N(s)\,ds$$

$$\int_0^T h(s)y(s)\,ds = a\int_0^T h(t)S(t)\,dt + \int_0^T h(t)N(t)\,dt.$$

Substituting these into the left side of (9.1.23), the main calculation involved is

$$E\left[\left(\int_0^T W_0(t)N(t)\,dt\right)\left(\int_0^T h(t)N(t)\,dt\right)\right],$$

which by the white noise properties given in Chapter 4

$$= d\int_0^T W_0(t)h(t)\,dt.$$

Hence we have that

$$\int_0^T \left(\sigma^2 S(t) - \sigma^2\left(\int_0^T S(t)W_0(t)\,dt\right)S(t) - dW_0(t)\right)h(t)\,dt = 0.$$

Hence because $h(\cdot)$ is arbitrary we obtain

$$\sigma^2\left(1 - \int_0^T S(t)W_0(t)\,dt\right)S(t) - dW_0(t) = 0, \qquad 0 \le t \le T.$$

Thus $W_0(\cdot)$ must be of the form

$$W_0(t) = \alpha S(t), \tag{9.1.24}$$

where α must satisfy

$$\sigma^2(1 - \alpha) = d\alpha$$

or

$$\alpha = \frac{\sigma^2}{\sigma^2 + d}$$

and for $\sigma^2 \to \infty$ (an instance of "maximum ignorance," see below)

$$\alpha = 1.$$

Regardless of the value of α, the main conclusion is (9.1.24) that $W_0(\cdot)$ is a multiple of $S(\cdot)$. If we are processing the data by a linear system with weighting pattern $W(\cdot)$ (see Chapter 4), the response at time T is

$$\int_0^T W(T - t) y(t)\, dt$$

and thus we must have

$$W_0(t) = W(T - t), \qquad 0 \le t \le T$$

or

$$W(t) = W_0(T - t) = \alpha S(T - t).$$

Note that the transfer function is given by

$$\psi(f) = \int_0^T e^{2\pi i f t} W(t)\, dt = \alpha \int_0^T e^{2\pi i f t} S(T - t)\, dt$$

$$= \alpha e^{2\pi i f T} \overline{\psi_S(f)}, \qquad (9.1.25)$$

where

$$\psi_S(f) = \int_0^T e^{2\pi i f t} S(t)\, dt.$$

The characterization of the optimal system transfer function by (9.1.25) as the conjugate (omitting constants) of the signal transform is the genesis of the "matched filter" principle in radar detection.

Returning now to the general best linear estimate problem, we note that only the second-order properties, means and covariances have been used. It is therefore of interest to see what we can say further when the distribution of $\begin{vmatrix} x \\ y \end{vmatrix}$ is Gaussian. Here we need first to introduce the notion of "conditional expectation."

9.2 CONDITIONAL EXPECTATION

Conditional expectation is at the heart of all filtering theory. We begin with a simplistic definition. With x and y as in Section 1, we define

$$E[x \,|\, y] = \text{"Conditional expectation of } x \text{ given } y\text{"}$$

$$= \int_{\mathbf{R}^m} x p(x \,|\, y) \,|dx|, \qquad (9.2.1)$$

where $p(x \,|\, y)$ is the conditional density†

$$= \frac{p(x, y)}{p(y)},$$

† The notation here is admittedly sloppy; we should use $p_{x,y}(x, y)$, $p_y(y)$, and so on, but this is cumbersome.

where $p(x, y)$ denotes the joint density and $p(y)$ the (marginal) density of y:

$$p(y) = \int_{\mathbf{R}^m} p(x, y) \, |dx|.$$

Here $|dx|$ denotes the volume element in $\mathbf{R}^m$, the m-dimensional Euclidean space. Note that we can express (9.2.1) as

$$E[x \mid y] = \frac{\displaystyle\int_{\mathbf{R}^m} x p(x, y) \, |dx|}{\displaystyle\int_{\mathbf{R}^m} p(x, y) \, |dx|}. \qquad (9.2.2)$$

It is important to note that

$$E[x \mid y]$$

is a function of the random variable y. With $|dy|$ denoting the volume element in $\mathbf{R}^m$, we obtain

$$E[\|E[x \mid y]\|] = \int_{\mathbf{R}^n} \left\| \int_{\mathbf{R}^m} x p(x \mid y) \, |dx| \right\| p(y) \, |dy|$$

$$\leq \int_{\mathbf{R}^n} \int_{\mathbf{R}^m} \|x\| \, p(x, y) \, |dx||dy|$$

$$= \int_{\mathbf{R}^m} \|x\| \, p(x) \, |dx| = E[\|x\|] < \infty.$$

Hence we may calculate

$$E[E[x \mid y]] = \int_{\mathbf{R}^n} p(y) \, |dy| \int_{\mathbf{R}^m} x p(x \mid y) \, |dx|$$

$$= \int_{\mathbf{R}^n} \int_{\mathbf{R}^m} x p(x, y) \, |dx||dy|$$

$$= \int_{\mathbf{R}^m} x p(x) \, |dx|$$

$$= E[x].$$

Even more importantly, for any 1×1 function $h(\cdot)$ defined on $\mathbf{R}^n$ such that

$$E[|h(y)|] < \infty \qquad (9.2.3)$$

we have that

$$E[E[x|y]h(y)] = \int_{\mathbf{R}^m} \int_{\mathbf{R}^n} h(y)p(y)\,|dy| \cdot xp(x|y)\,|dx|$$

$$= \int_{\mathbf{R}^m} \int_{\mathbf{R}^n} xh(y)p(x, y)\,|dx|\,|dy|$$

$$= E[xh(y)].$$

Or

$$E[(x - E[x|y])h(y)] = 0. \tag{9.2.4}$$

We can use (9.2.4) to obtain an alternate ("density-function" free) definition of conditional expectation—namely, suppose $f(\cdot)$ mapping $\mathbf{R}^n$ into $\mathbf{R}^m$ is such that

$$E[(x - f(y))h(y)] = 0 \tag{9.2.5}$$

for *every* $h(\cdot)$ satisfying (9.2.3). Then

$$f(y) = E[x|y].\dagger \tag{9.2.6}$$

Indeed, subtracting (9.2.4) from (9.2.5) we have that

$$E[(E[x|y] - f(y))h(y)] = 0 \tag{9.2.7}$$

for every $h(\cdot)$ satisfying (9.2.3). This is enough to prove (9.2.6), since $h(\cdot)$ being arbitrary in (9.2.7), it follows that

$$E[x|y] - f(y) = 0 \tag{9.2.7a}$$

for any $m \times q$ function $h(\cdot)$ satisfying

$$E[\|h(y)\|] < \infty,$$

and we may take

$$h(y) = (E[x|y] - f(y))^*.$$

Note that (9.2.5) is a stronger version of (9.1.15); the residual is "uncorrelated with any function of the data."

---

† A mathematical technicality here is that this need hold only "almost everywhere" (with respect to the distribution of y) and need not hold, for example, for points in R^n where the density function of y vanishes.

9.2.1   Variational Definition of Conditional Expectation

Drawing on (9.2.4) we can provide a "variational" definition of conditional expectation. A technicality here is that in the definition of conditional expectation it is only required that the first moments be finite. Here, however, we shall need to assume that

$$E[\|x\|^2] < \infty$$

(which will always be the case for us!).

Let us consider nonlinear (not-necessarily-linear) estimates $f(y)$, where $f(\cdot)$ maps $\mathbf{R}^n$ into R^m. The problem is now to find $f(\cdot)$ that minimizes the mean square error:

$$E[\|x - f(y)\|^2]. \tag{9.2.8}$$

Taking $f(y) = 0$, we see that the minimum must be finite. Now we can write

$$x - f(y) = x - E[x|y] + E[x|y] - f(y)$$

and by (9.2.7a)

$$E[(x - E[x|y])(E[x|y] - f(y))^*] = 0$$

and hence

$$E[(x - E[x|y]), (E[x|y] - f(y))] = 0.$$

Hence

$$E[\|x - f(y)\|^2] = E[\|x - E[x|y]\|^2] + E[\|E[x|y] - f(y)\|^2].$$

Hence

$$f(y) = E[x|y]$$

is the optimal estimate in the mean square.

As in the linear estimation case of Section 9.1, we may consider the problem of minimizing the covariance matrix of the error:

$$E[(x - f(y))(x - f(y))^*]$$

and the result is the same, the arguments going through virtually without change. In particular, by (9.2.7a) we obtain

$$E[(x - E[x|y])E[x|y]^*] = 0, \tag{9.2.9}$$

and hence the minimal error covariance matrix, which in the future we refer to as the "error covariance matrix," is given by

$$E[(x - E[x|y])(x - E[x|y])^*] = E[(x - E[x|y])x^*] \tag{9.2.10}$$

$$= E[xx^*] - E[E[x|y]E[x|y]^*]. \tag{9.2.11}$$

We close by noting some elementary properties of conditional expectation: For any $p \times m$ matrix L we have the following

(i)
$$E[Lx \mid y] = LE[x \mid y];$$

(ii)
$$E[f(x) \mid y] = E[f(x)]$$

if x is independent f y.

(iii)
$$E[\|E[x \mid y]\|^2] \leq E[\|x\|^2]$$
$$E[E[x \mid y]E[x \mid y]^*] \leq E[xx^*].$$

(iv) Let y denote the $(n + q) \times 1$ random variable.

$$y = \begin{vmatrix} y_1 \\ y_2 \end{vmatrix},$$

where y_1 is $n \times 1$ and y_2 is $q \times 1$. Then we often use the notation

$$E[x \mid y_1, y_2] \quad \text{for } E[x \mid y].$$

With this notation we note the important property:

$$E[E[x \mid y_1, y_2] \mid y_1] = E[x \mid y_1].$$

This follows from

$$E[E[x \mid y_1, y_2] \mid h(y_1)] = E[xh(y_1)]$$
$$= E[E[x \mid y_1]h(y_1)].$$

9.2.2   Conditional Expectation: The Gaussian Case

Let us examine now what simplification accrues when we specialize to the case when

$$\begin{vmatrix} x \\ y \end{vmatrix}$$

is Gaussian distributed. We have:

Theorem 9.2.1. Let the $(n + m) \times 1$ variable

$$\begin{vmatrix} x \\ y \end{vmatrix}$$

be Gaussian. Then

$$E[x \,|\, y] = E[x] + A_0 \tilde{y}, \tag{9.2.12}$$

where A_0 is any solution of the Wiener-Hopf equation (9.1.10)—A_0 satisfies (9.1.6).

Proof. Because of the Wiener-Hopf equation, we can verify that

$$E[(x - (E[x] + A_0 \tilde{y})) y^*] = 0. \tag{9.2.13}$$

Indeed, since

$$E[x - (E[x] + A_0 \tilde{y})] = 0$$

we have that (9.2.13)

$$= E[(x - (E[x] + A_0 \tilde{y})) \tilde{y}^*]$$

$$= R_{xy} - A_0 R_{yy}$$

$$= 0. \tag{9.2.14}$$

Now

$$\begin{vmatrix} x - (E[x] + A_0 \tilde{y}) \\ \tilde{y} \end{vmatrix} = \begin{vmatrix} \tilde{x} - A_0 \tilde{y} \\ \tilde{y} \end{vmatrix}$$

is Gaussian, being a linear transformation of the Gaussian

$$\begin{vmatrix} \tilde{x} \\ \tilde{y} \end{vmatrix}.$$

But by (9.2.14),

$$\tilde{x} - A_0 \tilde{y} \quad \text{and} \quad \tilde{y}$$

are uncorrelated. Hence (see Review Chapter) they are independent. Hence for any 1×1 function $h(\cdot)$ defined on $\mathbf{R}^n$ such that

$$E[|h(\tilde{y})|] < \infty$$

we have that

$$E[(\tilde{x} - A_0 \tilde{y}) h(\tilde{y})] = E[(\tilde{x} - A_0 \tilde{y})] E[h(\tilde{y})]$$

$$= 0.$$

Hence (9.2.5) is satisfied with

$$f(y) = E[x] + A_0 \tilde{y}$$

or

$$E[x|y] = E[x] + A_0 \tilde{y}$$

are required to be proved.

Note that

$$E[x] + A_0 \tilde{y} = A_0 y + c,$$

where c is given by (9.1.13). Thus we have shown that the best linear estimate of x in terms of y is also the conditional expectation of x given y when

$$\begin{vmatrix} x \\ y \end{vmatrix}$$

is Gaussian.

9.2.3   Conditional Density: The Gaussian Case

With $x \sim m \times 1$ and $y \sim n \times 1$ as before, let

$$\begin{vmatrix} x \\ y \end{vmatrix}$$

be Gaussian. We want to calculate the conditional density

$$p(x|y).$$

We could of course write down $p(x, y)$ and divide it by $p(y)$ and try to "simplify." It is more illuminating to proceed slightly differently.

We first calculate the mean and covariance of the conditional density $p(x|y)$. The mean

$$\int x p(x|y) \, |dx|$$

is in fact

$$= E[x|y]$$
$$= E[x] + A_0(y - E[y]).$$

The covariance matrix is by definition

$$= \int_{\mathbf{R}^m} (x - E[x|y])(x - E[x|y])^* p(x|y) \, |dx|$$
$$= E[(x - E[x|y])(x - E[x|y])^* | y]. \tag{9.2.15}$$

Now, as we have seen, the random variable

$$x - E[x|y]$$

is independent of y and hence by property (ii), it follows that (9.2.15)

$$= E[(x - E[x|y])(x - E[x|y])^*]$$
$$= \text{error covariance matrix } P$$
$$= R_{xx} - A_0 R_{yy} A_0^*$$
$$= R_{xx} - R_{xy} A_0^*$$
$$= R_{xx} - A_0 R_{yx}.$$

Let Λ denote the covariance matrix of

$$z = \begin{vmatrix} x \\ y \end{vmatrix}.$$

Then Λ is $(n + m) \times (n + m)$ and can be expressed as the compound matrix:

$$\Lambda = \begin{vmatrix} R_{xx} & R_{xy} \\ R_{yx} & R_{yy} \end{vmatrix},$$

where

$$R_{yx} = R_{xy}^*.$$

Of course the density function $p(x, y)$ is not definable unless Λ is nonsingular. We shall now prove that

$$|\Lambda| = |R_{yy}| \cdot |P|, \tag{9.2.16}$$

where "Det" denotes determinant. Note that we may perform "elementary operations" on a matrix without changing the determinant. Hence

$$|\Lambda| = \begin{vmatrix} R_{xx} & R_{xy} \\ R_{yx} & R_{yy} \end{vmatrix}_{det} = \begin{vmatrix} R_{xx} - A_0 R_{yx} & R_{xy} - A_0 R_{yy} \\ R_{yx} & R_{yy} \end{vmatrix}_{det},$$

which, using (9.2.15), (9.1.6),

$$= \begin{vmatrix} P & 0 \\ R_{yx} & R_{yy} \end{vmatrix}_{det}.$$

Note in particular that Λ is nonsingular if and only if P and R_{yy} are nonsingular. Hence P and R_{yy} are both singular under the assumption that Λ is nonsingular. In particular, then

$$A_0 = R_{xy} R_{yy}^{-1}. \tag{9.2.17}$$

The Gaussian density with mean $E[x|y]$ and covariance matrix P is given by

$$G[x|y] = \frac{1}{(\sqrt{2\pi})^m} \frac{1}{|P|^{1/2}} \exp\left(\frac{-1}{2}[P^{-1}(\tilde{x} - A_0\tilde{y}), (\tilde{x} - A_0\tilde{y})]\right), \quad (9.2.18)$$

where

$$\tilde{x} = x - \mu_x, \qquad \tilde{y} = y - \mu_y.$$

The density $p(x, y)$ is given by

$$p(x, y) = \frac{1}{(\sqrt{2\pi})^{m+n}} \frac{1}{|\Lambda|^{1/2}} \exp\left(\frac{-1}{2}[\Lambda^{-1}\tilde{z}, \tilde{z}]\right), \quad \tilde{z} = \left|\begin{matrix} \tilde{x} \\ \tilde{y} \end{matrix}\right|,$$

and the density $p(y)$ is expressed as

$$p(y) = \frac{1}{(2\sqrt{\pi})^n} \frac{1}{|R_{yy}|^{1/2}} \exp\left(\frac{-1}{2}[R_{yy}^{-1}\tilde{y}, \tilde{y}]\right).$$

Hence

$$\frac{p(x, y)}{p(y)} = \frac{1}{(\sqrt{2\pi})^m} \frac{1}{|P|^{1/2}} \exp\left(\frac{-1}{2}[(\Lambda^{-1}\tilde{z}, \tilde{z}) - (R_{yy}^{-1}\tilde{y}, \tilde{y})]\right). \quad (9.2.19)$$

Now we shall show that

$$\Lambda^{-1} = \left|\begin{matrix} P^{-1} & -P^{-1}A_0 \\ -A_0^* P^{-1} & R_{yy}^{-1} + A_0^* P^{-1}A_0 \end{matrix}\right|$$

by showing that

$$\left|\begin{matrix} R_{xx} & R_{xy} \\ R_{yx} & R_{yy} \end{matrix}\right| \left|\begin{matrix} P^{-1} & -P^{-1}A_0 \\ -A_0^* P^{-1} & R_{yy}^{-1} + A_0^* P^{-1}A_0 \end{matrix}\right| \quad (9.2.20)$$

is the $(n + m) \times (n + m)$ identity matrix. In fact, carrying out the indicated multiplication we obtain that (9.2.20)

$$= \left|\begin{matrix} (R_{xx} - R_{xy}A_0^*)P^{-1} & -R_{xx}P^{-1}A_0 + R_{xy}(R_{yy}^{-1} + A_0^*P^{-1}A_0) \\ (R_{yx} - R_{yy}A_0^*)P^{-1} & -R_{yx}P^{-1}A_0 + R_{yy}(A_0^*P^{-1}A_0 + R_{yy}^{-1}) \end{matrix}\right|.$$

But from (9.2.15)

$$(R_{xx} - R_{xy}A_0^*)P^{-1} = PP^{-1} = I_{m \times m}$$

and
$$R_{xy} - R_{yy}A_0^* = (R_{xy} - A_0 R_{yy})^* = 0.$$

Next
$$-R_{xx}P^{-1}A_0 + R_{xy}A_0^* P^{-1}A_0 + R_{xy}R_{yy}^{-1} = (-R_{xx} + R_{xy}A_0^*)P^{-1}A_0 + R_{xy}R_{yy}^{-1}$$
$$= -A_0 + R_{xy}R_{yy}^{-1}$$
$$= 0.$$

Similarly,
$$-R_{yx}^{-1}P^{-1}A_0 + R_{yy}A_0^* P^{-1}A_0 + R_{yy}R_{yy}^{-1} = (-R_{yx} + R_{yy}A_0^*)P^{-1}A_0 + I$$
$$= I,$$

since
$$(-R_{yx} + R_{yy}A_0^*) = (-R_{xy} + A_0 R_{yy})^* = 0.$$

Hence
$$[\Lambda^{-1}\tilde{z}, \tilde{z}] - [R_{yy}^{-1}\tilde{y}, \tilde{y}] = [P^{-1}\tilde{x}, \tilde{x}] - 2[P^{-1}A_0\tilde{y}, \tilde{x}]$$
$$+ [(R_{yy}^{-1} + A_0^* P^{-1}A_0)\tilde{y}, \tilde{y}] - [R_{yy}^{-1}\tilde{y}, \tilde{y}]$$
$$= [P^{-1}\tilde{x}, \tilde{x}] - 2[P^{-1}A_0\tilde{y}, \tilde{x}] + [A_0^* P^{-1}A_0\tilde{y}, \tilde{y}]$$
$$= [P^{-1}(\tilde{x} - A_0\tilde{y}), (\tilde{x} - A_0\tilde{y})],$$

showing that
$$p(x|y) = \frac{1}{(\sqrt{2\pi})^m} \cdot \frac{1}{|P|} \exp\left(\frac{-1}{2}[P^{-1}(\tilde{x} - A_0\tilde{y}), (\tilde{x} - A_0\tilde{y})]\right) \quad (9.2.21)$$

and hence is Gaussian, in particular. Note that the conditional density is well defined so long as P is nonsingular, even if Λ is singular.

Formula (9.2.16) is most useful when x is 1×1 so that P is 1×1 and hence
$$P = |P|.$$

Moreover, in that case the error covariance ($=$ mean square error) is given by
$$P = \frac{|\Lambda|}{|R_{yy}|}. \quad (9.2.16a)$$

9.2.4 Mutual Information

The concept of "mutual information" is an important one in communication theory. Given any two random variables x and y, the mutual information is defined by
$$I(x; y) = E\left[\log \frac{p(x, y)}{p(x)p(y)}\right]. \quad (9.2.22)$$

While this is not generally calculable, we can evaluate it explicitly in the Gaussian case in terms of the moments, exploiting our formula for the conditional expectation (9.2.18). Thus

$$
E\left[\log \frac{p(x, y)}{p(x)p(y)}\right] = E\left[\log \frac{p(x \mid y)}{p(x)}\right]
$$

$$
= E[\log p(x \mid y)] - E[\log p(x)]
$$

$$
= E\left[\frac{-1}{2} [P^{-1}(\tilde{x} - A_0 \tilde{y}), (\tilde{x} - A_0 \tilde{y})]\right] - \frac{1}{2} \log |P|
$$

$$
+ E\left[\frac{1}{2} [R_{xx}^{-1} \tilde{x}, \tilde{x}]\right] + \frac{1}{2} \log |R_{xx}|.
$$

Hence

$$
I(x; y) = \frac{1}{2} \log \frac{|R_{xx}|}{|P|}. \tag{9.2.23}
$$

But as we have seen

$$
|\Lambda| = |P| |R_{yy}|
$$

and hence (9.2.23) is also given by

$$
\frac{1}{2} \log \frac{|R_{xx}| |R_{yy}|}{|\Lambda|}. \tag{9.2.24}
$$

Note that (as we should expect)

$$
I(x; y) = \infty, \qquad \text{if } |P| = 0
$$

and (9.2.23) may be defined even if $|\Lambda|$ is zero.

9.3 GRAM-SCHMIDT ORTHOGONALIZATION AND COVARIANCE MATRIX FACTORIZATION

Let x be an $m \times 1$ Gaussian with zero mean, and let us use the notation

$$
x = \text{col}(x_1, \ldots, x_m).
$$

Let R_{xx} denote the covariance matrix of x. Then

$$
R_{xx} = \{\lambda_{ij}\}, \qquad 1 \le i, j \le m,
$$

where

$$
\lambda_{ij} = E[x_i x_j].
$$

Let us construct a new set of Gaussian variables $\{y_i\}$, $i = 1, \ldots, n$, as follows

$$y_1 = x_1,$$
$$y_2 = x_2 - E[x_2 | x_1],$$
$$y_j = x_j - E[x_j | x_1, \ldots, x_{j-1}], \qquad 2 < j < m,$$
$$y_m = x_m - E[x_m | x_1, \ldots, x_{m-1}].$$

Note that the $\{y_i\}$ are zero-mean Gaussians, but furthermore

$$E[y_i y_j] = 0, \qquad i \neq j. \tag{9.3.1}$$

Indeed, for any i, by construction we have

$$E[y_i x_j] = 0, \qquad j = 1, \ldots, i - 1;$$

and since y_j is a linear combination of $x_1, \ldots, x_{j-1}, x_j$, we obtain

$$E[y_i y_j] = 0, \qquad j = 1, \ldots, i - 1.$$

For $j > i$, the result follows from:

$$E[y_i y_j] = E[y_j y_i].$$

We say that y_i is "orthogonal" to y_j if

$$E[y_i y_j] = 0.$$

Hence the variables $\{y_i\}$ are mutually orthogonal. Equivalently, the covariance matrix of

$$Y = \text{col}\,[y_i, \ldots, y_m]$$

is diagonal:

$$E[YY^*] = D,$$

where

$$D = \{d_{ij}\}, \qquad d_{ij} = 0, i \neq j$$

and

$$d_{11} = E[x_1^2]$$
$$d_{ii} = E[(x_i - E[x_i | x_1, \ldots, x_{i-1}])^2], \qquad i \geq 2.$$

Note that we can express $\{y_i\}$ as

$$y_i = a_{i1}x_1 + a_{i2}x_2 + \cdots + a_{ii}x_i,$$

where
$$a_{ii} = 1.$$

Define the matrix L by
$$L = \{\ell_{ij}\}, \qquad 1 \le i, j \le m,$$

where
$$\ell_{ij} = a_{ij}, \qquad i \le j,$$
$$= 0, \qquad i > j.$$

Thus defined, L is an $m \times m$ "lower-triangular" matrix, and we then obtain

$$Y = Lx.$$

The determinant of a lower-triangular matrix is the product of the diagonal elements, and hence in our case

$$|L| = 1.$$

Hence L is nonsingular, and

$$x = L^{-1}Y,$$

where L^{-1} is also lower-triangular, with the diagonal elements also equal to unity. Furthermore, for any $p \times m$ matrix A, we have

$$Ax = (AL^{-1})Y,$$

so that any linear combination of the $\{x_i\}$ can also be expressed as a linear combination of the $\{y_i\}$. Moreover, we have that

$$R_{xx} = E[xx^*] = L^{-1}E[YY^*]L^{*-1}$$
$$= L^{-1}DL^{*-1} = (L^{-1}\sqrt{D})(L^{-1}\sqrt{D})^*. \qquad (9.3.2)$$

Thus we have factorized the covariance matrix R_{xx} as

$$R_{xx} = \mathscr{L}\mathscr{L}^*, \qquad (9.3.3)$$

where $\mathscr{L}$ is lower-triangular. Such a factorization is not unique, in the sense that we can find many lower-triangular matrices satisfying (9.3.3). Indeed we have only to replace $\mathscr{L}$ by

$$\mathscr{L}\mathscr{D},$$

where $\mathscr{D}$ is a diagonal matrix with entries which are $+1$ or -1. Thus we can have 2^m such matrices.

9.3.1 Simulating Gaussian Vectors

Finally, let Z be an $m \times 1$ Gaussian with zero mean and identity covariance matrix. Then

$$\mathscr{L}Z \tag{9.3.4}$$

is Gaussian and has R_{xx} for its covariance matrix. We note that (9.3.4) provides us with a "simulation" technique for constructing Gaussian vectors with prescribed covariance matrix, from a "random number" generator. Thus, let $\xi_1, \xi_2, \ldots, \xi_{2m}$ be mutually independent random variables uniformly distributed between 0 and 1. As we have seen in Chapter 7, Section 7.1, defining

$$z_i = (\sqrt{-2 \log (1 - \xi_i)}) \sin 2\pi\xi_{2i}$$

will yield

$$Z = \begin{vmatrix} z_1 \\ \vdots \\ z_m \end{vmatrix},$$

which is Gaussian with identity covariance matrix. Using the factorization (9.3.3) and taking

$$\mathscr{L} = L^{-1}\sqrt{D},$$

$$x = \mathscr{L}Z,$$

we can see that x is Gaussian with covariance R_{xx}. Note also that

$$|R_{xx}| = |D| = \prod_i d_{ii}. \tag{9.3.5}$$

Remark 1. Even if x has nonzero mean, we may define

$$y_i = x_i - E[x_i \mid x_{i-1}, \ldots, x_1], \qquad i = 1, \ldots, m,$$

so that

$$y_1 = x_1 - E[x_1]$$

and $\{y_i\}$ would again be zero mean Gaussian, orthogonal for $i \neq j$. Also

$$y_i = \tilde{x}_i - E[\tilde{x}_i \mid \tilde{x}_{i-1}, \ldots, \tilde{x}_1],$$

or we can write

$$Y = \text{col}\,\{y_i\} = L\tilde{x} = Lx - LE[x]$$

and

$$x = L^{-1}Y + E[x]. \tag{9.3.6}$$

Remark 2. We may extend the orthogonalization procedure to the case where the variables x_i are multidimensional. To be specific, let each x_i be $n \times 1$ with zero mean. In that case we have again

$$y_1 = x_1,$$
$$y_i = x_i - E[x_i, \ldots, x_{i-1}]. \tag{9.3.7}$$

The $\{y_i\}$ are again zero-mean orthogonal Gaussians:

$$E[y_i y_j^*] = 0, \quad i \neq j.$$

Let

$$E[y_i y_i^*] = D_i$$

and

$$y_i = \sum_{j=1}^{i} a_{ij} x_j, \tag{9.3.8}$$

where the a_{ij} are now $n \times n$ matrices, with

$$a_{ii} = I \quad (n \times n \text{ identity}).$$

Defining the compound matrix L by

$$L = \{\ell_{ij}\}.$$
$$\ell_{ij} = a_{ij}, \quad i \leq j,$$
$$= 0, \quad i > j.$$

we obtain a block lower-triangular matrix such that

$$|L| = 1.$$

Let D now define the compound matrix

$$D = \text{Diag}(D_{11}, \ldots, D_{mm}),$$

each D_{ii} being nonnegative definite, and again

$$R_{xx} = \mathscr{L}\mathscr{L}^*,$$

where

$$\mathscr{L} = L^{-1}\sqrt{D}$$

and

$$|R_{xx}| = \prod_i |D_{ii}|, \qquad (9.3.9)$$

generalizing (9.3.5). Moreover, if none of the D_{ii} is singular, we can express the density function of $x = \text{col}\,(x_1, \ldots, x_n)$ as

$$p(x) = \prod_i p_i(y_i), \qquad (9.3.10)$$

where

$$p_i(y_i) = \frac{1}{(\sqrt{2\pi})^i |D_{ii}|^{1/2}} \left(\exp \frac{-1}{2} [D_{ii}^{-1} y_i, y_i] \right)$$

and of course $\{y_i\}$ given by (9.3.7). Also, formula (9.3.7) continues to be valid in the general case where the $\{x_i\}$ are allowed to have nonzero means, and again (9.3.6) holds with

$$x = \text{col}\,[x_1, \ldots, x_m].$$

9.4   THE MAXIMUM LIKELIHOOD PRINCIPLE

The Maximum Likelihood Principle provides an oft-invoked tool for generating estimates as an alternative to conditional expectation, and unlike the latter it lends itself more easily to numerical implementation. With x and y as before, to estimate x from y we simply maximize the density

$$p(x, y)$$

with respect to x for fixed y. The point of maximum is the MLE (or Maximum Likelihood Estimate). The maximization problem is a numerical optimization problem with a large literature devoted to it (see, e.g., references 10 and 11). The only drawback is that there is no easy way to evaluate (any measure of) the corresponding error—such as the mean square error.

The case where $p(x, y)$ is Gaussian is of primary interest, since we can evaluate the solution analytically. It is more convenient to consider

$$\max_{x \in R^m} \log p(x, y). \qquad (9.4.1)$$

Now

$$\log p(x, y) = \log p(x|y) + \log p(y),$$

and hence it is enough to consider

$$\max_{x \in R^m} \log p(x|y). \qquad (9.4.2)$$

We have already evaluated $p(x \mid y)$ in (9.2.21) and we see that if we omit constants we need to maximize

$$\frac{-1}{2}[P^{-1}(\tilde{x} - A_0\tilde{y}), \tilde{x} - A_0\tilde{y}]$$

or minimize

$$[P^{-1}(\tilde{x} - A_0\tilde{y}), \tilde{x} - A_0\tilde{y}],$$

and the minimum is attained at

$$\tilde{x} - A_0\tilde{y} = 0$$

or

$$x = \mu_x + A_0(y - \mu_y).$$

But this is also the conditional expectation which is then the same as the MLE in the Gaussian case. Nevertheless we shall see that the direct maximization (9.4.1) is a convenient alternative to calculating the conditional expectation estimate, not invoking the Wiener-Hopf equation. Note in particular that the MLE actually minimizes the mean square error in the Gaussian case.

Since the gradient must vanish at the maximum, we may seek instead a "root of the gradient equation"

$$\nabla_x \log p(x, y) = 0 \qquad (9.4.3)$$

for fixed y. Note the appearance of the subscript x denoting the variable with respect to which the gradient is being taken.

9.5 ESTIMATION OF SIGNAL PARAMETERS IN ADDITIVE NOISE MODELS

Let us now specialize our estimation models closer to practice. We consider a communication channel with additive noise, where v_n represents the received signal at the nth sampling interval, so that we can write

$$v_n = s_n + N_n, \qquad (9.5.1)$$

where $\{N_n\}$ represents the channel noise and $\{s_n\}$ the transmitted signal at the nth sampling interval. The canonical problem is to estimate the signal s_n from the received waveform samples $\{v_n\}$. "Real-time" or "on-line" operation would mean that s_n would have to be estimated from v_k, $k \leq n$. Moreover, we would need to specify the class of signals being transmitted in some fashion. We may have one of two signal models: we have the "sure signal"—or "deterministic signal"—model when we assume that signal parameters are unknown but that signals are specifiable once the parameters are specified. Here we consider the general (Bayesian) case when the parameters are random, so that $\{s_n\}$ is a "trivial" random process. Or we may have the model where the signal is a nontrivial random process—a "stochastic signal." These, of course, are not necessarily mutually exclusive points of view, and further various shades in between are also often employed.

We begin with the "sure signal" case. Our signal model is the "linear" model

$$s_n = \sum_1^m \theta_k S_k(n),$$

where the $S_k(\cdot)$ are known (specified) but the parameters $\{\theta_k\}$, $k = 1, m$ are unknown. Suppose now we observe v_n, $n = 1, \ldots, p$. Then, for each p, letting

$$v = \begin{vmatrix} v_1 \\ \vdots \\ v_p \end{vmatrix}$$

we can write (9.5.1) as

$$v = \sum_1^m \theta_k S_k + N, \tag{9.5.2}$$

where v, S_k, and N are all $p \times 1$. The $\{S_k\}$ are known to the receiver but the parameters $\{\theta_k\}$, $k = 1, \ldots, m$, are not, and their values specify the transmitted waveform. We assume N is Gaussian with zero mean and covariance R_N. Although we have stated the problem in a signal transmission setting, such a model can occur in a variety of other applications; in any event, the model (9.5.2) can be considered divorced from any specific application.

To proceed with the estimation problem, it is convenient to let

$$\theta = \mathrm{col}\,(\theta_1, \ldots, \theta_m)$$

and write

$$\sum_1^m \theta_k S_k = L\theta,$$

where L is the $p \times m$ matrix defined by

$$L = \{\ell_{ij}\}, \qquad\qquad i = 1, \ldots, p; j = 1, \ldots, m$$

$$\ell_{ij} = i\text{th component of } S_j$$

$$= [S_j, e_i]$$

where $\{e_i\}$ are $p \times 1$ unit basis vectors:

$$e_i = \mathrm{col}\,[e_{i1}, e_{i2}, \ldots, e_{im}],$$

where

$$e_{ij} = \delta^i_j = 1, \qquad i = j,$$
$$= 0, \qquad i \neq j.$$

Hence (9.5.2) takes the canonical form

$$v = L\theta + N. \tag{9.5.3}$$

If

$$L\theta = 0 \qquad \text{for some } \theta \neq 0$$

or, equivalently,

$$\sum_1^m \theta_k S_k(n) = 0, \qquad n = 1, \dots, p$$

we cannot hope to estimate this value of θ from (9.5.3) We assume therefore that

$$L\theta = 0 \Rightarrow \theta = 0.$$

Or, equivalently, zero is not an eigenvalue of

$$L^*L.$$

9.5.1 The "Bayesian" Model

We assume next that θ is random. We have then a "Bayesian" model. More specifically we assume that it is Gaussian with mean μ and covariance matrix Λ. Then $p(\theta, v)$ is Gaussian and we can apply the estimation theory we have developed. The first simplification is to assume that $\mu = 0$; indeed since μ is known we may consider

$$v - L\mu \tag{9.5.4}$$

in place of v. To avoid further notation we shall continue to use v and take μ to be zero.

9.5.2 Conditional Expectation Estimate

The conditional expectation estimate is given by

$$\hat{\theta} = E[\theta|v] = A_0 v,$$

where A_0 satisfies the Wiener-Hopf equation

$$R_{\theta v} = A_0 R_{vv},$$

where we can readily calculate that

$$R_{\theta v} = E[\theta(L\theta + N)^*] = \Lambda L^*$$

$$R_{vv} = E[(L\theta + N)(L\theta + N)^*] = L\Lambda L^* + R_N.$$

We assume that R_{vv} is nonsingular, for which it is enough if either R_N or $L\Lambda L^*$ is nonsingular. Then

$$A_0 = \Lambda L^*(L\Lambda L^* + R_N)^{-1}$$

and

$$\hat{\theta} = \Lambda L^*(L\Lambda L^* + R_N)^{-1}v. \tag{9.5.5}$$

The corresponding (minimal) error covariance matrix P (see (9.2.15)) is given by

$$P = E[\theta\theta^*] - A_0 E[v\theta^*]$$

$$= \Lambda - \Lambda L^*(L\Lambda L^* + R_N)^{-1}L\Lambda. \tag{9.5.6}$$

9.5.3 Maximum Likelihood Estimate

Let us next consider the MLE, even though we know that is the same as the conditional expectation. Thus we consider

$$\max_{\theta \in R^m} \log p(\theta, v).$$

It is convenient to use

$$p(\theta, v) = p(v \mid \theta)p(\theta)$$

and since θ is independent of N, we can readily verify (see Problem 9.2) that

$$p(v \mid \theta) = P_N(v - L\theta), \tag{9.5.7}$$

where the subscript indicates that it is the density function of the noise vector N, which is well defined if we assume that R_N is nonsingular. Then we can calculate that the gradient (see Review Chapter for more on the notion of gradient)

$$\nabla_\theta \log p(\theta, v) = \nabla_\theta \log P_N(v - L\theta) + \nabla_\theta \log P(\theta)$$

$$= \frac{-1}{2}(\nabla_\theta[R_N^{-1}(v - L\theta), (v - L\theta)] + \nabla_\theta[\Lambda^{-1}\theta, \theta]). \tag{9.5.8}$$

Now

$$\frac{-1}{2}\frac{d}{d\lambda}[R_N^{-1}(v - L(\theta + \lambda h)), v - L(\theta + \lambda h)] = [R_N^{-1}(v - L(\theta + \lambda h)), Lh]$$

$$\frac{-1}{2}\frac{d}{d\lambda}[\Lambda^{-1}(\theta + \lambda h), (\theta + \lambda h)] = -[\Lambda^{-1}(\theta + \lambda h), h]$$

and hence (9.5.8)

$$= (L^*R_N^{-1}(v - L\theta) - \Lambda^{-1}\theta)^*.$$

The gradient vanishes at the point $\hat{\theta}$ given by

$$L^*R_N^{-1}(v - L\hat{\theta}) - \Lambda^{-1}\hat{\theta} = 0$$

or

$$\hat{\theta} = (L^*R_N^{-1}L + \Lambda^{-1})^{-1}L^*R_N^{-1}v, \qquad (9.5.9)$$

which appears different from (9.5.5). We realize that the difference comes from the fact that we have assumed that R_N is nonsingular. Equation (9.5.5) is more general in that we only require that $L\Lambda L^*$ or R_N be nonsingular. This is only of "mathematical technicality" significance, however, since the noise covariance R_N is rarely, if ever, singular. The more important difference is in the form of the estimate—(9.5.9) is simpler, and the factor, the matrix

$$(L^*R_N^{-1}L + \Lambda^{-1})^{-1}, \qquad (9.5.10)$$

is actually the error covariance matrix P, given by (9.5.6). Proving this is an algebraic exercise. We begin with the identity:

$$L^*(I + R_N^{-1}L\Lambda L^*) = (L^*R_N^{-1}L\Lambda + I)L^*.$$

Now

$$(I + R_N^{-1}L\Lambda L^*) \qquad \text{and} \qquad (I + L^*R_N^{-1}L\Lambda)$$

are both nonsingular (see Problem 14 on page 33). Hence

$$(L^*R_N^{-1}L\Lambda + I)^{-1}L^* = L^*(I + R_N^{-1}L\Lambda L^*)^{-1}.$$

Hence

$$\Lambda(L^*R_N^{-1}L\Lambda + I)^{-1}L^*R_N^{-1}L\Lambda = \Lambda L^*(I + R_N^{-1}L\Lambda L^*)^{-1}R_N^{-1}L\Lambda$$

$$= \Lambda L^*(R_N + L\Lambda L^*)^{-1}L\Lambda.$$

Hence

$$\Lambda - \Lambda L^*(R_N + L\Lambda L^*)^{-1}L\Lambda$$

$$= \Lambda - \Lambda(L^*R_N^{-1}L\Lambda - I)^{-1}L^*R_N^{-1}L\Lambda$$

$$= (I - (L^*R_N^{-1}L + \Lambda^{-1})^{-1}L^*R_N^{-1}L)\Lambda$$

$$= (L^*R_N^{-1}L + \Lambda^{-1})^{-1}(L^*R_N^{-1}L + \Lambda^{-1} - L^*R_N^{-1}L)\Lambda$$

$$= (L^*R_N^{-1}L + \Lambda^{-1})^{-1}$$

$$= P$$

as we needed to show. Thus we have

$$\hat{\theta} = PL^*R_N^{-1}v,$$

where P is the $p \times p$ error covariance matrix.

If R_N is nonsingular, for large Λ, or in the limiting case

$$\Lambda = +\infty$$

we have that

$$P(L^*R_N^{-1}L)^{-1}$$

and

$$\hat{\theta} = (L^*R_N^{-1}L)^{-1}L^*R_N^{-1}v. \tag{9.5.11}$$

Intuitively $\Lambda = \infty$ corresponds to "maximum ignorance"; to say that θ is Gaussian with infinite variance is tantamount to saying θ is simply an unknown parameter. The optimal estimate is then obtained by maximizing (9.5.7). The model is no longer Bayesian and we have a case of maximum-likelihood "parameter estimation."

If the noise $\{N_n\}$ is a white noise sequence with variance $= \sigma^2$, we have

$$\hat{\theta} = (L^*L + \sigma^2\Lambda^{-1})^{-1}L^*v \tag{9.5.12}$$

where

$$L^*v = \left\{ \sum_{i=1}^{p} \ell_{ij}v_i \right\}, \qquad j = 1, \ldots, m.$$

Moreover in this case, the error covariance matrix

$$= \sigma^2(L^*L + \sigma^2\Lambda^{-1})^{-1}.$$

In the "maximum ignorance" case, namely

$$\Lambda = \infty,$$

the corresponding error is

$$= \sigma^2(L^*L)^{-1} \tag{9.5.13}$$

and the estimate is

$$\hat{\theta} = (L^*L)^{-1}L^*v, \tag{9.5.14}$$

where it should be noted that the noise covariance does not enter.

9.5.4   Fit Error

Corresponding to the optimal estimate $\hat{\theta}$,

$$L\hat{\theta}$$

is the best estimate of the signal. While we cannot observe the estimate error we can observe the "fit error":

$$v - L\hat{\theta}. \tag{9.5.15}$$

We use the term "fit error" since we may think of $L\hat{\theta}$ as the best fit to v. For the case where $\{N_n\}$ is white noise and we consider maximum ignorance,

$$v - L\hat{\theta} = v - L(L^*L)^{-1}L^*v$$

$$= (L\theta + N) - L(L^*L)^{-1}L^*(L\theta + N)$$

$$= (I - L(L^*L)^{-1}L^*)N.$$

Hence the fit error covariance is given by

$$E[(v - L\hat{\theta})(v - L\hat{\theta})^*] = (I - L(L^*L)^{-1}L^*)\sigma^2(I - L(L^*L)^{-1}L^*)$$

$$= \sigma^2(I - L(L^*L)^{-1}L^*)$$

$$< \sigma^2 I.$$

Note that $L(L^*L)^{-1}L^*$ is an orthogonal projection of rank m. Hence

$$E[\|(v - L\hat{\theta})\|^2] = \sigma^2(p - m).$$

Hence the ratio

$$\frac{E[\|v - L\hat{\theta}\|^2]}{E[\|N\|^2]} = \frac{p - m}{p} < 1$$

but $\rightarrow 1$ as $p \rightarrow \infty$—as we take more data.

9.6   EXAMPLE: POLYNOMIAL IN NOISE: CURVE FITTING

A canonical application of the model (9.5.2) is to "curve fit" to data. Thus consider the case where we have a continuous-time "truth" model:

$$v(t) = S(t) + N(t), \qquad 0 \le t \le T$$

where $N(\cdot)$ is random noise (error). In the absence of further knowledge about the signal, we may assume that the signal is a lower-order polynomial in t, say of order m. This will lead us to the model (9.5.2).

To say that $S(\cdot)$ is a polynomial of order m is equivalent to

$$\frac{d^{m+1}S(t)}{d^{m+1}} \equiv 0, \qquad 0 < t < T.$$

Thus we have the representation

$$S(t) = Cx(t)$$
$$\dot{x}(t) = Ax(t),$$

where

$$x(t) = \begin{vmatrix} S(t) \\ S^{(1)}(t) \\ \vdots \\ S^{(m)}(t) \end{vmatrix}, \qquad C = |1 \quad 0 \quad \cdots \quad 0|_{1 \times (m+1)}$$

where

$$S^{(k)}(t) = \frac{d^k S(t)}{dt^k}$$

and A is a "companion" matrix $((m+1) \times (m+1))$:

$$A = \begin{vmatrix} 0 & 1 & 0 & 0 & \cdots & 0 \\ 0 & 0 & 1 & 0 & \cdots & 0 \\ \multicolumn{6}{c}{\cdots\cdots\cdots\cdots\cdots\cdots} \\ 0 & 0 & 0 & 0 & \cdots & 1 \\ 0 & 0 & 0 & 0 & \cdots & 0 \end{vmatrix} = \begin{vmatrix} 0_{m \times 1} & I_{m \times m} \\ 0_{1 \times 1} & 0_{1 \times m} \end{vmatrix}.$$

We note that A is "nilpotent," namely

$$A^{m+1} = 0$$

and hence

$$e^{A}t = \sum_{0}^{m} \frac{A^{k}t^{k}}{k!} = \begin{vmatrix} 1 & t & \dfrac{t^2}{2} & \cdots & & \dfrac{t^m}{m!} \\ 0 & 1 & t & \cdots & & \dfrac{t^{m-1}}{m-1!} \\ \multicolumn{6}{c}{\cdots\cdots\cdots\cdots\cdots\cdots\cdots} \\ 0 & 0 & 0 & \cdots & 0 & 1 & t \\ 0 & 0 & 0 & \cdots & 0 & 0 & 1 \end{vmatrix}.$$

Also

$$Ce^{At} = \left| 1, t, \frac{t^2}{2!}, \dots, \frac{t^m}{m!} \right|$$

and hence finally we have

$$S(t) = Ce^{At}x(0).$$

Estimating $S(t)$ is thus equivalent to estimating $x(0)$.

We shall assume digital processing so that the data are sampled periodically, with Δ as the sampling interval. Hence we have

$$v_n = s_n + N_n, \qquad\qquad n = 1, \dots, p$$
$$s_n = S(n\Delta) = Ce^{An\Delta}x(0).$$
$$N_n = N(n\Delta), \qquad\qquad n = 1, \dots, p; \ p\Delta = T$$

and we assume that $\{N_n\}$ is zero-mean Gaussian and denote the $p \times p$ covariance by R_N. Letting

$$v = \begin{vmatrix} v_1 \\ \vdots \\ v_p \end{vmatrix} \qquad \mathcal{N} = \begin{vmatrix} N_1 \\ \vdots \\ N_p \end{vmatrix}$$

$$L = \begin{vmatrix} Ce^{A\Delta} \\ \vdots \\ Ce^{Ap\Delta} \end{vmatrix}_{p \times (m+1)}$$

we arrive at the model

$$v = Lx_0 + \mathcal{N},$$

corresponding to (9.5.2), and we can apply our estimation theory. Here $L^{*}L$ is nonsingular as soon as $N > m + 1$ (the number of data point exceeds polynomial degree). In fact

$$L^*L = \sum_1^p e^{A^*k\Delta} C^* C e^{Ak\Delta} = \{\lambda_{ij}\}, \qquad i, j = 0, 1, \ldots, m,$$

where

$$\lambda_{ij} = \left(\sum_{n=1}^p n^{i+j} \right) \frac{\Delta^{i+j}}{i!j!}.$$

9.6.1   Maximum Likelihood Estimate: Maximum Ignorance

We may assume that R_N is nonsingular and $p > m + 1$. Usually $\{N_k\}$ is white noise with covariance we shall denote by "d." Then the MLE for maximum ignorance is given by (cf. 9.5.14))

$$\hat{x}_0 = (L^*L)^{-1} L^* v$$

or

$$\sum_{j=0}^m \left(\sum_{n=1}^p n^{k+j} \right) \frac{\Delta^{k+j} \theta_j}{k!j!} = \sum_{n=1}^p \frac{(n\Delta)^k v_n}{k!}, \qquad k = 0, 1, \ldots, m,$$

where

$$\hat{x}_0 = \begin{vmatrix} \theta_0 \\ \vdots \\ \theta_m \end{vmatrix}.$$

Since

$$S(t) = \sum_{j=0}^m \theta_j \frac{t^j}{j!}$$

we may obtain the polynomial coefficients

$$a_j = \frac{\theta_j}{j!}, \qquad j = 0, \ldots, m$$

solving

$$\sum_{j=0}^m \left(\sum_{n=1}^p n^{k+j} \right) \Delta^j a_j = \sum_{n=1}^p n^k v_n, \qquad k = 0, \ldots, m,$$

which is recognized as the least squares solution in curve fitting. The fit error is given by

$$E\|v - L(L^*L)^{-1} L^* v\|^2 = (p - (m + 1))d.$$

The corresponding error covariance matrix in estimating the coefficients $\{a_k\}$ of the polynomial is

$$dM(L^*L)^{-1} M^*,$$

where

$$M = \text{Diag } \{k!, k = 0, \ldots, m\}$$

and hence

$$= dQ^{-1},$$

where

$$Q = \{q_{ij}\}$$

$$q_{ij} = \left(\sum_{1}^{p} n^{i+j} \right) \Delta^{i+j}$$

since

$$\text{Tr } Q = \sum_{k=0}^{m} \left(\sum_{n=1}^{p} n^{2k} \right) \Delta^{2k}$$

and increases without bound as $p \to \infty$, for fixed Δ, and Q^{-1} converges to zero as $p \to \infty$, or as $T \to \infty$. For $m = 1$, we can calculate that

$$q_{00} = p$$

$$q_{10} = \frac{p(p+1)}{2} \Delta = q_{01}$$

$$q_{11} = \frac{p(p+1)(2p+1)}{6} \Delta^2.$$

9.7 A TRACKING PROBLEM: LINEAR MODEL

In this section we begin with an application and go on to consider its generalizations.

The application is a "tracking" problem: literally tracking a moving vehicle, from available sensor data.

Let $r(t)$ denote the (3×1 column vector) position (range) with respect to a known fixed or moving coordinate frame vector at time t. In the simplest model, we assume that the acceleration is zero (the vehicle is cruising) so that our dynamic model is

$$\ddot{r}(t) \equiv 0. \tag{9.7.1}$$

This is our basic "dynamic equation." The (relative) velocity

$$v(t) = \dot{r}(t)$$

is then of course constant. Of the variety of possible sensor information patterns we may consider, let us again for simplicity assume that the position vector is sensed at a fixed sampling rate so that we have the linear "observation model"

$$y(n\Delta) = r(n\Delta) + N_n, \tag{9.7.2}$$

where Δ is the sampling interval and $\{N_N\}$ represents the sensor error modelling as white Gaussian with zero mean and

$$E[N_n N_m^*] = \delta_m^n R_n,$$

where we assume that R_n is nonsingular for each n.

The "tracking problem" is to estimate the position and velocity at $(n+p)\Delta$:

$$r(n\Delta + p\Delta), \qquad v(n\Delta + p\Delta),$$

where p is a fixed nonnegative integer, based on the sensor data available up to the time $n\Delta$. This can also be expressed as

$$y(k\Delta), \qquad k = 1, 2, \ldots, n$$

("n" data "samples"). This is also called a "prediction" problem if $p > 0$.

To cast this into the framework of our estimation theory, let us note first that we can rewrite the dynamic equations in discrete time as

$$r_{n+1} = r_n + \Delta v_n$$

$$v_{n+1} = v_n,$$

$$y_n = r_n + N_n,$$

where we use the notation

$$r(n\Delta) = r_n$$

$$v(n\Delta) = v_n.$$

Writing

$$x_n = \begin{vmatrix} r_n \\ v_n \end{vmatrix},$$

where x_n is now 6×1, we can express (9.7.1), (9.7.2) in the "state-space" form:

$$\begin{aligned} x_{n+1} &= A x_n \\ y_n &= C x_n + N_n, \end{aligned} \qquad (9.7.3)$$

where

$$A = \begin{vmatrix} I_3 & \Delta I_3 \\ 0 & I_3 \end{vmatrix}, \qquad (9.7.4)$$

I_3 being the 3×3 identity matrix and C the 3×6 matrix

$$C = [I_3 \quad 0]. \qquad (9.7.4a)$$

Since

$$x_n = A^n x_0,$$

we have, substituting into (9.7.3),

$$y_n = CA^n x_0 + N_n,$$

and of course what we need to estimate is

$$x_{n+p} = A^{n+p} x_0$$

based on

$$y_1, \ldots, y_n.$$

Let us assume that x_0 is Gaussian with zero mean and covariance Λ and is independent of the sensor noise. Then for the model (9.7.3) with A and C arbitrary, the optimal mean square estimate $\hat{x}_{n+p}$ is given by

$$\hat{x}_{n+p} = E[A^{n+p} x_0 | y_1, \ldots, y_n].$$

But

$$E[A^{n+p} x_0 | y_1, \ldots, y_n] = A^{n+p} E[x_0 | y_1, \ldots, y_n],$$

and we note that it is enough to estimate x_0 as the state at $t = 0$ (in fact equivalent, if A is nonsingular, as in our case). Let us use the notation

$$\hat{x}_{0|n} = E[x_0 | y_1, \ldots, y_n].$$

The similarity to the signal parameter estimation problem of Section 9.5 is again apparent, and we can in fact use (9.5.9). Hence we obtain

$$\hat{x}_{0|n} = (\mathcal{R}_n + \Lambda^{-1})^{-1} \sum_1^n A^{*k} C^* R_k^{-1} y_k, \tag{9.7.5}$$

where

$$\mathcal{R}_n = \sum_1^n A^{*k} C^* R_k^{-1} C A^k. \tag{9.7.6}$$

The corresponding error covariance (cf. (9.5.10)) is given by

$$P_n = (\mathcal{R}_n + \Lambda^{-1})^{-1}. \tag{9.7.7}$$

R may be singular for small values of n.

Getting back to the tracking problem, for A given by (9.7.4) and C by (9.7.4a), we can calculate $\mathcal{R}_n$, taking

$$R_k = dI_3, \qquad \Lambda = \text{Diag}(\lambda_1 I_3, \lambda_2 I_3).$$

We have

$$\mathcal{R}_n = \frac{1}{d} \begin{vmatrix} \left(n + \dfrac{1}{\lambda_1}\right) I_3 & \dfrac{n(n+1)\Delta}{2} I_3 \\ \dfrac{n(n+1)\Delta}{2} I_3 & \left(\dfrac{\Delta^2 n(n+1)(2n+1)}{6} + \dfrac{1}{\lambda_2}\right) I_3 \end{vmatrix}.$$

Hence for $\Lambda = \infty$, $n > 1$, we obtain

$$
P_n = d \begin{vmatrix} \dfrac{4n+2}{n^2-n} I_3 & \dfrac{-6}{n(n-1)\Delta} I_3 \\[3mm] \dfrac{-6}{n(n-1)\Delta} I_3 & \dfrac{-12}{n(n^2-1)\Delta^2} I_3 \end{vmatrix}.
$$

Finally,

$$
\hat{x}_{0|n} = \begin{vmatrix} \hat{r}_{0|n} \\ \hat{v}_{0|n} \end{vmatrix}
$$

$$
= \begin{vmatrix} \dfrac{4n+2}{n^3-n} I_3 & \dfrac{-6}{(n^2-1)\Delta} I_3 \\[3mm] \dfrac{-6}{(n^2-1)\Delta} I_3 & \dfrac{-12}{n(n^2-1)\Delta^2} I_3 \end{vmatrix} \begin{vmatrix} \sum_1^n y_k \\ \Delta \sum_1^n k y_k \end{vmatrix},
$$

where the noise covariance does not appear and

$$
\hat{x}_{n+p} = \begin{vmatrix} \hat{r}_{0|n} + (n+p)\Delta \hat{v}_{0|n} \\ \hat{v}_{0|n} \end{vmatrix}.
$$

The error covariance matrix (for $\Lambda = \infty$)

$$
E[(x_{n+p} - \hat{x}_{n+p})(x_{n+p} - \hat{x}_{n+p})^*] = A^{n+p} P_n (A^*)^{n+p}.
$$

We omit the calculations except to note that every entry in this matrix goes to zero as $n \to \infty$—the position error as $1/n^2$ (independent of Δ) and the velocity error as $1/n^4\Delta$.

9.7.1   Miscellaneous Ad Hoc Estimates

Of course one may consider other ad hoc estimates, drawing on experience or intuition. In our example, for instance, we may estimate the velocity as

$$
\hat{v} = \frac{1}{\Delta}(y_n - y_{n-1})
$$

and then take

$$
\hat{r}_{n+p} = y_n + p\Delta\hat{v},
$$
$$
\hat{v}_{n+p} = \hat{v}.
$$

If our sensor model is correct, we know that these estimates will have higher error. However, they have some advantages: They are simpler and do not require knowledge of P_k or Λ. And of course they give the correct answer if there is no noise! While these are simple and obvious facts in our present (elementary) case, the situation is far more complicated in a real tracking problem.

9.7.2 Time-Varying Model

Still retaining the discrete-time linear model, we may easily extend our results for (9.7.3) to the time-varying case:

$$
\begin{aligned}
x_n &= A_{n-1}x_{n-1}\\
y_n &= C_n x_n + N_n,
\end{aligned}
\tag{9.7.8}
$$

where $\{A_k\}$ and $\{C_k\}$ are specified, yielding

$$
\hat{x}_{0|n} = (\mathscr{R}_n + \Lambda^{-1})^{-1}\sum_1^n \phi_k^* C_k^* R_k^{-1} y_k,
\tag{9.7.9}
$$

$$
\mathscr{R}_n = \sum_1^n \phi_k^* C_k^* R_k^{-1} C_k \phi_k,
\tag{9.7.6a}
$$

$$
\phi_k = A_{k-1}\phi_{k-1}, \qquad \phi_0 = \text{identity},
$$

$$
\hat{x}_{n+p} = \phi_{n+p}\hat{x}_{0|n}.
\tag{9.7.10}
$$

The error covariance matrix for $p \geq 0$:

$$
\begin{aligned}
P_{n+p} &= E[(x_{n+p} - \hat{x}_{n+p})(x_{n+p} - \hat{x}_{n+p})^*]\\
&= \phi_{n+p}(\Lambda^{-1} + \mathscr{R}_n)^{-1}\phi_{n+p}^*
\tag{9.7.11}\\
&= \phi_{n+p}\phi_n^{-1}P_n\phi_n^{*-1}\phi_{n+p}^*,
\tag{9.7.12}
\end{aligned}
$$

where

$$
P_n = (\mathscr{R}_n + \Lambda^{-1})^{-1}.
$$

Note that $\mathscr{R}_n$ is monotone nondecreasing, namely

$$
\mathscr{R}_{n+1} \geq \mathscr{R}_n
$$

while the error P_n is monotone nonincreasing, namely

$$
P_{n+1} \leq P_n.
$$

9.7.3 On-Line ("Recursive") Estimation

We note that (9.7.10) is a "batch" formula: It processes the entire available data at each sample time, and it requires therefore that the time history be stored.

If memory is limited we may go to an "on-line" or updating version which requires that we store only the previous estimate. In other words, we must express

$$\hat{x}_{n+p}$$

in terms of $\hat{x}_{n-1+p}$ and y_n. We shall do this essentially by brute force, starting from (9.7.9). Thus we calculate

$$\hat{x}_{0|n+1} = (\mathcal{R}_{n+1} + \Lambda^{-1})^{-1}\left(\sum_1^{n+1} \phi_k^* C_k^* R_k^{-1} y_k\right),$$

and splitting the second factor as

$$\sum_1^{n+1} \phi_k^* C_k^* R_k^{-1} y_k = \sum_1^{n} \phi_k^* C_k^* R_k^{-1} y_k + \phi_{n+1}^* C_{n+1}^* R_{n+1}^{-1} y_{n+1}$$

and using (9.7.9) we have

$$\hat{x}_{0|n+1} = (\mathcal{R}_{n+1} + \Lambda^{-1})^{-1}(\mathcal{R}_n + \Lambda^{-1})\hat{x}_{0|n}$$
$$+ (\mathcal{R}_{n+1} + \Lambda^{-1})^{-1}\phi_{n+1}^* C_{n+1}^* R_{n+1}^{-1} y_{n+1}.$$

Again,

$$\mathcal{R}_n + \Lambda^{-1} = \mathcal{R}_{n+1} + \Lambda^{-1} - \phi_{n+1}^* C_{n+1}^* R_{n+1}^{-1} C_{n+1} \phi_{n+1},$$

hence

$$\hat{x}_{0|n+1} = \hat{x}_{0|n} + P_{n+1}\phi_{n+1}^* C_{n+1}^* R_{n+1}^{-1}(y_{n+1} - C_{n+1}\phi_{n+1}\hat{x}_{0|n})$$
$$\hat{x}_{0|0} = 0, \tag{9.7.13}$$

where

$$P_n = (\mathcal{R}_n + \Lambda^{-1})^{-1}$$

and is of course the error covariance matrix

$$E[x_0 - \hat{x}_{0|n})(x_0 - \hat{x}_{0|n})^*]$$

and itself satisfies a recursive equation

$$P_{n+1} - P_n = P_{n+1}(\mathcal{R}_n + \Lambda^{-1} - \mathcal{R}_{n+1} - \Lambda^{-1})P_n \tag{9.7.14}$$

or

$$P_{n+1} = P_n(I + \phi_{n+1}^* C_{n+1}^* R_{n+1}^{-1} C_{n+1} \phi_{n+1} P_n)^{-1}$$
$$P_0 = \Lambda. \tag{9.7.15}$$

Thus (9.7.13) and (9.7.15) together provide us with our recursive or updating relations.

We may extend this procedure to derive recursive equations for $\hat{x}_{n+p}$. We shall give one version which exploits the recursive relations for $p = 0$. Thus we have

$$\hat{x}_{n+p} = \phi_{n+p}\phi_n^{-1}\hat{x}_{0|n} = A_{n+p-1} \cdots A_n\hat{x}_{0|n}, \qquad p \geq 1$$

and use (9.7.9) to write

$$\hat{x}_{n+1} = \phi_{n+1}\hat{x}_{0|n+1}$$

and use the recursive equation for $\hat{x}_{0|n}$, yielding

$$\hat{x}_{n+1} = \phi_{n+1}\hat{x}_{0|n+1}$$

or

$$\hat{x}_{n+1} = A_n\hat{x}_n + \mathscr{P}_{n+1}C_{n+1}^*R_{n+1}^{-1}(y_{n+1} - C_{n+1}A_n\hat{x}^n)$$
$$\hat{x}_0 = 0 \tag{9.7.16}$$

where $\mathscr{P}_n$ is again the error covariance matrix

$$\mathscr{P}_n = E[(x_n - \hat{x}_n)(x_n - \hat{x}_n)^*] = \phi_n P_n \phi_n^*$$

and satisfies the "error propagation" equation:

$$\mathscr{P}_{n+1} = \phi_{n+1}P_{n+1}\phi_{n+1}^* = A_n\mathscr{P}_nA_n^* + \phi_{n+1}(P_{n+1} - P_n)\phi_{n+1}^*$$

which, from (9.7.14),

$$= A_n\mathscr{P}_nA_n^* + \phi_{n+1}P_{n+1}\phi_{n+1}^*C_{n+1}^*R_{n+1}^{-1}C_{n+1}\phi_{n+1}P_n\phi_{n+1}^*$$

or

$$\mathscr{P}_{n+1} = A_n\mathscr{P}_nA_n^* - \mathscr{P}_{n+1}C_{n+1}^*R_{n+1}^{-1}C_{n+1}A_n\mathscr{P}_nA_n^* \tag{9.7.17}$$

or

$$\mathscr{P}_{n+1} = (I + C_{n+1}^*R_{n+1}^{-1}C_{n+1}A_n\mathscr{P}_nA_n^*)^{-1}A_n\mathscr{P}_nA_n^*$$
$$\mathscr{P}_0 = \Lambda. \tag{9.7.18}$$

Thus recursive equations for $\hat{x}_n$ are given by (9.7.16), (9.7.17), or (9.7.18). Note the "feedback" form of (9.7.16). We call $\mathscr{P}_{n+1}C_{n+1}^*R_{n+1}^{-1}$ the feedback gain matrix.

9.7.4 Example: System Identification

As an example leading to a time-varying model (9.7.8), let us consider a (linear) "system identification" problem. Thus we are given the state-space model:

$$z_n = Az_{n-1} + Bu_{n-1}, \tag{9.7.19}$$

where the system matrix A is known—say $p \times p$ but B is unknown and is to be identified by exciting the system with the known input $\{u_n\}$ and observing the noisy output

$$y_n = Cz_n + N_n, \qquad n \geq 1,$$

where $\{N_n\}$ is white Gaussian with unit covariance matrix and C is known. To simplify the notation we shall assume that u_k is one-dimensional so that B is $p \times 1$. We can cast this in the form (9.7.8) by "enhancing the state space." Thus let

$$x_n = \begin{vmatrix} z_n \\ B_n \end{vmatrix},$$

where z_n satisfies (9.7.19) and we impose the equation

$$B_n = B_{n-1}.$$

Then we have the linear state-space model:

$$\begin{aligned} x_n &= A_{n-1}x_{n-1} \\ y_n &= |C \quad 0|x_n + N_n \end{aligned} \tag{9.7.20}$$

where

$$A_n = \begin{vmatrix} A & U_{n-1} \\ 0 & I_p \end{vmatrix}$$

$$U_n = u_n I_p$$

$$I_p = p \times p \text{ unit matrix.}$$

Our estimate is given by

$$\hat{B}_n = E[B \mid y_1, \ldots, y_n] = |0 \quad I_p|\hat{x}_{0|n} = |0 \quad I_p|\hat{x}_n, \tag{9.7.21}$$

where $\hat{x}_{0|n}$ (or $\hat{x}_n$) can be calculated using the batch or recursive algorithm.

Of main interest to us is the maximum-ignorance case $\Lambda = +\infty$ and thus we need to examine $\mathscr{R}_n$. Again we are primarily interested in what happens as $n \to \infty$. As we have noted, $\mathscr{R}_n$ is monotone-nondecreasing and hence let

$$\mathscr{R}_\infty = \lim_n \mathscr{R}_n.$$

Suppose $\mathscr{R}_\infty$ is singular. Then for some nonzero x,

$$0 = [\mathscr{R}_\infty x, x] \geq [\mathscr{R}_n x, x]$$

and hence

$$\mathscr{R}_n x = 0, \qquad \text{for every } n.$$

But from (9.7.6a) it follows that

$$|C \quad 0|\phi_k x = 0, \qquad \text{for every } k,$$

where we can readily calculate that

$$\phi_k = \begin{vmatrix} A_k & \sum_0^{k-1} A^{k-1-j}Bu_j \\ 0 & I_p \end{vmatrix};$$

and letting

$$x = \begin{vmatrix} z \\ B \end{vmatrix}$$

we have that for every k

$$CA^k z + \sum_0^{k-1} CA^{k-1-j}Bu_j = 0. \qquad (9.7.22)$$

Hence we study conditions under which this cannot happen. First we require that $(C - A)$ be observable so that

$$CA^k z = 0, \qquad \text{for every } k \rightarrow z = 0. \qquad (9.7.23)$$

Next we confine attention to the case where A is stable (see Chapter 3). In that case

$$\sum_1^\infty \|CA^k B\|^2 < \infty. \qquad (9.7.24)$$

Note that as a result in (9.7.22) the first term goes to zero; and hence we put conditions on the input that the second (summation) term does not. The input $\{u_k\}$ is one (long) sample of a stationary random process which is ergodic. Thus we assume that the time-average limit is given by

$$\lim_{N \to \infty} \frac{1}{N} \sum_{k=1}^N u_{k+m} u_k = R(m)$$

and we assume that

$$R(0) > 0. \qquad (9.7.25)$$

In that case, let

$$s_k = \sum_0^k CA^{k-j}Bu_j = \sum_0^k CA^j Bu_{k-j}.$$

We can calculate that

$$\lim_{N \to \infty} \frac{1}{N} \sum_1^N \|s_k\|^2 = \text{Tr} \int_{-1/2}^{1/2} \psi(\lambda)\psi(\lambda)^* p_u(\lambda)\, d\lambda$$

$$= \int_{-1/2}^{1/2} \|\psi(\lambda)\|^2 p_u(\lambda)\, d\lambda,$$

where

$$\psi(\lambda) = \sum_0^\infty CA^k B e^{-2\pi i k \lambda}$$

and $p_u(\lambda)$ is the spectral density of $\{u_k\}$:

$$R(k) = \int_{-1/2}^{1/2} e^{2\pi i k \lambda} p_u(\lambda)\, d\lambda.$$

Hence (9.7.22) holds for every k if and only if

$$\|\psi(\lambda)\|^2 p_u(\lambda) = 0, \qquad -1/2 \le \lambda \le 1/2 \qquad (9.7.26)$$

(where we should add (the mathematical technicality) "almost everywhere" on the right side). This condition is violated, for example, if $p_u(\lambda)$ is nonzero, on some nonzero sub-interval of $[-1/2 \le \lambda \le 1/2]$, for which in turn it is enough if (9.7.25) holds (assuming $p_u(\cdot)$ is continuous except perhaps at a finite number of points). It should be noted that

$$\psi(\lambda) = C(I - Ae^{-2\pi i \lambda})^{-1} B$$

and hence if C is singular, given any λ, we can find B such that

$$(I - Ae^{-2\pi i \lambda})B$$

is in the null space of C. Of course things become much simpler if C is nonsingular. In this case $\psi(\lambda)$ is never zero and hence we may allow $p_u(\lambda)$ to be a δ-function and take

$$u_k = \cos 2\pi \lambda_0 k, \qquad -1/2 < \lambda_0 < 1/2,$$

for example.

We refer to (9.7.25) as an "identifiability" condition, and it can be shown that as a consequence the error in estimating B goes to zero asymptotically as $n \to \infty$. Here, however, we shall merely illustrate the general idea of proof by simplifying to the case $p = 1$, $C = 1$ (we are in the case where C is nonsingular!). In that case the error for $\Lambda = \infty$ is

$$= |0 \quad I_p| \mathscr{R}_n^{-1} \begin{vmatrix} 0 \\ I_p \end{vmatrix} = \frac{\sum_1^n \rho^{2k}}{\left(\sum_1^n \rho^{2k} - \dfrac{\sum_1^n \rho^k s_k}{\sum_1^n s_k^2} \right) \left(\sum_1^n s_k^2 \right)}, \qquad (9.7.27)$$

where

$$s_k = \sum_0^{k-1} \rho^{k-1-j} u_j.$$

Here we may take any bounded sequence $\{u_k\}$ such that

$$\lim_N \frac{1}{N} \sum_1^N u_k^2 > 0,$$

so that

$$\sum_1^n s_k^2 \to \infty, \qquad \text{as } n \to \infty.$$

On the other hand (and this is the main point)

$$\sum_1^n \rho^k s_k = \sum_1^n \rho^k \left(\sum_0^{k-1} \rho^{k-1-j} u_j \right) = \rho \sum_0^{n-1} \rho^k \left(\sum_0^k \rho^{k-j} u_j \right)$$

$$= \rho \sum_{j=0}^{n-1} \left(\sum_{k=j}^{n-1} \rho^{2k-j} \right) u_j$$

$$= \rho \sum_0^{n-1} \frac{\rho^j (1 - \rho^{2(n-j)}) u_j}{1 - \rho^2}$$

is clearly bounded if $\{u_j\}$ is bounded, and hence the first factor in parentheses in the denominator in (9.7.27) goes to

$$\sum_1^\infty \rho^{2k} = \frac{1}{1 - \rho^2}$$

and hence (9.7.27) goes to zero.

9.8   LINEAR FILTERING: GENERAL CASE

In this section we consider the general case of linear filtering where the signal is a nontrivial process. We may then formulate the filtering problem as follows. Let

$$z_n = \begin{vmatrix} s_n \\ v_n \end{vmatrix}, \qquad n \geq 0$$

be a given $(r + q) \times 1$ random process where $\{s_n\}$ is $r \times 1$ and $\{v_n\}$ is $q \times 1$. We want to "filter" the signal $\{s_n\}$ from the "noisy" observed data $\{v_n\}$—more specifically, for each n estimate s_n (or more generally s_{n+p}, where p is a fixed integer) from all the data available up to time n—that is, from $v_{n-k}, k = 0, \ldots, n$. The optimal mean-square estimate we know is given by the conditional expectation:

$$\hat{s}_{n+p|n} = E[s_{n+p} | v_{n-k}, \quad k = 0, \ldots, n]. \tag{9.8.1}$$

Since we are only interested in the optimal linear filter we may assume that $\{z_i\}$ is a Gaussian process, with zero mean. The filter then being linear, we need to find $\{W_{n+p,k}\}, 0 \leq k \leq n$, such that

$$\hat{s}_{n+p|n} = \sum_0^n W_{n+p,k} v_{n-k}. \tag{9.8.2}$$

As we have seen, the error must be uncorrelated with the data (or, equivalently, the Wiener-Hopf equation must hold). Hence

$$E[(s_{n+p} - \hat{s}_{n+p|n}) v_{n-j}^*] = 0, \qquad j = 0, 1, \ldots, n$$

or the $W_{n+p,k}$ must satisfy

$$E[s_{n+p} v_{n-j}^*] = \sum_1^n W_{n+p,k} E[v_{n-k} v_{n-j}^*], \qquad j = 0, 1, \ldots, n. \tag{9.8.3}$$

This is a linear equation which always has a solution, unique or not, as we have seen. But, even given today's high-speed large-memory digital computing, solving (9.8.3) for each n is an impossible task, not to speak of the difficulty in storing all the moment matrices required. Hence simplifications need to be in the model.

9.8.1   Stationary Processes: Steady-State Response

The first simplification is to assume that the process $\{z_i\}$ is stationary, so that in particular we may then further limit attention to the steady-state response only.

9.8.2 Martingales

Note now that because of assumed stationarity the processes are defined for $n \in I$.

To study the steady-state response we fix n and study first the sequence

$$\zeta_N = E[s_{n+p} | v_{n-k}, k = 0, \dots, N], \qquad N \geq 0.$$

We shall show first that ζ_N converges in the mean square as $N \to \infty$. We begin by showing that for any $k \geq 0$,

$$E[\zeta_{N+k} | \zeta_N] = \zeta_N. \tag{9.8.4}$$

A sequence of random variables with this property is called a "Martingale," and here we have in addition that

$$E[\zeta_N \zeta_N^*] \leq E[s_N s_N^*] < \infty. \tag{9.8.5}$$

We have thus a "square integrable" Martingale. To prove (9.8.4), we invoke (9.2.5). Thus for any $h(\cdot)$ as therein,

$$E[(\zeta_{N-k} - \zeta_N)h(\zeta_N)] = E[((s_{n+p} - \zeta_N) - (s_{n+p} - \zeta_{N+k}))h(\zeta_N)]$$
$$= 0$$

using the principle that the residual is uncorrelated with any function of the data. Next, from (9.2.11) it follows that

$$0 \leq E[(\zeta_{N+k} - \zeta_N)(\zeta_{N+k} - \zeta_N)^*]$$
$$= E[\zeta_{N+k} \zeta_{N+k}^*] - E[\zeta_N \zeta_N^*]. \tag{9.8.6}$$

This could also have been deduced from the fact that

$$E[s_{n+p} s_{n+p}^*] - E[\zeta_N \zeta_N^*] = E[(s_{n+p} - \zeta_N)(s_{n+p} - \zeta_N)^*],$$

which, by the mean square optimality of ζ_{N+1}

$$\leq E[(s_{n+p} - \zeta_{N+1})(s_{n+p} - \zeta_{N+1})^*]$$
$$= E[s_{n+p} s_{n+p}^*] - E[\zeta_N \zeta_N^*].$$

Hence the sequence

$$E[\zeta_N \zeta_N^*]$$

is monotonic nondecreasing but must converge to a finite matrix because of (9.8.5).

From (9.8.6) it follows further that $\{\zeta_N\}$ is a Cauchy sequence (in the mean square sense) and hence converges in the mean square sense. (We have incidentally proved that a square integrable Martingale converges in the mean square.) We denote the limit by $\hat{s}_{n+p}$ and use the notation

$$\hat{s}_{n+p} = E[s_{n+p} | v_{n-k}, \quad k \geq 0]. \tag{9.8.7}$$

Note in particular that for any $k \geq 0$ we have

$$E[(s_{n+p} - \hat{s}_{n+p})v_{n-k}^*] = \lim_N E[(s_{n+p} - \zeta_N)v_{n-k}^*]$$

$$= 0, \tag{9.8.8}$$

or, in other words, the residual is again uncorrelated with the data.

Next we shall show that if we can find a solution of the Wiener-Hopf equation

$$E[s_{n+p}v_{n-k}^*] = \sum_0^\infty W_j E[v_{n-j}v_{n-k}^*] \tag{9.8.9}$$

where

$$\sum_0^\infty \|W_j\|^2 < \infty, \tag{9.8.10}$$

then

$$\hat{s}_{n+p} = \sum_0^\infty W_k v_{n-k}, \tag{9.8.11}$$

with the infinite series on the right converging in the mean square sense (cf. Chapter 3). For this, note first that (9.8.9) yields

$$E\left[\left(s_{n+p} - \sum_0^\infty W_j v_{n-j}\right)v_{n-k}^*\right] = 0, \qquad k \geq 0.$$

From (9.8.8) it would seem intuitively obvious that

$$\sum_0^\infty W_j v_{n-j} = \hat{s}_{n+p}. \tag{9.8.12}$$

But this still requires a formal proof. We proceed as follows.

$$E\left(\left\|s_{n+p} - \sum_0^\infty W_k v_{n-k}\right\|^2\right) = E\left[\left\|s_{n+p} - \zeta_N + \zeta_N - \sum_0^\infty W_k v_{n-k}\right\|^2\right]$$

$$= E[\|s_{n+p} - \zeta_N\|^2] + E\left[\left\|\zeta_N - \sum_0^\infty W_k v_{n-k}\right\|^2\right]$$

$$+ 2 \operatorname{Tr} E\left[(s_{n+p} - \zeta_N)\left(\zeta_N - \sum_0^\infty W_k v_{n-k}\right)^*\right],$$

where the third term

$$= 2 \operatorname{Tr} \, E \left[(s_{n+p} - \zeta_N) \left(\sum_{N+1}^{\infty} W_k v_{n-k} \right)^* \right]$$

goes to zero as $N \to \infty$, since

$$E[\|s_{n+p} - \zeta_N\|^2] \le E[\|s_{n+p}\|^2] < \infty$$

and

$$E \left[\left\| \sum_{N+1}^{\infty} W_k v_{n-k} \right\|^2 \right] \to 0.$$

Hence

$$E \left[\left\| s_{n+p} - \sum_{0}^{\infty} W_k v_{n-k} \right\|^2 \right]$$

$$= E[\|s_{n+p} - \hat{s}_{n+p}\|^2] + E \left[\left\| \hat{s}_{n+p} - \sum_{0}^{\infty} W_k v_{n-k} \right\|^2 \right]. \quad (9.8.13)$$

Next

$$E[\|s_{n+p} - \zeta_N\|^2]$$

$$= E \left[\left\| s_{n+p} - \sum_{0}^{\infty} W_k v_{n-k} \right\|^2 \right] + E \left[\left\| \zeta_N - \sum_{0}^{\infty} W_k v_{n-k} \right\|^2 \right]$$

$$+ 2 \operatorname{Tr} \, E \left[\left(s_{n+p} - \sum_{0}^{\infty} W_k v_{n-k} \right) \left(\zeta_N - \sum_{0}^{\infty} W_k v_{n-k} \right)^* \right].$$

But

$$E \left[\left(s_{n+p} - \sum_{0}^{\infty} W_k v_{n-k} \right) \zeta_N^* \right] = 0$$

by the Wiener-Hopf equation, and so also is

$$E \left[\left(s_{n+p} - \sum_{0}^{\infty} W_k v_{n-k} \right) \sum_{0}^{\infty} v_{n-k}^* W_k^* \right].$$

Hence letting $n \to \infty$

$$E[\|s_{n+p} - \hat{s}_{n+p}\|^2] = E \left[\left\| s_{n+p} - \sum_{0}^{\infty} W_k v_{n-k} \right\|^2 \right] + E \left[\left\| \hat{s}_{n+p} - \sum_{0}^{\infty} W_k v_{n-k} \right\|^2 \right].$$

Combining this with (9.8.13), we see that (9.8.12) follows.

9.8.3   Linear Smoothing: Interpolation

To proceed further, we now specialize to $p < 0$—in which case we have an "interpolation" or "smoothing" problem. Of particular interest is the "two-sided interpolation" problem:

$$E[s_{n-N}|v_{n-k}, \quad 0 \le k \le 2N] = \sum_{0}^{2N} W_{N,k}v_{n-k}, \qquad (9.8.14)$$

which we can express as

$$= \sum_{-N}^{N} W_{N,N+j}v_{n-j-N}.$$

We are estimating the signal at the midpoint from data on both sides. By the stationarity of the process we obtain

$$E[s_n|v_j, n - N \le j \le n + N] = \sum_{-N}^{N} W_{N,N+j}v_{n-j}. \qquad (9.8.15)$$

In this form, however, we can consider the steady-state or limiting case, as $N \to \infty$, for each n. Thus (9.8.15) defines a square integrable Martingale sequence in N. Denoting (9.8.15) by

$$\hat{s}_{n|N}$$

let

$$\hat{s}_n = \lim_{N \to \infty} \hat{s}_{n|N} = E[s_n|v_{n-k}, k \in \mathbf{I}].$$

We note that we can express $\hat{s}_n$ in the form

$$\hat{s}_n = \sum_{-\infty}^{\infty} W_k v_{n-k} \qquad (9.8.16)$$

provided that we can solve the Wiener-Hopf equation

$$E[s_n v_{n-k}^*] = \sum_{-\infty}^{\infty} W_j E[v_{n-j}v_{n-k}^*], \qquad k \in \mathbf{I}, \qquad (9.8.17)$$

where

$$\sum_{-\infty}^{\infty} \|W_j\|^2 < \infty. \qquad (9.8.18)$$

We call the linear time-invariant system with weighting pattern $\{W_k\}$ the linear "steady-state smoother." Note that it is *not* "physically realizable." We refer to it as the "infinite-delay" filter, since the delay in (9.8.14) or (9.8.15) is N

(estimating s_n from data up to $n+N$) and we are allowing N to go to infinity—see Chapter 3.

We shall now show that we can obtain an explicit solution for the steady-state smoother defined by the Wiener-Hopf equation (9.8.17), which, because the processes are stationary, takes the form

$$R_{sv}(k) = \sum_{-\infty}^{\infty} W_j R_{vv}(k-j), \qquad k \in I, \tag{9.8.19}$$

where

$$R_{sv}(k) = E[s_n v_{n-k}^*]$$
$$R_{vv}(k-j) = E[v_{n-j} v_{n-k}^*].$$

In fact,

$$\sum_{-\infty}^{\infty} W_j R_{vv}(k-j) = \sum_{-\infty}^{\infty} W_j \int_{-1/2}^{1/2} e^{2\pi i(k-j)\lambda} P_{vv}(\lambda)\, d\lambda$$
$$= \int_{-1/2}^{1/2} \left(\sum_{-\infty}^{\infty} W_j e^{-2\pi i j\lambda} P_{vv}(\lambda) \right) e^{2\pi i k\lambda}\, d\lambda$$
$$= \int_{-1/2}^{1/2} \psi(\lambda) P_{vv}(\lambda) e^{2\pi i k\lambda}\, d\lambda,$$

where $\psi(\lambda)$ is the smoother transfer function:

$$\psi(\lambda) = \sum_{-\infty}^{\infty} e^{-2\pi i k\lambda} W_k.$$

Hence by (9.8.19)

$$\psi(\lambda) P_{vv}(\lambda) = P_{sv}(\lambda), \tag{9.8.20}$$

where

$$R_{sv}(k) = \int_{-1/2}^{1/2} e^{2\pi i\lambda k} P_{sv}(\lambda)\, d\lambda.$$

Hence, assuming that $P_{vv}(\lambda)$ is nonsingular for every λ, $-1/2 \le \lambda \le 1/2$, we obtain

$$\psi(\lambda) = P_{sv}(\lambda) P_{vv}(\lambda)^{-1} \tag{9.8.21}$$

and the corresponding error covariance matrix

$$= R_{ss}(0) - \sum_{-\infty}^{\infty} W_k E[v_{n-k} s_n^*]$$
$$= \int_{-1/2}^{1/2} P_{ss}(\lambda)\, d\lambda - \int_{-1/2}^{1/2} \psi(\lambda) P_{vs}(\lambda)\, d\lambda$$
$$= \int_{-1/2}^{1/2} [P_{ss}(\lambda) - P_{sv}(\lambda) P_{vv}(\lambda)^{-1} P_{vs}(\lambda)]\, d\lambda. \tag{9.8.22}$$

Note that the spectral density of the process

$$\left| \begin{matrix} s_n \\ v_n \end{matrix} \right|$$

is given by

$$P(\lambda) = \left| \begin{matrix} P_{ss}(\lambda) & P_{sv}(\lambda) \\ P_{vs}(\lambda) & P_{vv}(\lambda) \end{matrix} \right|.$$

To show that $\{W_k\}$ satisfies condition (9.8.18) we need to show that

$$\int_{-1/2}^{1/2} \text{Tr } \psi(\lambda)\psi(\lambda)^* \, d\lambda < \infty$$

for which by (9.8.21) it is enough if the smallest eigenvalue of $P_{vv}(\lambda)$ is

$$\geq \mu > 0, \qquad -1/2 \leq \lambda \leq 1/2.$$

Remark. The importance of (9.8.22) is that it provides a lower bound to all filtering error. If this error is too large, no filter will be acceptable.

9.8.4   Example: Signal in Additive White Noise

As an example let us consider the case

$$v_n = s_n + N_n,$$

where $\{N_n\}$ is white Gaussian with nonsingular covariance D and is independent of the signal process. Then

$$P_{vv}(\lambda) = P_{ss}(\lambda) + D$$

and the smallest eigenvalue of $P_{vv}(\lambda)$ is

$$\geq \min d_i \qquad D = \text{Diag } \{d_i\}.$$

Hence condition (9.8.18) is satisfied, and further since now

$$P_{sv}(\lambda) = P_{ss}(\lambda)$$

we have, using (9.8.21)

$$\psi(\lambda) = P_{ss}(\lambda)[P_{ss}(\lambda) + D]^{-1} \tag{9.8.21a}$$

and the minimal error-covariance matrix is given by

$$\int_{-1/2}^{1/2} (P_{ss}(\lambda) - P_{ss}(\lambda)(D + P_{ss}(\lambda))^{-1}P_{ss}(\lambda))\, d\lambda$$

$$= \int_{-1/2}^{1/2} (I + P_{ss}(\lambda)D^{-1})^{-1}P_{ss}(\lambda)\, d\lambda. \quad (9.8.22a)$$

We can actually evaluate the error covariance explicitly for the one-dimensional case where

$$P_{ss}(\lambda) = \frac{1 - \rho^2}{1 + \rho^2 - 2\rho \cos 2\pi\lambda}$$

and take $D = d$. We obtain

$$(1 - \rho^2)\int_{-1/2}^{1/2} \frac{d\lambda}{\left(\dfrac{1-\rho^2}{d} + 1 + \rho^2 - 2\rho \cos 2\pi\lambda\right)} = \frac{1 - \rho^2}{\sqrt{\left(1 + \rho^2 + \dfrac{1-\rho^2}{d}\right)^2 - 4\rho^2}}.$$

9.8.5 Wiener Filters/Kalman Filters

Let $p = 0$. Then the steady-state filter is defined by

$$\hat{s}_n = E[s_n | v_{n-k}, k \geq 0]$$

being the limit of the Martingale

$$E[s_n | v_{n-k}, 0 \leq k \leq N]$$

as $N \to \infty$. Also as we have seen, we can express $\hat{s}_n$ as

$$\hat{s}_n = \sum_0^\infty W_j v_{n-j} \quad (9.8.23)$$

provided that we can find a solution of the Wiener-Hopf equation

$$E[s_n v_{n-k}^*] = \sum_0^\infty W_j E[v_{n-j} v_{n-k}^*] \quad (9.8.24)$$

or, equivalently,

$$R_{sv}(k) = \sum_0^\infty W_j R_{vv}(k - j) \quad (9.8.24a)$$

with

$$\sum_0^\infty \|W_j\|^2 < \infty.$$

Note in particular the $\{W_k\}$ given by (9.8.24a) do not depend on n. Thus we see that $\hat{s}_n$ given by (9.8.23) is the output of linear time-invariant systems with weighting pattern $\{W_k\}$ satisfying (9.8.24a), which is physically realizable. Such a system is usually referred to as a "Wiener" filter after Wiener [1], even though Kolmogorov [2] had also obtained a solution at about the same time. Note that because of the physical realizability condition the right side is not a convolution and thus not immediately amenable to transform techniques. Unfortunately the solution procedure will take us too far afield to go into here!

Actually a better form of solution more suitable to digital computer usage was found by Kalman [4], taking advantage of state-space models for the processes. The loss in generality due to this restriction (the spectral densities must be rational) is far outweighed by the fact that the solution can be expressed in recursive form (analogous to (9.7.16), which is in fact a special case of the Kalman filter). Again, we cannot go into this theory here either!

When $p > 0$, we have a "predictor" and the considerations are similar, except that the prediction problem has content even when

$$v_n = s_n. \tag{9.8.25}$$

Indeed a central result due to Szëgo (see, reference 3) in prediction theory is the formula for the one-step prediction error in the one-dimensional case

$$\log_e (E[s_{n+1} - E[s_{n+1} \,|\, s_{n-k}, k \ge 0])^2]) = \int_{-1/2}^{1/2} \log_e P_{ss}(\lambda)\, d\lambda. \tag{9.8.26}$$

Note that this formula does not require that we know the weighting pattern corresponding to the predictor. The prediction error is zero, if the right-hand side is $-\infty$.

As a simple example of this formula let us consider the 1×1 process $\{s_n\}$ with

$$E[s_m s_{m+n}] = \rho^{|n|}, \qquad 0 < \rho < 1.$$

In that case

$$E[(s_{n+1} - \rho s_n)s_{n-k}] = \rho^{k+1} - \rho\rho^k = 0 \qquad \text{for } k \ge 0.$$

Hence

$$E[s_{n+1} \,|\, s_{n-k}, k \ge 0] = \rho s_n.$$

The mean square one-step prediction error

$$= E[(s_{n+1} - \rho s_n)^2] = 1 - \rho^2.$$

But

$$P_{ss}(\lambda) = \frac{1 - \rho^2}{1 + \rho^2 - 2\rho \cos 2\pi\lambda}, \qquad -1/2 \le \lambda \le 1/2,$$

and it is known from the theory of functions of a complex variable (Jensen's theorem: see, e.g., reference 7, p. 256) that

$$\int_{-1/2}^{1/2} \log_e (1 + \rho^2 - 2\rho \cos 2\pi\lambda)\, d\lambda = \log_e 1 = 0,$$

verifying (9.8.26) in turn.

PROBLEMS

9.1. In the notation of Section 9.1, show that

$$E[h(y)\,|\,y] = h(y).$$

9.2. In the notation of Section 9.2, let

$$\left|\begin{matrix} x \\ y \end{matrix}\right|,$$

where x is $p \times 1$, and y is $n \times 1$, be Gaussian. Let L be any nonsingular $p \times p$ matrix, and let M be any nonsingular $n \times n$ matrix. Show that the mutual information is given by

$$I(Lx; My) = I(x; y).$$

Hint: Using (9.2.23):

$$I(Lx; My) = I(x; My)$$

$$I(x; My) = I(My; x) = I(y; x).$$

9.3. If S and N are independent $n \times 1$ Gaussian variables and

$$Y = S + N,$$

show that

$$P(Y\,|\,S) = P_N(Y - S),$$

where $P_N(\cdot)$ is the density function of N. Calculate

$$I[S; Y], \qquad I[N; Y],$$

and

$$E[E[N\,|\,Y]E[S\,|\,Y]^*].$$

9.4. Let R denote the $n \times n$ matrix defined by

$$R = \{r_{i,j}\}, \qquad 1 \le i, j \le n$$

$$r_{ij} = \frac{-1}{n-1}.$$

Show that R is a covariance matrix. Factorize R as

$$R = \mathcal{L}\mathcal{L}^*,$$

where $\mathcal{L}$ is lower triangular. What is the rank of $\mathcal{L}$?

9.5. With the notation in Section 9.3, show that (use 9.2.16a):

$$d_{ii} = E[y_i^2] = \frac{|R_i|_{\det}}{|R_{i-1}|_{\det}},$$

where R_i is the covariance matrix of $x_1, \ldots, x_i$, and

$$R_i = 1 \qquad \text{for } i = 0.$$

Show that if

$$|R_i|_{\det} = 0,$$

then

$$|R_{i+k}|_{\det} = 0 \qquad \text{for } k \ge 0.$$

9.6. (A "signal design" problem.) In the notation of Section 9.5, and (9.5.2), let $m = 1$ and write

$$v = \theta S + N,$$

where the covariance R_N is not assumed to be nonsingular. Suppose now that one could choose any S in $\mathbf{R}^n$ subject only to the condition that

$$\|S\|^2 = 1.$$

How would you choose S in order to minimize the mean square estimation error? What is the corresponding estimate?

9.7. With the continuous-time model of Section 9.1,

$$v(t) = S(t) + N(t), \qquad 0 \le t \le T,$$

show how to estimate the derivative $\dfrac{d}{dt} S(t)$, given $v(t)$, $0 \le t \le T$, assuming digital processing. How does the error depend on the degree of the polynomial? On the sampling period Δ?

How would you similarly estimate the integral

$$\int_0^T S(t)\, dt?$$

9.8. Use the covariance factorization technique of Section 9.3 to construct $N \geq 4$ consecutive samples of a zero-mean stationary Gaussian process with covariance given in the second part of Problem 2.8.

9.9. (Hadamard's Theorem) In the notation of Section 9.2:

$$\Lambda = \begin{vmatrix} R_{xx} & R_{xy} \\ R_{yx} & R_{yy} \end{vmatrix}.$$

Show that

$$\det \Lambda \leq (\det R_{xx})(\det R_{yy}).$$

9.10. Let

$$y_n = S_n + N_n + \mu, \qquad n \in \mathbf{I},$$

where $\{S_n\}$ is a 1×1 stationary Gaussian process with mean zero and $\{N_n\}$ is white Gaussian with variance d, independent of the process $\{S_n\}$. It is desired to estimate μ from N consecutive samples y_n, $n = 1, \ldots, N$. Assuming that μ is Gaussian mean zero and variance λ, independent of $\{S_n\}$ and $\{N_n\}$, calculate

$$E[\mu \mid y_i, i \leq N]$$

corresponding to "maximum ignorance." Find the error and its limit as $N \to \infty$. Also determine what happens as $d \to 0$.

9.11. Let $\{x_n\}$ denote a $p \times 1$ discrete-time process, $n \geq 0$. We call it a Markov process if, for every n,

$$E[x_n \mid x_k, 0 \leq k \leq n-1] = E[x_n \mid x_{n-1}].$$

Show that the state process in a time-varying Kalman model (3.5.3) is Markov (under assumption (3.2.1)). In fact,

$$E[x_n \mid x_k, 0 \leq k \leq n-1] = A_{n-1} x_{n-1}$$

as follows readily from (3.5.2). Moreover, the corresponding error covariance matrix is

$$B_k D_k B_k^*.$$

Note also that in the time-invariant case (model given by (3.2.23), with A stable) we have

$$E[x_n \mid x_k, k \le n - 1] = A x_{n-1}$$

The one-step prediction error

$$= BDB^*,$$

where D is the covariance matrix of N_k.

Conversely, given that $\{x_n\}$ is a zero-mean Gaussian Markov process, show that the process is defined by state-space model (3.5.3).

Hint: Let

$$A_{n-1} x_{n-1} = E[x_n \mid x_{n-k}, k \ge 1]$$
$$\zeta_{n-1} = x_n - A_{n-1} x_{n-1}.$$

Show that ζ_n is a white noise sequence and that x_n is independent of ζ_k, $k \ge n$.

9.12. (The Sampling Principle as Optimal Smoother) Let the process $x(t)$, $-\infty < t < \infty$, be as defined in Theorem 6.1 of Chapter 6, band-limited to $(-W, W)$. Show that the right side of (6.1) is the optimal smoother based on periodic samples—that is,

$$\hat{x}(t) = E\left[x(t) \mid x\left(\frac{n}{2W}\right), n \in I \right] = \sum_{-\infty}^{\infty} a_n(t) x\left(\frac{n}{2W}\right)$$

or, equivalently $\{a_n(t)\}$ given by (6.2) satisfy for each t, $-\infty < t < \infty$, the Wiener-Hopf equation

$$R\left(t - \frac{m}{2W}\right) = \sum_{-\infty}^{\infty} a_n(t) R\left(\frac{n-m}{2W}\right), \qquad m \in I$$

and (6.1) follows from the fact that the corresponding error covariance is zero.

9.13. Let $x(t)$, $-\infty < t < \infty$, be a stationary process with mean zero, not necessarily band-limited. Let $R(\cdot)$ denote the stationary covariance function and $P(\cdot)$ the spectral density. Show that the smoothing/interpolation problem

$$\hat{x}(t) = E\left[x(t) \mid x\left(\frac{m}{2W}\right), m \in I \right]$$

has the solution

$$\hat{x}(t) = \sum_{-\infty}^{\infty} A_n(t) x\left(\frac{n}{2W}\right),$$

where

$$A_n(t) = \left(\frac{1}{2W}\right) \int_{-W}^{W} A(t, f) e^{-2\pi i n f / 2W} \, df$$

where

$$A(t, f) = \left(\sum_{k=-\infty}^{\infty} e^{2\pi i t (f + 2kW)} P(f + 2kW) \right) \left(\sum_{-\infty}^{\infty} P(f + 2kW) \right)^{-1}.$$

Call the corresponding interpolation error covariance the "optimal filtering" error, and (6.20) the "aliasing error." Show that the

aliasing error = optimal filtering error

$$+ \int_{-W}^{W} (s(0, f) - s(t, f)) s(0, f)^{-1} (s(0, f) - s(t, f))^* \, df$$

where

$$s(t, f) = \sum_{-\infty}^{\infty} e^{4\pi i k W t} P(f + 2kW).$$

Calculate the covariance function of the process $\hat{x}(\cdot)$—is it stationary?
Hint: The Wiener-Hopf equation becomes:

$$\int_{-\infty}^{\infty} \left(e^{2\pi i f t} I - \sum_{-\infty}^{\infty} A_n(t) e^{2\pi i f n / 2W} \right) P(f) e^{-2\pi i f m / 2W} \, df = 0.$$

Since for each t,

$$A(t, f) = \sum_{-\infty}^{\infty} A_n(t) e^{2\pi i f n / 2W}$$

is periodic in f with period $2W$, we have

$$\int_{-W}^{W} \left(\sum_{k=-\infty}^{\infty} (e^{2\pi i t (f + 2kW)} I - A(t, f)) P(f + 2kW) \right) e^{-2\pi i m f / 2W} = 0,$$

$$m \in \mathbf{I}$$

or,

$$\sum_{k=-\infty}^{\infty} (e^{2\pi i t (f + 2kW)} P(f + 2kW) = A(t, f) \sum_{-\infty}^{\infty} P(f + 2kW), \qquad -W < f < W$$

from which the result follows. The error

$$= R(0) - \sum_{-\infty}^{\infty} A_n(t) R\left(\frac{n}{2W} - t\right)$$

$$= \int_{-\infty}^{\infty} (I - A(y, f)e^{-2\pi i f t})P(f)\, df$$

$$= \int_{-W}^{W} (S(0, f) - S(t, f)S(0, f)^{-1}S(t, f)^*)\, df.$$

Aliasing error — optimal filtering error

$$= \sum\sum (a_n(t)I - A_n(t)) R\left(\frac{n - m}{2W}\right)(a_m(t)I - A_n(t))^*$$

$$= \int_{-W}^{W} (e^{2\pi i f t}I - A(t, f))s(0, f)(e^{2\pi i f t}I - A(t, f))^*\, df.$$

9.14. Specialize $P(\cdot)$ in Problem 9.13 to

$$P(f) = D, \qquad |f| < 3W$$
$$\quad\; = 0, \qquad |f| > 3W.$$

Show that in this case

$$A_n(t) = \frac{a_n(t)(1 + 2\cos 4\pi t W)I}{3}$$

with $a_n(t)$ given by (6.2). Calculate the corresponding error covariance. Compare the error with the choice

$$A_n(t) = a_n(t)I,$$

as in Theorem 6.2.
    Answer: Error covariance

$$= (2WD)\left(\frac{1}{3}\right)(8 - 4\cos 4\pi t W - 4\cos^2 4\pi t W)$$

(6.20) yields: $4(1 - \cos 4\pi i W)(2WD)$.

9.15. Let

$$v_n = s_n + N_n$$

where $\{N_n\}$ is white Gaussian with nonsingular covariance matrix D and is independent of the stationary zero mean Gaussian signal process $\{s_n\}$. Let

$$\hat{s}_{n|N} = E[s_n \,|\, v_{n-k}, 0 \le k \le N$$

$$= \sum_0^N W_k v_{n-k}.$$

Show that

$$W_0 = PD^{-1}$$

where P is the error-covariance matrix. Show that this holds also for the Martingale cases:

$$\hat{s}_n = E[s_n \,|\, v_{n-k}, k \le 0]$$

and

$$\hat{s}_n = E[s_n \,|\, v_{n-k}, k \in \mathbf{I}].$$

NOTES AND COMMENTS

The linear estimation theory presented should be considered standard fare and is scattered in many of the references listed, including texts on numerical analysis where it touches on least squares and matrix factorization. Within our limited scope we could not of course go into the Wiener/Kolmogorov filtering theory—an authoritative treatment for the one-dimensional case is reference 3. The multidimensional case is largely nowadays treated using the Kalman state-space theory—the pioneering work being reference 4. Our tracking example is a simple special case of this much more comprehensive theory.

REFERENCES

Classic Treatises

1. N. Wiener. *Extrapolation, Interpolation and Smoothing of Stationary Time Series.* John Wiley and Sons, 1949. [Unpublished work, 1942]

2. A. N. Kolmogorov. "Interpolation und Extrapolation von Stationären zufälligen Folgen," *Bulletin of the Academy of Sciences of the USSR*, Series Math 5 (1941), pp. 3–14. (English translation: Rand Corporation, RM 3090-PR.)

3. J. L. Doob. Chapter 12 of *Stochastic Processes*. John Wiley and Sons, 1953.

4. R. E. Kalman. "A New Approach to Linear Filtering and Prediction Problems," *Transactions of the ASME: Journal of Basic Engineering*, vol. 82, series D, no. 1 (1960), pp. 35–45.

5. A. M. Iaglom. *An Introduction to the Theory of Random Functions*. Prentice-Hall, 1962.

6. W. B. Davenport and R. L. Root. *An Introduction to the Theory of Random Signals in Noise*. McGraw-Hill, 1958.

7. E. Hille. *Analytic Function Theory*, vol. 1. Ginn and Co., 1973.

Recent Publications

8. A. Papoulis. *Probability, Random Variables and Stochastic Processes.* McGraw-Hill, 1984.

9. A. V. Balakrishnan. *Kalman Filtering Theory.* Optimization Software Publications, 1987.

10. B. T. Polyak. *Introduction to Optimization.* Optimization Software Publications, 1987.

11. Yu. G. Evtushenko. *Numerical Optimization Techniques.* Optimization Software Publications, 1985.

INDEX